Oxford Graduate Texts in Mathematics

Series Editors
R. Cohen   S. K. Donaldson   S. Hildebrandt
T. J. Lyons   M. J. Taylor

T0177497

OXFORD GRADUATE TEXTS IN MATHEMATICS

Books in the series

1. Keith Hannabuss: *An Introduction to Quantum Theory*
2. Reinhold Meise and Dietmar Vogt: *Introduction to Functional Analysis*
3. James G. Oxley: *Matroid Theory*
4. N. J. Hitchin, G. B. Segal, and R. S. Ward: *Integrable Systems: Twistors, Loop Groups, and Riemann Surfaces*
5. Wulf Rossmann: *Lie groups: An Introduction through Linear Groups*
6. Qing Liu: *Algebraic Geometry and Arithmetic Curves*
7. Martin R. Bridson and Simon M. Salamon (eds): *Invitations to Geometry and Topology*
8. Shmuel Kantorovitz: *Introduction to Modern Analysis*
9. Terry Lawson: *Topology: A Geometric Approach*
10. Meinolf Geck: *An Introduction to Algebraic Geometry and Algebraic Groups*
11. Alastair Fletcher and Vladimir Markovic: *Quasiconformal Maps and Teichmüller Theory*
12. Dominic Joyce: *Riemannian Holonomy Groups and Calibrated Geometry*
13. Fernando Villegas: *Experimental Number Theory*
14. Péter Medvegyev: *Stochastic Integration Theory*
15. Martin A. Guest: *From Quantum Cohomology to Integrable Systems*
16. Alan D. Rendall: *Partial Differential Equations in General Relativity*
17. Yves Félix, John Oprea and Daniel Tanré: *Algebraic Models in Geometry*
18. Jie Xiong: *Introduction to Stochastic Filtering Theory*
19. Maciej Dunajski: *Solitons, Instantons, and Twistors*
20. Graham R. Allan: *Introduction to Banach Spaces and Algebras*
21. James Oxley: *Matroid Theory, Second Edition*
22. Simon Donaldson: *Riemann Surfaces*

Riemann Surfaces

Simon Donaldson

Royal Society Research Professor
Imperial College London

OXFORD
UNIVERSITY PRESS

OXFORD
UNIVERSITY PRESS

Great Clarendon Street, Oxford OX2 6DP

Oxford University Press is a department of the University of Oxford.
It furthers the University's objective of excellence in research, scholarship,
and education by publishing worldwide in

Oxford New York

Auckland   Cape Town Dar es Salaam Hong Kong   Karachi
Kuala Lumpur Madrid Melbourne Mexico City   Nairobi
New Delhi Shanghai Taipei Toronto
With offices in
Argentina Austria Brazil Chile Czech Republic France Greece
Guatemala Hungary Italy Japan South Korea Poland Portugal
Singapore Switzerland Thailand Turkey Ukraine Vietnam

Oxford is a registered trade mark of Oxford University Press
in the UK and in certain other countries

Published in the United States
by Oxford University Press Inc., New York

ISBN 978-0-19-960674-0

Printed and bound by CPI Group (UK) Ltd, Croydon, CR0 4YY

Preface

The origins of this book go back to 1981 and a 'junior seminar' in Oxford, in which the author tried to present parts of Riemann surface theory to a small group of fellow graduate students. Of course, those presentations were part and parcel of his own attempt to understand the theory—an attempt which has continued over the three succeeding decades and certainly has a long way to go yet. Moving on, the core of this book was developed from graduate courses given around about 2000 in Oxford, Stanford and Imperial College, London.

The theory of Riemann surfaces occupies a very special place in mathematics. It is a culmination of much of traditional calculus, making surprising connections with geometry and arithmetic. It is an extremely 'useful' part of mathematics, knowledge of which is needed by specialists in many other fields. It provides a model for a large number of more recent developments in areas including manifold topology, global analysis, algebraic geometry, Riemannian geometry and diverse topics in mathematical physics.

On the other hand, there are a great many excellent books about Riemann surface theory, so one has to ask what can be the justification for writing a new one. The approach and emphasis in this book are a little different from most texts. We emphasise connections with topology and manifold theory. Our main technical tools are differential forms, and we view Riemann surface theory to a large extent as a prototype for more advanced results in global analysis, such as Hodge theory and $\bar{\partial}$-methods in complex analysis. The emphasis on topology is focused particularly on the *mapping class group* and the concept of 'monodromy' in various settings. Our presentation here has been influenced by current developments in symplectic topology and the role of 'generalised Dehn twists' in the work of Arnold and Seidel, for example. On the other hand, we want to keep in touch with the traditional roots of Riemann surface theory, including special functions, algebraic manipulation in concrete examples and so on.

It may be helpful to explain the structure of the book. The core is contained in Chapters 8, 9 and 10 (Part III), where we prove the fundamental analytical results. These are what we call the 'Main Theorem' for compact Riemann surfaces—which implies the existence of meromorphic functions and the Uniformisation Theorem. Our approach emphasises real-variable and partial-differential-equation techniques and Hilbert space methods. Thus a reader who has studied the Main Theorem will be well on the way to understanding the general theory of elliptic operators over compact manifolds. Our proof of the Uniformisation Theorem, which avoids the more usual Perron method, is designed to appear as a natural extension of the Main Theorem, although the lack of compactness requires more subtle analysis. To arrive at this core goal we need to develop some background, and that is the purpose of Parts I and II. Part I is supposed to be slightly light-hearted and to set up the interplay between complex analysis and topology, with the latter treated informally. Part II could be seen as a rapid 'first course' in Riemann surface theory, including elliptic curves. The lecture course which this book grew out of was originally intended for students who had taken standard courses on differentiable manifolds and algebraic topology. In subsequent years it turned out that many of the audience did not have this background, so the material in Chapter 5 was included to give a very swift treatment of the standard material (tangent spaces, differential forms etc.) in the case of surfaces. We also give an extremely sketchy outline of the relevant algebraic topology. Realistically, Chapter 5 will probably be too condensed as a source for students who have not seen these ideas at all, and should be backed up by reading a text such as Spivak (1979).

Having proved the central, deep results in Part III, the remainder of the book tries to illustrate various facets of the more advanced theory. Chapter 11 is concerned with what Mumford (1975) calls 'the AMAZING SYNTHESIS which surely overwhelmed each one of us as graduate students'. Following that, we concentrate particularly on moduli theory and the connection with the mapping class group (although there is now a splendid and much more comprehensive text (Hubbard 2006) covering this area).

The mathematical level varies quite a lot as the book progresses, and there is probably no single reader, or no single reader at a single time, who will want to study all of it in detail. The treatment in Parts I, II and III is aimed at beginning graduate students with a reasonable background in general topology and real and complex analysis. The proofs are intended to be, for the most part, written in full detail. At the same time there are various topics, algebraic curves for example, which are close to but not exactly part of the main thrust. In the case of some topics, our treatment is quite brief (for example, we do not mention Bézout's Theorem) and the reader may want to consult a more thorough treatment such as Kirwan (1992). In Part IV, we gradually expect more background knowledge, the picture is painted with a broader brush and some proofs are only sketched in. For example, we start to use notions such as fibrations and higher-dimensional complex manifolds without explicitly defining them. By the end of the book, where we consider Picard–Fuchs equations, the material really outruns our technical resources (for a proper treatment one would start again with a thorough understanding of direct image sheaves and so on), and our discussion becomes largely descriptive.

There are exercises at the end of each chapter. It would be better if there were more. The subject is one in which familiarity with examples, and the ability to perform concrete calculations, is essential. Some of the harder questions may not be completely realistic, but it is hoped that they will at least stimulate the reader to think profitably about the material.

There is probably more material in the book than can be covered in a single lecture course. With a well-prepared audience, in about 30 lectures, it seems possible to cover most of Parts I, II and III in detail and then outline a selection from Part IV. On the other hand, if one has to develop foundational material such as Chapter 5 in detail, then it is probably realistic to get to Chapter 9.

Thanks are due to many people. The most thorough version of the course was given in Stanford (in 1997) and Young-Eun Choi produced a superb set of lecture notes, without which the book would probably never have been written. The staff of Oxford University Press have been enormously patient as many deadlines have slipped by, and again the book would probably never have appeared without their gentle but persistent reminders. Joel Fine, Richard Thomas and Ivan Smith have read parts of the manuscript, catching errors and making many suggestions for improvements. Till Brönnle has done meticulous and sterling work in preparing the diagrams and LATEX files.

I thank my wife, Nora, and my children for their support over the many years of this project. The book is dedicated to them.

Contents

Part I

Preliminaries

1 Holomorphic functions

1.1 Simple examples: algebraic functions

This is an introductory chapter in which we recall some examples of holomorphic functions in complex analysis. We emphasise the idea of 'analytic continuation', which is a fundamental motivation for Riemann surface theory.

One encounters holomorphic functions in various ways. One way is through power series, say $f(z) = \sum a_n z^n$. It often happens that a function which is initially defined on some open set $U \subset \mathbf{C}$ turns out to have natural extensions to larger open sets. For example, the power series

$$f(z) = 1 + z + z^2 + z^3 + \ldots$$

converges only for $|z| < 1$, but, writing $f(z) = 1/(1 - z)$, we see that the function actually extends to $\mathbf{C} \setminus \{1\}$. A more subtle example is the gamma function. For $\mathrm{Re}(z) > 0$, we write

$$\Gamma(z) = \int_0^\infty t^{z-1} e^{-t}\, dt.$$

The integral is convergent and defines a holomorphic function of z on this half-plane. Integration by parts shows that $\Gamma(n) = (n - 1)!$ when n is a positive integer. It is clear that $\Gamma(z)$ tends to infinity as z tends to 0, but let us examine this more carefully by writing

$$\Gamma(z) = \int_0^1 t^{z-1} e^{-t}\, dt + \int_1^\infty t^{z-1} e^{-t}\, dt.$$

The second integral is defined for all z, and holomorphic in z. We write the first integral as

$$\int_0^1 t^{z-1}(e^t - 1)\, dt + \int_0^1 t^{z-1}\, dt.$$

Now the term

$$\int_0^1 t^{z-1}(e^t - 1)\, dt$$

is defined, and holomorphic in z, for $\text{Re}(z) > -1$. The other integral we can evaluate explicitly:

$$\int_0^1 t^{z-1}\, dt = \frac{1}{z}.$$

So we conclude that, for $\text{Re}(z) > 0$,

$$\Gamma(z) = \frac{1}{z} + \Gamma_1(z)$$

say, where Γ_1 extends to a holomorphic function on the larger half-plane $\{z : \text{Re}(z) > -1\}$. So Γ has a meromorphic extension to the larger half-plane. Repeating the procedure, by considering

$$e^{-t} - \left(1 - t + \frac{t^2}{2!} - \frac{t^3}{3!} + \cdots + \frac{(-t)^k}{k!}\right),$$

we get a meromorphic extension to $\text{Re}(z) > -(k+1)$, and thence to the whole of $\mathbf{C}$.

It often happens that, when extending a function, one encounters 'multiple-valued functions'. For example,

$$f(z) = 1 + \frac{z}{2} - \frac{z^2}{2^2 2!} + \frac{3z^3}{2^3 3!} - \frac{5 \cdot 3z^5}{2^4 4!} + \cdots$$

is a perfectly good holomorphic function on the disc $|z| < 1$, which we recognise as $\sqrt{1+z}$. This cannot be extended holomorphically to $z = -1$ but, more, if we try to extend the function to $\mathbf{C} \setminus \{-1\}$ we find that, going once around the origin, the function switches to the other branch of the square root. Particularly important examples of this phenomena occur for 'algebraic functions'.

Let $P(z, w)$ be a polynomial in two complex variables. We want to think of the equation $P(z, w) = 0$ as defining w 'implicitly' as a function of z. For example, if $P(z, w) = w^2 - (1 + z)$, then we would get the function $w = \sqrt{1+z}$ above. Or, if $P(z, w) = z^3 + w^2 - 1$, we would get the function $w = \sqrt{1 - z^3}$. Of course, this does not make sense, precisely because of the problem of multiple values. The next theorem, which will be fundamental later, expresses precisely the way in which such functions are defined.

Let $X \subset \mathbf{C}^2$ be the set of points (z, w) with $P(z, w) = 0$. We define the partial derivatives

$$\frac{\partial P}{\partial z}, \frac{\partial P}{\partial w}$$

in the obvious way. They are again polynomial functions of the two variables z, w.

Theorem 1. *Suppose (z_0, w_0) is a point in X and $\partial P/\partial w$ does not vanish at (z_0, w_0). Then there is a disc D_1 centred at z_0 in C, a disc D_2 centred at w_0 and a holomorphic map $\phi : D_1 \to D_2$ with $\phi(z_0) = w_0$ such that*

$$X \cap (D_1 \times D_2) = \{(z, \phi(z)) : z \in D_1\}.$$

Of course, the set appearing in this statement is the 'graph' of the map ϕ. Essentially, the theorem says that $w = \phi(z)$ gives the unique local solution to the equation $P(z, w) = 0$, where 'local' means 'close to (z_0, w_0)'.

Remark Theorem 1 holds for any holomorphic function $P(z, w)$ of two variables, and not just for polynomials.

To prove the theorem, recall that if f is a holomorphic function on an open set containing the closure of a disc D which does not vanish on the boundary ∂D, then the number of solutions of the equation $f(w) = 0$ in D, counted with multiplicity, is given by the contour integral

$$\frac{1}{2\pi i} \int_{\partial D} \frac{f'(w)}{f(w)} dw.$$

If there is only one solution, w_1, it is given by another contour integral,

$$w_1 = \frac{1}{2\pi i} \int_{\partial D} \frac{w f'(w)}{f(w)} dw.$$

We apply these formulae to the family of functions of the variable w $f_z(w) = P(z, w)$, where we regard z as a parameter. First, take $z = z_0$. Then the hypothesis that $\partial P/\partial w \neq 0$ means that f_{z_0}' does not vanish at w_0. Thus we can find a small disc D_2 centred at w_0 so that f_{z_0} has no other zeros in the closure of D_2. Since the boundary of D_2 is compact, there is some $\delta > 0$ such that $|f_{z_0}| > 2\delta$ on ∂D_2. By continuity, this means that if z is sufficiently close to z_0, we still have $|f_z| > \delta$, say, on ∂D_2. Thus we can apply the formula above for the number of roots of the equation $f_z(w) = 0$. When $z = z_0$, this number must be 1 so, by continuity, the same is true for z close to z_0. Then we define $\phi(z)$ to be this unique root. The second formula shows that

$$\phi(z) = \int_{\partial D_2} \frac{w}{P} \frac{\partial P}{\partial w} dw,$$

which is clearly holomorphic in the variable z. This completes the proof.

1.2 Analytic continuation: differential equations

Next we want to give a precise meaning to possibly many-valued extensions of a holomorphic function.

Definition 1. *Let ϕ be a holomorphic function defined on a neighbourhood of a point $z_0 \in \mathbf{C}$. Let $\gamma : [0, 1] \to \mathbf{C}$ be a continuous map with $\gamma(0) = z_0$. An analytic continuation of ϕ along γ consists of a family of holomorphic functions ϕ_t for $t \in [0, 1]$, where ϕ_t is defined on a neighbourhood U_t of $\gamma(t)$ such that*

- *$\phi_0 = \phi$ on some neighbourhood of z_0;*
- *for each $t_0 \in [0, 1]$ there is a $\delta_{t_0} > 0$ such that if $|t - t_0| < \delta_{t_0}$, the functions ϕ_t and ϕ_{t_0} are equal on their common domain of definition $U_t \cap U_{t_0}$.*

For example, suppose that $z_0 = 0$ and ϕ is the function defined by the power series above, giving a branch of $\sqrt{1 + z}$. Let γ be the path which traces out the circle centred at -1,

$$\gamma(t) = -1 + e^{2\pi i t}.$$

The reader will see how to construct an analytic continuation of ϕ along this path with ϕ_1 equal to the other branch of the square root,

$$\phi_1 = -\phi,$$

on a suitable neighbourhood of 0.

Alongside the algebraic equations discussed above, another very important way in which holomorphic functions arise is as solutions to *differential equations*. Of course, the simplest example here is the differential equation

$$\frac{du}{dz} = g,$$

where g is a given function, whose solution is the indefinite integral of g and is given by a contour integral. We know that we may again encounter multi-valued functions, for example when g is the function $1/z$. In our current language, the contour integral along a (smooth) path γ furnishes an analytic continuation of a local solution along γ.

We consider next second-order linear, homogeneous equations of the form

$$u'' + Pu' + Qu = 0, \tag{1.1}$$

where P and Q are given functions of z, and u is to be found. Suppose first that P and Q are holomorphic near $z = 0$. So, they have power series expansions

$$P(z) = \sum p_n z^n, \quad Q(z) = \sum q_n z^n,$$

valid in some common region $|z| < R$. We seek a solution to the equation in the form $u(z) = \sum u_n z^n$. Equating terms, we get, for each $n \geq 0$,

$$(n + 2)(n + 1)u_{n+2} + \sum_{i \geq 0}(n + 1 - i)p_i u_{n+1-i} + \sum_{j \geq 0} q_j u_{n-j} = 0.$$

Both the sums are finite and contain only terms u_i for $i < n + 2$, so this gives a recursion formula. Given any choice of u_0, u_1, there is a unique way to define all the u_i satisfying the equations, and the resulting power series converges in $|z| < R$ (Exercise 4 below). We conclude that the solutions of our equation on the disc $|z| < R$ form a two-dimensional complex vector space.

Now suppose that P, Q are holomorphic on some open set $\Omega \subset \mathbf{C}$ and let $\gamma : [0, 1] \to \Omega$ be a path in Ω.

Proposition 1. *If u is a solution to equation (1.1) on a neighbourhood of $\gamma(0)$, then u has an analytic continuation along γ, through solutions of the equation.*

We leave the proof as an exercise. (In fact, one can generalise the result to the case when P, Q are themselves defined initially in a neighbourhood of $\gamma(0)$ and have analytic continuations along γ.)

This leads to the notion of the 'monodromy' of solutions to a differential equation. Suppose γ is a loop in Ω, with $\gamma(0) = \gamma(1)$. Let u_1, u_2 be a basis for the solutions of the differential equation on a small neighbourhood of $\gamma(0)$. Analytic continuation of u_1, u_2 along γ yields another pair of solutions $\tilde{u}_1, \tilde{u}_2$, say. These are linear combinations of the original pair

$$\tilde{u}_1 = au_1 + bu_2, \quad \tilde{u}_2 = cu_1 + du_2.$$

In more invariant language, let V be the two-dimensional vector space of local solutions; then analytic continuation along γ gives a linear map

$$M_\gamma : V \to V.$$

Interesting examples arise when the complement of Ω is a discrete subset of $\mathbf{C}$ and P, Q are meromorphic.

Definition 2. *A point $z_0 \in \mathbf{C}$ is a regular singular point of the equation $u'' + Pu' + Qu = 0$ if P has at worst a simple pole at z_0 and Q has at worst a double pole.*

Consider a model case,

$$u'' + \frac{A}{z}u' + \frac{B}{z^2}u = 0,$$

where A, B are complex constants. We try a solution $u = z^\alpha$ defined on a cut plane, say. This satisfies the equation if

$$\alpha(\alpha - 1) + A\alpha + B = 0.$$

If this quadratic equation (the *indicial equation*) has two distinct roots α_1, α_2 we get two solutions to our equation $z^{\alpha_1}, z^{\alpha_2}$ in the cut plane. If there is a double root α, the second solution is $z^\alpha \log z$. The general case is similar.

Proposition 2. *If* $P(z) = A/z + P_0(z)$ *and* $Q(z) = B/z^2 + C/z + Q_0(z)$, *where* P_0, Q_0 *are holomorphic near* 0, *and if the indicial equation has roots* α_1, α_2 *with* $\alpha_1 - \alpha_2 \notin \mathbf{Z}$, *then there are solutions*

$$u_1(z) = z^{\alpha_1} w_1(z), \quad u_2(z) = z^{\alpha_2} w_2(z)$$

to the equation $u'' + Pu' + Qu = 0$, *where* w_1, w_2 *are holomorphic in a neighbourhood of* 0.

The proof goes by power series expansion, as before. If $\alpha_1 - \alpha_2 \in \mathbf{Z}$, the second solution may involve logarithmic terms. Expressed in terms of monodromy, if z_0 is a regular singular point of the equation $u'' + Pu' + Q = 0$ and γ is a loop around z_0, then in the case when $\alpha_1 - \alpha_2 \notin \mathbf{Z}$, the monodromy around γ is (in a suitable basis) the diagonal matrix with entries $e^{2\pi i \alpha_1}, e^{2\pi i \alpha_2}$. In the other case, we may get a non-trivial Jordan form.

This discussion can be generalised to higher-order equations of the form

$$u^{(r)} + P_1 u^{(r-1)} + \cdots + P_r u' + P_{r+1} = 0.$$

A point $z_0 \in \mathbf{C}$ is a regular singular point if P_j has at worst a pole of order j there. We get an indicial equation of degree r, and everything goes through as before. If the P_j are defined for $|z| > R$, we have the notion of a 'regular singular point at infinity'. This is when P_j is $O(z^{-j})$ for large z. When this holds, one can construct power solutions for large z of the form $u = z^\alpha (1 + a_1 z^{-1} + a_2 z^{-2} \ldots)$.

A very important example, which we will return to later, is the *hypergeometric equation*,

$$z(1 - z)u'' + (c - (a + b + 1)z)u' - abu = 0,$$

which has regular singular points at $0, 1, \infty$. Here a, b, c are fixed parameters. It is not too hard to show that any second-order equation with three regular singular points can be transformed into a hypergeometric equation by a combination of Möbius maps $\tilde{z} = (az + b)/(cz + d)$ changing the independent variable, and multiplication operations $\tilde{u} = L(z)u(z)$ for fixed functions L. (For example, we could replace the dependent variable u by $z^2 u$, in which case $L(z) = z^2$.)

Exercises

1. Show that the gamma function has a simple pole at the point $z = -k$ for positive integers k, with residue $(-1)^k / k!$.

2. The Riemann zeta function is defined for $\mathrm{Re}(s) > 1$ by

$$\zeta(s) = \sum_{n=1}^{\infty} n^{-s}.$$

By writing

$$\zeta(s)\Gamma(s) = \int_0^{\infty} \frac{x^{s-1}e^{-x}}{1 - e^{-x}}dx,$$

obtain a meromorphic continuation of ζ over **C**.

3. Let $P(z, w) = w^3 - (z^2 + 1)$ and let $w = \phi_0(z)$ be the local solution of $P(z, \phi_0(z))$, around $z = 0, w = 1$. Find two paths beginning and ending at $z = 0$ along which ϕ_0 can be analytically continued and which lead to the two different other solutions around $z = 0$.

4. Show that if P, Q are holomorphic in $|z| < R$, then the power series solutions of the differential equation $u'' + Pu' + Qu = 0$ converge for $|z| < R$.

5. Show that the indicial equation of the hypergeometric equation at $z = 0$ has roots $0, c$ and that the solution corresponding to the root 0 is the hypergeometric function

$$F(z) = 1 + \frac{ab}{c}z + \frac{a(a+1)b(b+1)}{c(c+1)2!}z^2 + \frac{a(a+1)a+2)b(b+1)(b+2)}{c(c+1)(c+2)3!}z^3$$
$$+ \cdots,$$

in $|z| < 1$.

6. Prove the assertion that any second-order ordinary differential equation with three regular singular points can be transformed to a hypergeometric equation.

2 Surface topology

2.1 Classification of surfaces

In this second introductory chapter, we change direction completely. We discuss the topological classification of surfaces, and outline one approach to a proof. Our treatment here is almost entirely informal; we do not even define precisely what we mean by a 'surface'. (Definitions will be found in the following chapter.) However, with the aid of some more sophisticated technical language, it not too hard to turn our informal account into a precise proof. The reasons for including this material here are, first, that it gives a counterweight to the previous chapter: the two together illustrate two themes—complex analysis and topology—which run through the study of Riemann surfaces. And, second, that we are able to introduce some more advanced ideas that will be taken up later in the book.

The statement of the classification of closed surfaces is probably well known to many readers. We write down two families of surfaces Σ_g, Ξ_h for integers $g \geq 0, h \geq 1$.

The surface Σ_0 is the 2-sphere S^2. The surface Σ_1 is the 2-torus T^2. For $g \geq 2$, we define the surface Σ_g by taking the 'connected sum' of g copies of the torus. In general, if X and Y are (connected) surfaces, the connected sum $X \sharp Y$ is a surface constructed as follows (Figure 2.1). We choose small discs D_X in X and D_Y in Y and cut them out to get a pair of 'surfaces-with-boundaries', coresponding to the circle boundaries of D_X and D_Y. Then we glue these boundary circles together to form $X \sharp Y$.

One can show that this resulting surface is (up to natural equivalence) independent of the choices of discs. Also, the operation $\sharp$ is commutative and associative, up to natural equivalence. Now we define inductively, for $g \geq 2$,

$$\Sigma_g = \Sigma_{g-1} \sharp \Sigma_1 = \Sigma_{g-1} \sharp T^2$$

(Figure 2.2), which we can write as

$$\Sigma_g = T^2 \sharp \ldots \sharp T^2.$$

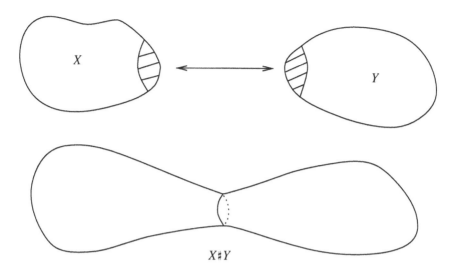

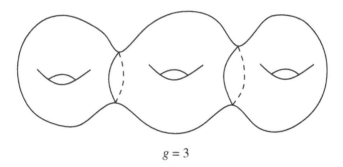

Figure 2.1 *Connected sum*

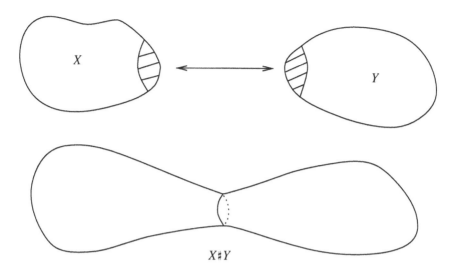

$g = 3$

Figure 2.2 *Genus-three surface*

There are many other representations of these surfaces, topologically equivalent. For example, we can think of Σ_g as being obtained by deleting $2g$ discs from the 2-sphere and adding g cylinders to form g 'handles'. Or we can start with a disc and add g ribbons in the manner shown in Figure 2.3. The boundary of the resulting surface-with-boundary is a circle, and we form Σ_g by attaching a disc to this boundary to get a closed surface (i.e. a surface with no boundary).

The surface Σ_1 is the real projective plane $\mathbf{R}P^2$. We can form it by starting with a Möbius band and attaching a disc to the boundary circle. We cannot do this within ordinary three-dimensional space without introducing self-intersections: more formally, $\mathbf{R}P^2$ cannot be embedded in $\mathbf{R}^3$. But we can still perform the construction to make a topological space and, if we like, we can

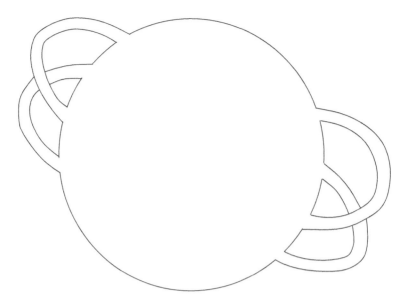

Figure 2.3   *Disc with handles*

think of embedding this in some $\mathbf{R}^n$ for larger n. Again, there are many other models possible. Notice that we can think of our Möbius band as a disc with a twisted ribbon attached (Figure 2.4). Then the construction falls into the same pattern as our third representation of Σ_g.

Now we make the family of surfaces Ξ_h by taking connected sums of copies of $\Xi_1 = \mathbf{R}P^2$:

$$\Xi_h = \mathbf{R}P^2 \sharp \ldots \sharp \mathbf{R}P^2.$$

Now let S be any closed, connected, surface. (More precisely, we mean compact and without boundary, so, for example, $\mathbf{R}^2$ would not count as closed.) We say that S is orientable if it does not contain any Möbius band, and non-orientable if it does. This leads to the following statement.

Classification Theorem.   *If S is orientable, it is equivalent to one (and precisely one) of the Σ_g. If S is not orientable, it is equivalent to one (and precisely one) of the Ξ_h.*

(We emphasise again that this statement has quite a different status from the other 'Theorems' in this book, since we have not even defined the terms precisely.)

To see an example of this, consider the 'Klein bottle' K. This can be pictured in $\mathbf{R}^3$ as shown in Figure 2.5, except that there is a circle of self-intersection. So, we should think of pushing the handle of the surface into a fourth dimension, where it passes through the 'side', just as we can take a curve in the plane as

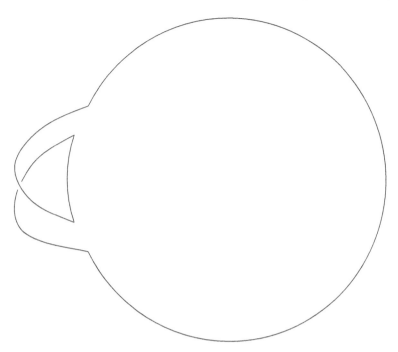

Figure 2.4 *Disc with twisted ribbon—a Möbius band*

Figure 2.5 *Klein bottle*

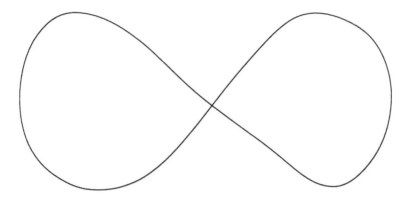

Figure 2.6 *Self-intersecting curve*

pictured in Figure 2.6 and remove the intersection point by lifting one branch into three dimensions.

Now it is true, but perhaps not immediately obvious, that K is equivalent to Ξ_2. To see this, cut the picture in Figure 2.5 down the vertical plane of symmetry. Then you can see that K is formed by gluing two Möbius bands together along their boundaries. By the definition of the connected sum, this shows that $K = \mathbf{R}P^2 \sharp \mathbf{R}P^2$, since the complement of a disc in $\mathbf{R}P^2$ is a Möbius band.

Now we will outline a proof of the classification theorem. The proof uses ideas that (when developed in a rigorous way, of course) go under the name of 'Morse theory'. A detailed technical account is given in the book by Hirsch (1994). The idea is that, given our closed surface S, we choose a typical real-valued function on S. Here 'typical' means, more precisely, that f is what is called a Morse function. What this requires is that if we introduce a choice of a gradient vector field of f on S, then there are only a finite number of points P_i, called critical points, in S where v (see below) vanishes, and near any of one of these points P_i we can parametrise the surface by two real numbers u, v such that the function is given by one of three local models:

- $u^2 + v^2 + \text{constant}$;
- $u^2 - v^2 + \text{constant}$;
- $-u^2 - v^2 + \text{constant}$.

The critical point P_i is said to have *index* 0, 1 or 2, respectively, in these three cases.

For example, if S is a typical surface in $\mathbf{R}^3$, we can take the function f to be the restriction of the z co-ordinate (say): the 'height' function on S. The vector field v is the projection of the unit vertical vector to the tangent spaces of S, and the critical points are the points where the tangent space is horizontal. The points of index 0 and 2 are local minima and maxima respectively, and the critical points of index 1 are 'saddle points' (Figure 2.7).

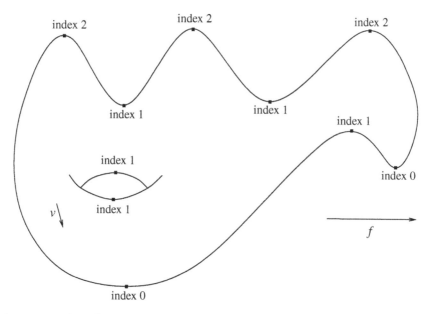

Figure 2.7   *Indices of Morse function*

Now, for each $t \in \mathbf{R}$, we can consider the subset $S_t = \{x \in S : f(x) \leq t\}$. There are a finite number of special cases, the 'critical values' $f(P_i)$ where P_i is a critical point. If f is a sufficiently typical function, then the values $f(P_i)$ will all be different, so to each critical value we can associate just one critical point. If t is not a critical value, then S_t is a surface-with- boundary. Morover, if t varies over an interval not containing any critical values, then the surfaces-with-boundaries S_t are equivalent for different parameters t in this interval. To see this, if $t_1 < t_2$ and $[t_1, t_2]$ does not contain any critical value, then one deforms S_{t_2} into S_{t_1} by pushing down the gradient vector field v (Figure 2.8).

Now consider the exceptional case when t is a critical value, t_0 say. The set S_{t_0} is no longer a surface. However, we can analyse the difference between $S_{t+\epsilon}$ and $S_{t-\epsilon}$ for small positive ϵ. The crucial thing is that this analysis is concentrated around the corresponding critical point, and the change in the surface follows one of three standard local models, depending on the index.

- *Index 0.* Near the critical point, $S_{t_0 \pm \epsilon}$ corresponds to $\{u^2 + v^2 \leq \pm \epsilon\}$, which is empty in one case and a disc in the other case. In other words, $S_{t_0 + \epsilon}$ is obtained from $S_{t_0 - \epsilon}$ by adding a disc as a new connected component.
- *Index 2.* This is the reverse of the index-0 case. The surface $S_{t_0 + \epsilon}$ is formed by attaching a disc to a boundary component of $S_{t_0 - \epsilon}$.
- *Index 1.* This is a little more subtle. The local picture is to consider

$$\{u^2 - v^2 \leq \pm \epsilon\},$$

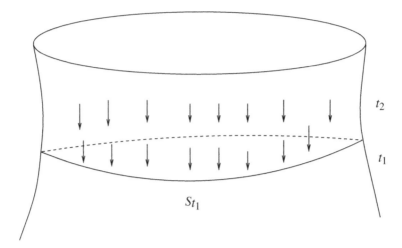

Figure 2.8 S_t

as shown in Figure 2.9. This is equivalent to adding a strip to the boundary (Figure 2.10). Thus $S_{t_0+\epsilon}$ is formed by adding a strip to the boundary of $S_{t_0-\epsilon}$.

We can now see a strategy to prove the Classification Theorem. What we should do is to prove a more general theorem, classifying surfaces-with-boundaries (not necessarily connected, and including the case of an empty boundary). Suppose we have any class of model surfaces-with-boundaries which is closed under the three operations associated to the critical points of index 0, 1 and 2 explained above. Then it follows that our original closed surface S must lie in this class, since it is obtained by a sequence of these operations.

For $r \geq 0$, let $\Sigma_{g,r}$ be the surface-with-boundary (possibly empty) obtained by removing r disjoint discs from Σ_g. Similarly for $\Xi_{h,r}$. Now we aim to prove that the class of disjoint unions of copies of the $\Sigma_{g,r}$ and $\Xi_{h,r}$ (for any collection of g's h's and r's) is closed under the three operations above. That is, if X is in our class (a disjoint union of copies of $\Sigma_{g,r}$ and $\Xi_{h,r}$) and we obtain X' by performing one of the operations, then X' is also in our class. This does require a little thought, and consideration of various cases.

- *Index 0*. This is obvious, since $\Sigma_{1,1}$ is the disc, so we can include a new disc component in our class.
  Index 2. This is also obvious. We cap off some boundary component with a disc, turning some $\Sigma_{g,r}$ into a $\Sigma_{g,r-1}$ or a $\Xi_{h,r}$ into a $\Xi_{h,r-1}$.
- *Index 1*. This requires more work. There are various cases to consider.
  Case 1. The ends of the attaching strip lie on different components of X.
  Case 2. The ends of the attaching strip lie on the same component of X.

Figure 2.9 *The case of index 1*

Now Case 2 subdivides into two subcases.

Case 2(i). The ends of the strip lie on the same boundary component of a component of X.

Case 2(ii). The ends of the strip lie on different boundary components of one common component of X.

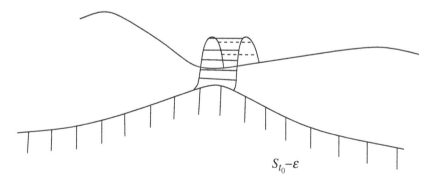

$$S_{t_0} - \varepsilon$$

Figure 2.10 *Attaching a strip*

Further, each of these cases 2(i) and 2(ii) subdivide further because there are two ways we can do the attaching, differing by a twist, in just the same fashion as when we form a Möbius band from a disc above. (The reader may like to think through why we do not need to make this distinction in Case 1.)

Now let us get to work.

Case 1. Let the relevant components of X be A and B, say. Then we can write $A = A' \setminus \text{disc}$, $B = B' \setminus \text{disc}$ for some A', B', and '$\setminus$ disc' means the operation of removing a disc, so, of course, the boundaries of the indicated discs contain the attaching regions. Then we can see that the manifold we get when we attach a strip is $(A' \natural B') \setminus \text{disc}$ (Figure 2.11).

Case 2(i). Let the component of X to which we attach the strip be $A = A' \setminus \text{disc}$. Then, in the untwisted case, we get, after the strip attachment, $A' \setminus \text{disc} \setminus \text{disc}$ (Figure 2.12). In the twisted case, we get $(A' \natural \mathbf{R}P^2) \setminus \text{disc}$ (Figure 2.13).

Case 2(ii). Here the distinction between the 'twisted' and 'untwisted' attachments is more subtle. Suppose again that the component of X where we attach the disc has the form $A = A' \setminus \text{disc} \setminus \text{disc}$, with the two indicated discs

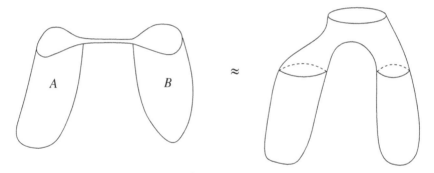

Figure 2.11 *Attaching a strip to different components*

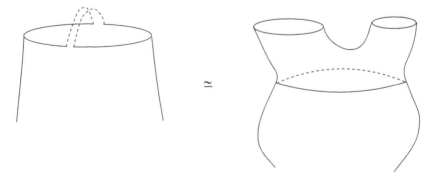

Figure 2.12 *Attaching a strip to the same boundary component*

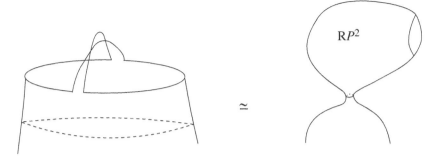

Figure 2.13 *Attaching a twisted strip*

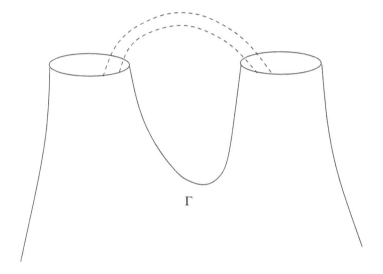

Figure 2.14 *Joining the attaching regions*

corresponding to the attaching regions. Choose a path Γ in A between points in the two attaching regions (Figure 2.14).

Then we say the twisted case is when the union of the attached strip and a strip about Γ in A form a Möbius band, and that the untwisted case is when the union forms an ordinary band. A little thought shows that the operation takes $A' \setminus \text{disc} \setminus \text{disc}$ to $A' \sharp T^2 \setminus \text{disc}$ in the untwisted case and to $A' \sharp K$ in the twisted case. However, we have seen that K is equivalent to Ξ_2, so we can write the new surface as $A' \sharp \mathbf{R}P^2 \sharp \mathbf{R}P^2 \setminus \text{disc}$.

At this point we have the proof of a result, although not quite the one we want. Our argument shows that any connected surface S is equivalent to a connected sum $\Sigma_g \sharp \Xi_h$. This holds because the class of disjoint unions of surfaces-with-boundaries of the form $\Sigma_g \sharp \Xi_h \setminus r$ discs is closed under the operations above.

To complete the proof of the Classification Theorem stated above, we need the following lemma.

Lemma 1. *The surfaces $T^2 \sharp \mathbf{R}P^2$ and $\mathbf{R}P^2 \sharp \mathbf{R}P^2 \sharp \mathbf{R}P^2$ are equivalent.*

Given this, we see that if $h > 1$, the surface $\Sigma_g \sharp \Xi_h$ is equivalent to Ξ_{h+2g}, so the result we obtained above implies the stronger form.

To prove the lemma, it suffices to show that $K \sharp \mathbf{R}P^2$ is equivalent to $T^2 \sharp \mathbf{R}P^2$, since we know that K is equivalent to $\mathbf{R}P^2 \sharp \mathbf{R}P^2$. $\mathbf{R}P^2 \setminus$ disc and $T^2 \setminus$ disc are pictured together in Figure 2.15. From this, one can see easily that $(T^2 \sharp \mathbf{R}P^2) \setminus$ disc is pictured as in Figure 2.16. Similarly, a little thought shows that $K \setminus$ disc

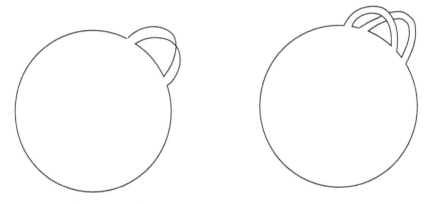

Figure 2.15 $\mathbf{R}P^2 \setminus disc$ and $T^2 \setminus disc$

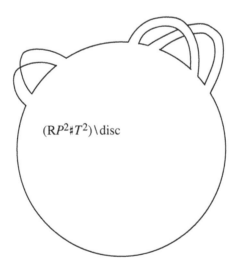

$(RP^2 \sharp T^2) \setminus \text{disc}$

Figure 2.16 $(T^2 \sharp \mathbf{R}P^2) \setminus disc$

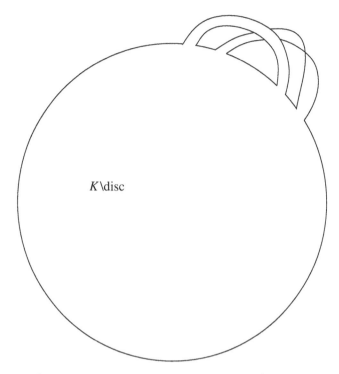

$K \setminus$disc

Figure 2.17 $K \setminus$ disc

is as in Figure 2.17. (That is, it is similar to the torus case but with one strip twisted.) So $(K \sharp \mathbf{R}P^2) \setminus$ disc is as pictured in Figure 2.18.

We can deform this picture into that shown in Figure 2.16 by sliding handles around the boundary as shown in Figures 2.19 and 2.20. When we attach the disc, this gives an equivalence between $K \sharp \mathbf{R}P^2$ and $T^2 \sharp \mathbf{R}P^2$, as desired.

2.2   Discussion: the mapping class group

This topological classification of surfaces has been known for many years and, while our discussion above is completely informal, a fully rigorous proof is not really difficult by modern standards. From this, one might be tempted to think that the subject of surface topology is a closed, fully understood area. One might be further tempted to think that the analogous classification problem in higher dimensions—the topological classification of manifolds—should not be too much harder. However, the second of these notions is certainly false, and the first is false if one broadens the concept of surface topology slightly. Moreover, these two issues are tightly connected, as we will now explain.

Figure 2.18 $(K \sharp \mathbf{R}P^2) \setminus disc$

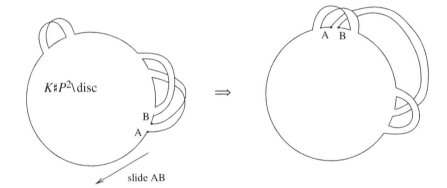

Figure 2.19   *Sliding handles (i)*

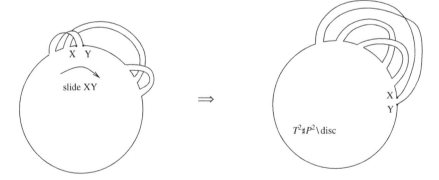

Figure 2.20   *Sliding handles (ii)*

Suppose one tries to implement the same strategy to classify three-dimensional manifolds. It is not hard to show that any closed 3-manifold can be built up from standard pieces in a similar fashion to what we have discussed above. More precisely, any closed 3-manifold has a *Heegard decomposition*. This is defined as follows. Take the standard picture of the surface Σ_g in $\mathbf{R}^3$ and let N_g be the three-dimensional region enclosed by the surface. So N_g is a 3-manifold with boundary Σ_g. Now let N'_g be another copy of N_g with boundary Σ'_g, i.e. another copy of Σ_g. Let $\phi : \Sigma_g \to \Sigma'_g$ be a homeomorphism. Then we can obtain a 3-manifold M_ϕ by gluing N_g to N'_g along their boundaries by ϕ. More precisely, we define M_ϕ to be the quotient of the disjoint union $N_g \cup N'_g$ by the equivalence relation which identifies $x \in \Sigma_g \subset N_g$ with $\phi(x) \in \Sigma'_g \subset N'_g$. Then a Heegard decomposition of a 3-manifold M is a homeomorphism $M \cong M_\phi$ for some ϕ, and, as we have said, any M arises in this way, determined up to equivalence by a ϕ. Of course, if we fix some standard identification between Σ_g and Σ'_g as a reference, then we can regard ϕ as a homeomorphism from Σ_g to itself.

Now the point is that the apparent simplicity of this description of 3-manifolds is illusory, because the set of self-homeomorphisms of a surface Σ_g is *enormously complicated* (at least once $g \geq 2$). These self-homeomorphisms obviously form a group, and there is a natural notion of equivalence (isotopy) such that the set of equivalence classes of self-homeomorphisms modulo isotopy forms a countable discrete group MC_g, called the 'mapping class group' of genus g. The complication which we refer to really resides in the complexity of this group. Looking back at the classification of surfaces from this perspective, we can see that what made the argument run so smoothly in that case is that the analogous group associated to the one-dimensional manifold—the circle— is very simple. The group of self-homeomorphisms of the circle modulo isotopy has just two elements, realised by the identity and a reflection map. This means that when we talked about 'attaching a disc to a circle boundary', say, the meaning was essentially unambiguous. (However, there is an issue lurking here because we do have two distinct ways of attaching surfaces along circle boundaries, and we should really have kept track of this throughout our discussion above. In the end it turns out that this does not matter, because it happens that for any surface-with-boundary there is a self-homeomorphism of the surface inducing the non-trivial map on any given boundary component. This is much the same issue as that which the reader was invited to consider when attaching twisted or untwisted bands in Case 1 above.)

Expressing our main point in another way, the complexity of surface topology arises not from the relatively easy fact that any orientable surface S is equivalent to some Σ_g but from the fact that there is a vast set of essentially different choices of equivalence between Σ_g and S, any two differing by an element of the mapping class group. One can make an analogy with the

Figure 2.21   *Dehn twist*

theory of finite-dimensional vector spaces over a field k. The classification is very simple: any one is isomorphic to a standard model k^d, where d is the dimension. However, there is a large choice in the isomorphism to the standard model, corresponding to the symmetry group $GL(d, k)$, which is a relatively complicated non-abelian group.

To illustrate these remarks, we introduce 'Dehn twists'. Let S be an orientable surface and let $C \subset S$ be an embedded circle. Since S is orientable, there is a neighbourhood N of C which we can identify with a standard band or cylinder $S^1 \times [-1, 1]$. We define a homeomorphism $\phi_0 : S^1 \times [-1, 1] \to S^1 \times [-1, 1]$ as follows. Regard S^1 as the unit circle in $\mathbf{C}$ and fix a function $f(t)$ on $[-1, 1]$ which is equal to 0 for $t \leq -1/2$, say, and to 2π for $t \geq 1/2$, say. Then set

$$\phi_0(e^{i\theta}, t) = (e^{i(\theta + f(t))}, t).$$

The choice of f means that ϕ_0 is the identity near the boundary of the cylinder. Now, if we identify the cylinder with the neighbourhood N in S, we can regard ϕ_0 as a homeomorphism from N to N, equal to the identity near the boundary. We define $\phi : S \to S$ by

$$\phi(x) = x \text{ if } x \notin N, \quad \phi(x) = \phi_0(x) \text{ if } x \in N.$$

The fact that ϕ_0 is the identity near the boundary means that ϕ is a homeomorphism from S to itself, the 'Dehn twist' around C (Figure 2.21).

Of course, the construction depends on various particular choices, namely the function f, the neighbourhood N and the identification of N with the cylinder, but up to isotopy the map ϕ is independent of these choices, and we get a well-defined element of the mapping class group. We will see later that the mapping class group and, in particular, Dehn twists arise naturally in questions of complex analysis and geometry.

Exercises

1. For each g, find an equation $f(x, y, z) = 0$ whose solution set in $\mathbf{R}^3$ is a closed surface of genus g.

2. Let Σ be a closed orientable surface and let C_1, C_2 be two embedded circles in Σ, neither of which separates Σ into two connected components. Show (or convince yourself) that there is a homeomorphism from Σ to itself which takes C_1 to C_2.

3. Identify the torus with $S^1 \times S^1$, where S^1 is the unit circle in $\mathbf{C}$. If p, q are coprime integers, we define the (p, q) curve in the torus to be the image of the map $i_{p,q} : S^1 \to S^1 \times S^1$ defined by $i_{p,q}(z) = (z^p, z^q)$. Show that $i_{p,q}$ is an embedding. Show that a suitable Dehn twist takes the (p, q) curve to a curve which can be deformed through homeomorphisms to the $(p - q, q)$ curve. Hence show that there is a composite of Dehn twists and inverse Dehn twists which maps the (p, q) curve to a curve which can be deformed to the $(1, 0)$ curve.

Part II
Basic theory

3 Basic definitions

3.1 Riemann surfaces and holomorphic maps

Definition 3. *A Riemann surface is given by the following data:*

- *A Hausdorff topological space X.*
- *A collection of open sets $U_\alpha \subset X$, where α ranges over some index set, which cover X (i.e. $X = \bigcup_\alpha U_\alpha$).*
- *For each α, a homeomorphism*

$$\psi_\alpha : U_\alpha \to \tilde{U}_\alpha,$$

where $\tilde{U}_\alpha$ is an open set in $\mathbf{C}$, with the property that for all α, β the composite map $\psi_\alpha \circ \psi_\beta^{-1}$ is HOLOMORPHIC on its domain of definition.

The maps ψ_α are referred to as 'charts', 'co-ordinate charts' or just 'local co-ordinates', and the entire collection of data $(U_\alpha, \tilde{U}_\alpha, \psi_\alpha)$ is called an 'atlas' of charts.

The reader who has never encountered this kind of notion before may find the definition hard to digest, so a few remarks are in order.

First, we define ψ_β^{-1} to be the obvious homeomorphism from $\tilde{U}_\beta$ to U_β, so $\psi_\alpha \circ \psi_\beta^{-1}$ is well defined as a map

$$\psi_\alpha \circ \psi_\beta^{-1} : V_{\alpha,\beta} \to V_{\beta,\alpha},$$

where $V_{\alpha,\beta} = \psi_\beta(U_\alpha \cap U_\beta)$ and $V_{\beta,\alpha} = \psi_\alpha(U_\alpha \cap U_\beta)$. Since $V_{\alpha,\beta}$ and $V_{\beta,\alpha}$ are open sets in $\mathbf{C}$, the notion of a holomorphic map, as specified in the definition, makes sense. Notice that, interchanging α and β, it is a consequence of the definition that $\psi_\alpha \circ \psi_\beta^{-1}$ is a homeomorphism from $V_{\alpha,\beta}$ to $V_{\beta,\alpha}$ with a holomorphic inverse.

The second remark is that in practice, when working with Riemann surfaces, one rarely sees this bulky collection of data explicitly. Suppose we have a point p in X. This lies in at least one U_α, so we choose one. The map ψ_α is just a complex-valued function defined on a neighbourhood of p, and we normally denote this by a symbol such as z. Then, in making calculations near to p, we label points by the corresponding value of the variable z in $\mathbf{C}$, so we are effectively working in the traditional notation of complex analysis. On the other hand, we might have chosen a different co-ordinate chart, ψ_β, which we might call w. Thus the map $\psi_\alpha \circ \psi_\beta^{-1}$ expresses, in more classical notation, z as a holomorphic function of w. The key feature of Riemann surface theory is that we have to study the behaviour of our calculations and constructions under such a holomorphic change of variables to obtain results which are independent of the choice of co-ordinate chart.

The third remark has more mathematical content. The main ideas embodied in the definition are not specific to the particular case at hand. If we take the same definition but replace the word HOLOMORPHIC by another appropriate condition (****) on maps between open sets in $\mathbf{C}$, then we get a definition of another kind of mathematical object. The main instances we need are the following:

- Taking (****) to be SMOOTH, we get the definition of a *smooth surface*. Here a smooth map between open sets in $\mathbf{C}$ is one which is infinitely often differentiable (C^∞) in the sense of two real variables, identifying $\mathbf{C}$ with $\mathbf{R}^2$.
- Taking (****) to be SMOOTH WITH POSITIVE JACOBIAN, we get the definition of an *oriented smooth surface*. Here the Jacobian is, as usual, the determinant of the 2×2 matrix of first derivatives of the map.

But there are many other interesting possibilities. For example, we could take (****) to be SMOOTH WITH JACOBIAN 1, to get the notion of a 'surface with an area form'. More generally, there is no need to restrict ourselves to two real dimensions. If we modify our definition to allow $\tilde{U}_\alpha$ to be open sets in $\mathbf{R}^n$ (for fixed n) and if we fix a suitable condition on maps between open sets in $\mathbf{R}^n$, we get the definition of a corresponding class of *n-dimensional manifolds*. Taking smooth maps, we get the notion of an *n*-dimensional smooth manifold. So a smooth surface is the same as a two-dimensional smooth manifold. A slightly more sophisticated possibility is this: if $n = 2m$ and we identify $\mathbf{R}^{2m}$ with $\mathbf{C}^m$, we may consider the condition that a map between open sets in $\mathbf{C}^m$ should be *holomorphic* in the sense of several complex variables. Then we get the notion of an *m*-dimensional *complex manifold*. So a Riemann surface is the same thing as a one-dimensional complex manifold.

Carrying on with the theory, we consider maps between Riemann surfaces.

Definition 4. *Let X be a Riemann surface with an atlas $(U_\alpha, \tilde{U}_\alpha, \psi_\alpha)$ and let Y be another Riemann surface, with an atlas $(V_i, \tilde{V}_i, \phi_i)$. A map $f : X \to Y$ is called holomorphic if for each α and i, the composite $\phi_i \circ f \circ \psi_\alpha^{-1}$ is holomorphic on its domain of definition.*

Here $\phi_i \circ f \circ \psi_\alpha^{-1}$ is a map from $\psi_\alpha(U_\alpha \cap f^{-1}(V_i))$ to $\tilde{V}_i$; these are open sets in $\mathbf{C}$, so the condition of being holomorphic is the usual one of complex analysis.

Again, this definition has obvious variants, which we will not spell out in detail. For example, we can get the definition of a smooth map between smooth surfaces, a smooth function on a smooth surface, a smooth map from $\mathbf{R}$ to a surface etc.

Now, of course, we say that two Riemann surfaces X and Y are *equivalent* if there is a holomorphic bijection $f : X \to Y$ with a holomorphic inverse. We will often treat equivalent Riemann surfaces as identical. For example, this allows us to remove the dependence of the definition of a Riemann surface on the particular choice of atlas. If we have one atlas $(U_\alpha, \tilde{U}_\alpha, \psi_\alpha)$, we can always concoct another one, for example by adding in some extra charts. Strictly, according to our definition, this gives a different Riemann surface, with the same underlying set X. However, the surfaces that arise are all equivalent, since the identity map gives a holomorphic equivalence between them.

We mention one more technical point. In the definition of a smooth surface (or, in general, of a manifold), one often imposes an extra 'countability' condition, which can be stated in various equivalent forms. One form is that there is an atlas consisting of a countable set of charts. This rules out certain rather pathological examples. In the case of Riemann surfaces it turns out that, for rather deep reasons (Rado's Theorem), this hypothesis is unnecessary—it is a consequence of the definition as we have stated it. See the discussion in Chapter 10.

3.2   Examples

3.2.1   First examples

First, any open set in $\mathbf{C}$ is naturally a Riemann surface. Familiar examples are the unit disc $D = \{z : |z| < 1\}$ and the upper half-plane $H = \{w \in \mathbf{C} : \mathrm{Im}(w) > 0\}$. These Riemann surfaces are equivalent, via the well-known map

$$z = \frac{w - i}{w + i}.$$

Next we consider the *Riemann sphere* S^2. As a set, this is $\mathbf{C}$ with one extra point ∞. The topology is that of the 'one-point compactification', i.e. open sets in S^2 are either open sets in $\mathbf{C}$ or unions

$$\{\infty\} \cup (\mathbf{C} \setminus K),$$

where K is a compact subset of **C**. Alternatively (as our notation has already suggested), we can define the topology by an obvious identification of $\mathbf{C} \cup \{\infty\}$ with the unit sphere in $\mathbf{R}^3$. We make S^2 into a Riemann surface with an atlas of two charts:

$$U_0 = \{z \in \mathbf{C} : |z| < 2\}, \ U_1 = \{z \in \mathbf{C} : |z| > 1/2\} \cup \{\infty\}.$$

We take $\tilde{U}_0 = \tilde{U}_1 = U_0$ and we let $\psi_0 : U_0 \to \tilde{U}_0$ be the identity map. We define ψ_1 by $\psi_1(\infty) = 0$ and $\psi_1(z) = 1/z$ for $z \in \mathbf{C}, |z| > 1/2$. Then the maps $\psi_0 \circ \psi_1^{-1}$ and $\psi_1 \circ \psi_0^{-1}$ are each the map $z \mapsto 1/z$ from the annulus $\{z : 1/2 < |z| < 2\}$ to itself, so these are both holomorphic and the condition of the definition is realised.

This example is perhaps confusing in its simplicity, but we have spelled it out in detail to illustrate how the definition works. Notice that the Riemann sphere is an example of a *compact* Riemann surface.

We insert here a short digression from our main path, which will not be taken up again until Chapter 13. This needs a little classical differential geometry. Let S be a smooth oriented surface in $\mathbf{R}^3$. Then S can be considered as a Riemann surface in the following way. We consider *isothermal co-ordinates* on an open set $U \subset S$. By definition, these constitute a chart $\psi : U \to \tilde{U} \subset \mathbf{R}^2$ such that the derivative of ψ^{-1} (which is a map from $\tilde{U} \subset \mathbf{R}^2$ to $\mathbf{R}^3$) takes orthogonal vectors in $\mathbf{R}^2$ to orthogonal vectors in $\mathbf{R}^3$. It is a fact that S can be covered by the domains of such isothermal co-ordinate systems. We will prove this, in a more general form, in Chapter 13. Given this fact, it is straightforward to see that the collection of isothermal co-ordinate systems and the chosen orientation define a Riemann surface structure on S. The comparatively difficult fact here is the existence of isothermal co-ordinates for general surfaces. But, for specific examples, we may be able to find these explicitly. In particular, for the standard sphere in $\mathbf{R}^3$, isothermal co-ordinates are furnished by projection from any point in the sphere to the tangent space of the antipodal point. In this way, one sees that the Riemann surface structure on the sphere in $\mathbf{R}^3$ is isomorphic to that of the Riemann sphere $\mathbf{C} \cup \{\infty\}$ defined above.

3.2.2   Algebraic curves

This is a much more extended example, in which we cover some important theory. We begin with 'affine curves'.

Let $P(z, w)$ be a polynomial in two complex variables. Define X, as a topological space, to be the set of zeros of P in $\mathbf{C}^2$,

$$X = \{(z, w) \in \mathbf{C}^2 : P(z, w) = 0\}.$$

Let us suppose that P has the following property. For each point (z_0, w_0) of X, at least one of the partial derivatives P_z, P_w does not vanish. Then we can make X into a Riemann surface in the following way. Suppose (z_0, w_0) is a point of X where P_w does not vanish. Then, according to Theorem 1, we can find small discs D_1, D_2 and a holomorphic map $\phi : D_1 \to D_2$ such that $X \cap (D_1 \times D_2)$ is the graph of ϕ—points of the form $(z, \phi(z))$. We make a co-ordinate chart with $U_\alpha = X \cap (D_1 \times D_2)$, with $\tilde{U}_\alpha = D_1$ and with ψ_α the restriction of the projection from $D_1 \times D_2$ to D_1. Symmetrically, if (z_1, w_1) is a point of X where P_z does not vanish, we can find discs B_1, B_2, say, and a holomorphic map $g : B_2 \to B_1$ describing $X \cap (B_1 \times B_2)$ as the set of points of the form $(g(w), w)$. Clearly, we can then find a collection of charts, either of the first kind or of the second kind, which cover X. We have to check that the 'overlap maps' between the charts are holomorphic. Now, between charts of the first kind, the overlap map will be the identity map on a suitable intersection of discs in $\mathbf{C}$. Likewise for the two charts of the second kind. Between a chart of the first kind, say $U_\alpha, \tilde{U}_\alpha, \psi_\alpha$ as above, and a chart of the second kind, the overlap map will be given by the composite

$$z \mapsto (z, \phi(z)) \mapsto \phi(z),$$

i.e. by the holomorphic map ϕ.

The preceding discussion is crucial in understanding the historical roots and the significance of the notion of a Riemann surface. Our Theorem 1 made precise the idea of an algebraic function, defined locally, and leads to the question of understanding how the different local pictures fit together. Now we can say, roughly speaking, that this is encoded in the topology of the Riemann surface $X \subset \mathbf{C}^2$, which is described locally by the branches of the algebraic function.

The addition of the point at infinity turns the non-compact Riemann surface $\mathbf{C}$ into the compact Riemann sphere. We extend this idea to the algebraic curves considered above, defining *projective curves*.

Recall that the complex projective space $\mathbf{CP}^n$ is the quotient of $\mathbf{C}^{n+1} \setminus \{0\}$ by the equivalence relation which identifies vectors $v, \lambda v$ in $\mathbf{C}^{n+1} \setminus \{0\}$ for any λ in $\mathbf{C} \setminus \{0\}$. A point in $\mathbf{CP}^n$ can be represented by homogeneous coordinates $[Z_0, \ldots, Z_n]$, with the understanding that $[\lambda Z_0, \ldots, \lambda Z_n]$ represents the same point. It is an exercise for the reader to prove that $\mathbf{CP}^n$ is compact, in its natural topology.

Let U_0 be the subset of $\mathbf{CP}^n$ consisting of points with the co-ordinate $Z_0 \neq 0$. Since, in this case,

$$[Z_0, Z_1, \ldots, Z_n] = [1, Z_1/Z_0, \ldots, Z_n/Z_0],$$

we can identify U_0 with $\mathbf{C}^n$. That is, a point $(z_1, \ldots, z_n)$ in $\mathbf{C}^n$ is identified with the point $[1, z_1, \ldots, z_n]$ in $\mathbf{CP}^n$. The complement of U_0 in $\mathbf{CP}^n$ is a copy of $\mathbf{CP}^{n-1}$, the 'hyperplane at infinity'. For example, when $n = 1$, $\mathbf{CP}^{n-1}$ is a single point, so $\mathbf{CP}^1 = \mathbf{C} \cup \{\infty\}$ and $\mathbf{CP}^1$ can be canonically identified with the Riemann sphere. For general n and any $i \leq n$, we can define a subset $U_i \subset \mathbf{CP}^n$ as the set of points where the co-ordinate Z_i does not vanish. Then

$$\mathbf{CP}^n = U_0 \cup U_1 \cdots \cup U_n$$

and each U_i is a copy of $\mathbf{C}^n$. When making calculations around a point in $\mathbf{CP}^n$, we can always choose a U_i containing that point, and then perform our calculations in $\mathbf{C}^n$. (In fact, $\mathbf{CP}^n$ is an n-dimensional complex manifold, with charts furnished by the U_i, although we will not make explicit use of this notion.)

Now let p be a *homogeneous polynomial* in the variables $Z_0, \ldots, Z_n$. This means that p can be written as

$$p(Z_0, \ldots, Z_n) = \sum_{i_0, \ldots, i_n} a_{i_0 \ldots i_n} Z_0^{i_0} \ldots Z_n^{i_n},$$

where for each term in the sum, $i_0 + \cdots + i_n = d$ for some fixed integer d, the *degree* of p. Equivalently,

$$p(\lambda Z_0, \ldots, \lambda Z_n) = \lambda^d p(Z_0, \ldots, Z_n).$$

Thus the equation $p(Z_0, \ldots, Z_n) = 0$ defines a subset of $\mathbf{CP}^n$ in the obvious way. For example, if p is the polynomial $p(Z_0, \ldots, Z_n) = Z_0$, then this zero set is just the hyperplane at infinity considered above. (More generally, if p has degree 1, then the zero set is a copy of $\mathbf{CP}^{n-1}$ in $\mathbf{CP}^n$.) The upshot of this is that if we have a collection of homogeneous polynomials p_1, p_r (of any degrees $d_1, \ldots, d_r$), then we can define a subset V of $\mathbf{CP}^n$ as the intersection of the zero sets of the p_i. Such a set in $\mathbf{CP}^n$ is called a *projective algebraic variety*, and their study is the field of projective algebraic geometry. Notice that V is compact, as a closed subset of the compact space $\mathbf{CP}^n$.

To make things more concrete, we will now suppose that $n = 2$ and consider a single homogeneous polynomial $p(Z_0, Z_1, Z_2)$ of degree d. Assume that Z_0 does not divide p. We denote the zero set of p in $\mathbf{CP}^2$ by $\bar{X}$. Let P be the corresponding inhomogeneous polynomial in two variables z, w:

$$P(z, w) = p(1, z, w).$$

By definition, the intersection of $\bar{X}$ with $U_0 = \mathbf{C}^2$ is the zero set X of P that we considered before: an affine algebraic curve. Thus

$$\bar{X} = X \cup (\bar{X} \cap L_\infty),$$

where $L_\infty = \mathbf{CP}^2 \setminus U_0$ is the line at infinity. Then $\bar{X} \cap L_\infty$ will be a finite set of points given by the zeros of $p(0, Z_1, Z_2)$, and $\bar{X}$ is a *compactification* of X obtained by adjoining this finite set.

Now suppose that the polynomial $P(z, w)$ obtained from p satisfies the condition of the previous subsection: that P_z, P_w do not both vanish at any point of X. Then we have made X into a Riemann surface. We can repeat the discussion, replacing Z_0 by Z_1 and Z_2. If the partial derivatives of the corresponding inhomogeneous polynomials satisfy the relevant non-vanishing conditions, then we make $\bar{X} \cap U_1$ and $\bar{X} \cap U_2$ into Riemann surfaces. It is easy to check that the three Riemann surface structures are equivalent on their common regions of definition $\bar{X} \cap U_i \cap U_j$, and thus we make $\bar{X}$ into a *compact* Riemann surface.

To state the result more neatly, we have the following proposition.

Proposition 3. *Suppose $p(Z_0, Z_1, Z_2)$ is a homogeneous polynomial of degree $d \geq 1$ and the only solution of the equations $\partial p / \partial Z_0 = \partial p / \partial Z_1 = \partial p / \partial Z_2 = 0$ is $Z_0 = Z_1 = Z_2 = 0$. Then the solutions of the equation $p = 0$ in $\mathbf{CP}^2$ form a compact Riemann surface.*

To see this, begin with *Euler's identity*

$$\sum_{i=0}^{2} Z_i \frac{\partial p}{\partial Z_i} = dp.$$

This shows first that p cannot be divisible by any Z_i. For if $p = Z_0 q$, say, then there is a non-zero point $(0, Z_1, Z_2)$ where q vanishes, and this contradicts the hypothesis. Now consider any point $[Z_0, Z_1, Z_2] \in \mathbf{CP}^2$ which lies in the zero set of p. At least one of the co-ordinate functions is not zero, so, without loss of generality, we can suppose $Z_0 = 1$ (using the symmetry under permuting co-ordinates). By hypothesis, at this point at least one of the partial derivatives of p does not vanish and, by Euler's identity, it follows that we cannot have $\partial p / \partial Z_1, \partial p / \partial Z_2$ both zero but $\partial p / \partial Z_0$ non-zero. Thus, without loss of generality, we have $\partial p / \partial Z_2$ non-zero. But this means that if we transfer to affine notation with $P(z, w) = p(z, w, 1)$, the partial derivative of P with respect to w does not vanish at the point in question, and we can fit into the preceding discussion.

When the criterion above on a homogeneous polynomial of degree d is satisfied, we call the zero set in $\mathbf{CP}^2$ a *smooth plane curve of degree d*.

It is not hard to extend this discussion to Riemann surfaces obtained as algebraic varieties in $\mathbf{CP}^n$ for larger n, but we will not go through this here.

To see how this construction works, consider the polynomial

$$P(z, w) = z^3 - zw^2 + 10z^2 + 3w + 16.$$

This has real coefficients so, to aid our geometric intuition, we can consider the corresponding *real algebraic curve* $X_{\mathbf{R}}$ in $\mathbf{R}^2$, which we can sketch. It takes some labour to work out an accurate picture of this, but one thing we can read off easily is the asymptotic behaviour. Informally, when z, w are large, the leading, cubic, terms in P should dominate the other terms, so we expect that the curve has asymptotes given by the zeros of $z^3 - zw^2$. Factorising this as

$$z^3 - zw^2 = z(z - w)(z + w),$$

we expect that the curve has asymptotic lines $z = 0, z = \pm w$, and this is indeed the case. Now consider the homogeneous polynomial of degree 3 corresponding to P,

$$p(Z_0, Z_1, Z_2) = Z_1^3 - Z_1 Z_2^2 + 10 Z_1^2 Z_0 + 3 Z_2 Z_0^2 + 16 Z_0^3.$$

This defines a subset $\overline{X}$ of $\mathbf{CP}^2$ as above, and one can check that the condition of Proposition 3 above is satisfied, so $\overline{X}$ is a Riemann surface. Now $\overline{X}$ meets the line at infinity L_∞ at the points $[0, Z_1, Z_2]$ which satisfy $P(0, Z_1, Z_2) = 0$. But

$$p(0, Z_1, Z_2) = Z_1^3 - Z_1 Z_2^2.$$

So $\overline{X} \cap L_\infty$ consists of three points $Z_1 = 0, Z_1 = \pm Z_2$, which of course correspond exactly to the asymptotic lines we saw in the affine picture. (One can carry over the entire projective-space construction to the case of real coefficients, getting a real projective curve $\overline{X}_\mathbf{R} \subset \overline{X}$, and in this case all the points of $\overline{X} \cap L_\infty$ lie in $\overline{X}_\mathbf{R}$ and so were apparent as asymptotes of our sketch of $X_\mathbf{R}$.)

What we see from this example—and which, of course, holds more generally—is that the projective-space construction gives us a systematic way to discuss the asymptotic phenomena of affine curves.

3.2.3 Quotients

We begin with a very simple case. Consider $2\pi \mathbf{Z}$ as a subgroup of $\mathbf{C}$ under addition and form the quotient set $\mathbf{C}/2\pi \mathbf{Z}$. First, this has a standard quotient topology, in which it is clearly homeomorphic to a cylinder $S^1 \times \mathbf{R}$. Second, we can make $\mathbf{C}/2\pi \mathbf{Z}$ into a Riemann surface in a very simple way. For each point z in $\mathbf{C}$, we consider the disc D_z centred on z and with radius $1/2$. Clearly, if z_1, z_2 are two points in D_z and if

$$z_1 = z_2 + 2\pi n$$

for $n \in \mathbf{Z}$, then we must have $n = 0$ and $z_1 = z_2$ (since $1/2 < \pi$). What this means is that the projection map $\pi : \mathbf{C} \to \mathbf{C}/2\pi \mathbf{Z}$ maps D_z bijectively to the quotient space. We use this to construct a chart about $\pi(z) \in \mathbf{C}/2\pi \mathbf{Z}$, taking $U = \pi(D_z)$ and $\tilde{U} = D_z$ and taking ψ_z as the local inverse of π. Then we cover $\mathbf{C}/2\pi \mathbf{Z}$ by some collection of charts of this form. The overlap maps between the charts will just have the shape

$$z \mapsto z + 2\pi n,$$

for appropriate $n \in \mathbf{Z}$, and these are certainly holomorphic. In fact, the Riemann surface $\mathbf{C}/2\pi\mathbf{Z}$ that we construct this way is equivalent to $\mathbf{C} \setminus \{0\}$, with the equivalence induced by the map $z \mapsto e^{iz}$.

Now let Λ be a *lattice* in $\mathbf{C}$; that is, a discrete additive subgroup. To be concrete, we can consider the lattice

$$\Lambda = \mathbf{Z} \oplus \mathbf{Z}\lambda,$$

where λ is some fixed complex number with positive imaginary part. We can repeat the discussion above without essential change; all we need to do is to choose the radius r of D_z so that

$$2r < \min_{n,m} |n + \lambda m|,$$

where n, m run over the integers, not both zero. Since $|n + \lambda m| \geq \mathrm{Im}(\lambda)$ if $m \neq 0$, and $|n + \lambda m| \geq 1$ if $m = 0$ and $n \neq 0$, it suffices to take

$$2r < \min(1, \mathrm{Im}(\lambda)).$$

In this way we see that $\mathbf{C}/\Lambda$ is a Riemann surface, clearly homeomorphic to the torus $S^1 \times S^1$ and, in particular, *compact*.

The construction of the Riemann surface structures in the examples above is rather trivial, but we have gone through it at some length because precisely the same ideas apply more generally. Suppose a group Γ acts on a Riemann surface X by holomorphic automorphisms. Suppose that the following two conditions hold:

- Around each point p of X, we can find an open neighbourhood N such that if $q_1, q_2 \in N$ and $\gamma \in \Gamma$ with $\gamma(q_1) = q_2$, then we must have $\gamma = 1$ and $q_1 = q_2$.
- Suppose p, q are points in X which do not lie in the same orbit of Γ. Then there are neighbourhoods N_1, N_2 of p, q, respectively, such that ΓN_1 is disjoint from N_2.

Then we can go through exactly the same construction to endow the quotient set X/Γ with a Riemann surface structure. Notice that the first condition above implies that Γ acts *freely* on X.

The important case of this quotient construction is when X is the upper half-plane H. Of course, it is the same to consider the unit disc D. The holomorphic automorphisms of H are given by Möbius maps

$$z \mapsto \frac{az + b}{cz + d},$$

where a, b, c, d are real and $ad - bc = 1$. In other words,

$$\mathrm{Aut}(H) = PSL(2, \mathbf{R}),$$

the quotient of the group $SL(2, R)$ of 2×2 real matrices of determinant 1 by the subgroup $\{\pm 1\}$. So if we have a suitable subgroup $\Gamma \subset PSL(2, \mathbf{R})$, we can construct a Riemann surface H/Γ. Taking this over to D, the holomorphic automorphisms are Möbius maps of the form

$$z \mapsto \frac{\alpha z + \beta}{\overline{\beta} z + \overline{\alpha}},$$

where $|\alpha|^2 - |\beta|^2 = 1$. Now we have the following proposition.

Proposition 4. *Suppose that Γ is a discrete subgroup of* $\mathrm{Aut}(H)$ *or* $\mathrm{Aut}(D)$ *which acts freely on H or D, respectively. Then the conditions above are satisfied.*

By a *discrete* subgroup we simply mean that there is a neighbourhood of the identity consisting only of the identity, for the topology induced by the natural topology on $PSL(2, \mathbf{C})$.

To prove this proposition, we work with the disc D, which of course implies the other case. Note that the proof is much more transparent when one has developed the theory of hyperbolic geometry, which we will do in Chapter 11, so the reader may prefer to skip the proof for the time being and return to it later.

To begin with, for the generalisation below, we will not use the assumption that Γ acts freely. The stabiliser of 0 in $\mathrm{Aut}(D)$ is the circle, acting in the obvious way. This is just the case $\beta = 0$ in the general description above. If Γ is discrete, its intersection with this circle is a discrete subgroup of the circle, which must be a finite cyclic group. So the stabiliser of 0 in Γ is a finite cyclic group Z, say. It follows easily from the discreteness of Γ that we can choose a $k < 1$ such that if $\gamma \in \Gamma$ is $\gamma(z) = (\alpha z + \beta)/(\overline{\beta} z + \overline{\alpha})$, where $|\alpha|^2 - |\beta|^2 = 1$, and if $|\beta| \leq k|\beta|$, then γ is in Z. Now suppose η is a point in the disc with $|\eta| \leq \epsilon$ and that $|\gamma(\eta)| \leq c$. Then

$$|\alpha \eta + \beta| \leq c|\overline{\beta} \eta + \overline{\alpha}|.$$

Applying the triangle inequality twice, we get

$$|\beta| \leq c(|\alpha| + |\beta|\epsilon) + \epsilon|\alpha|.$$

Hence

$$|\beta| \leq \frac{\epsilon + c}{1 - c\epsilon}|\alpha|.$$

Now choose ϵ so small that

$$\frac{2\epsilon}{1 - \epsilon^2} < k,$$

and let N be the disc of radius ϵ about 0. It follows that if $w_1, w_2 \in N$ and $\gamma \in \Gamma$ with $\gamma(w_1) = w_2$, then γ must lie in Z. So if Γ acts freely, as we now assume, γ must be the identity and the first condition is satisfied. This is for the particular point $0 \in D$, but since $\mathrm{Aut}(D)$ acts transitively on D, the argument applies in the general case.

Now we turn to the second condition. Again, there is no loss in taking one of the points to be 0. Let q be any point in the disc not in the same Γ-orbit as 0. We take $\tilde{N}_1, \tilde{N}_2$ to be the ϵ discs around 0 and q, respectively. If $\gamma \in \Gamma$ maps a point η in $\tilde{N}_1$ to $\tilde{N}_2$, then $|\gamma(\eta)| \leq c$ with $c = |q| + \epsilon$ and so, by the above, $|\beta| \leq k'|\alpha|$ with $k' = (\epsilon + c)/(1 - c\epsilon)$. We choose ϵ small enough that $k' < 1$. The set of points $(\alpha, \beta) \in \mathbf{C}^2$ with $|\alpha|^2 - |\beta|^2 = 1$ and $|\alpha| \leq k'|\beta|$ is compact. Since Γ is discrete, it follows that with these fixed $\tilde{N}_1, \tilde{N}_2$ there are only *finitely many* $\gamma \in \Gamma$ such that $\gamma(\tilde{N}_1) \cap \tilde{N}_2$ is non-empty. Now it is clear that we can find smaller neighbourhoods $N_i \subset \tilde{N}_i$ having the desired property.

Suppose that a Möbius map in $PSL(2, \mathbf{R})$, as above, has a fixed point in the upper half-plane. This is the same as saying that there is a solution z of the equation

$$z = \frac{az + b}{cz + d}$$

with $\mathrm{Im}(z) > 0$. If $c \neq 0$, the equation is equivalent to the quadratic equation

$$cz^2 + (d - a)z - b = 0,$$

so the condition is the same as saying that the discriminant is negative. Using the hypothesis $ad - bc = 1$, this discriminant is

$$(d - a)^2 + 4bc = (a + d)^2 - 4,$$

so we need $|a + d| < 2$. It is obvious that there is no fixed point in the case when $c = 0$, unless of course the map is the identity.

Now we introduce some famous examples, very important in number theory. Fix a prime p and let $\tilde{\Gamma}_p$ be the set of 2×2 matrices M with integer entries, with determinant 1 and with $M = \pm 1$ modulo p. Dividing by ± 1, we get a subgroup $\Gamma_p \subset PSL(2, \mathbf{R})$. For simplicity, we discuss here the case $p \geq 5$. Then it is clear that if a and d are equal to 1 modulo p, then the sum $a + d$ must have modulus more than 2. Thus Γ_p acts freely and we obtain a Riemann surface $X_p = H/\Gamma_p$.

We will finish this chapter with a refinement of the quotient construction above. For *any* discrete subgroup $\Gamma \subset \mathrm{Aut}(D)$ (or $\mathrm{Aut}(H)$), the quotient has a natural Riemann surface structure. That is, the condition that Γ acts freely can be dropped. This is more surprising, but little extra work is required for the proof. What our analysis above gives is that if p is a point in D, then the stabiliser Stab_p of p in Γ is a finite cyclic group, and there is a Stab_p-invariant

neighbourhood N of p which can be identified with a disc in such a way that
the action of Stab_p is the standard action of the cyclic subgroup of S^1. Then
the quotient space D/Γ is covered by open sets which are either copies of N
(for the points with a trivial stabiliser) or copies of proper quotients of N. But
the map $z \mapsto z^r$ induces an identification between the quotient of the disc of
radius ρ by the action of the cyclic group of order r and the disc of radius ρ^r.
In this way we identify all members of this cover with discs, and these form an
atlas of charts, making D/Γ into a Riemann surface.

Exercises

1. Show that a Riemann surface is naturally an oriented smooth surface.
2. Suppose that $F = F(x, y, z)$ is a smooth function on $\mathbf{R}^3$ with $F(0,0,0) = 0$
   and that the partial derivative $F_z = \partial F/\partial z$ does not vanish at $(0,0,0)$.
   Then one can show that there is a smooth function $f(x, y)$, defined on
   a disc $D = \{(x, y) \in \mathbf{R}^2 : x^2 + y^2 < r^2\}$ and taking values in an interval
   $I = (-\epsilon, \epsilon) \subset \mathbf{R}$, such that the intersection of $F^{-1}(0)$ with the cylinder
   $D \times I$ is the graph of f. (This result is analogous to Theorem 1, and is
   another instance of the implicit function theorem.) Using this, prove the
   following result. Let F be a smooth function on $\mathbf{R}^3$ and let $S = F^{-1}(0)$.
   Suppose that at each point of X at least one of the partial derivatives
   F_x, F_y, F_z does not vanish. Then X is naturally an oriented smooth surface.
3. (For those with some knowledge of differential geometry.) Construct
   isothermal co-ordinates on a general surface of revolution in $\mathbf{R}^3$.
4. Show that the set of points $(z, w) \in \mathbf{C}^2$ with $w^2 = \sin z$ is naturally a
   Riemann surface.
5. Prove Euler's formula.
6. Let A be a matrix in $GL(3, \mathbf{C})$. Show that A induces a map $\underline{A}$ from $\mathbf{CP}^2$ to
   itself such that if p is a homogeneous polynomial with zero set $Z_p \subset \mathbf{CP}^2$,
   then $\underline{A}(Z_p)$ is the zero set of another homogeneous polynomial p_A. Show
   that if p satisfies the criterion for Z_p to be a smooth plane curve, then so
   does p_A, and that the map $\underline{A} : Z_p \to \underline{A}(Z_p)$ is an isomorphism of Riemann
   surfaces. (*Remark*: All this just expresses the invariance of the theory under
   projective transformations. From another point of view, we could have
   started with an abstract three-dimensional vector space, and the choice of
   co-ordinates Z_0, Z_1, Z_2 corresponds to a choice of basis in this space.)
7. Suppose the homogeneous polynomial p has degree 2, and so can be
   regarded as a quadratic form on $\mathbf{C}^3$. Show that the criterion for the zero set
   to be a smooth plane curve is satisfied if and only if this quadratic form is
   non-degenerate. Using the classification of forms, or otherwise, show that
   the Riemann surface defined by p is equivalent to the Riemann sphere.

8.(a) Use the Liouville Theorem to show that Aut $\mathbf{C}$ consists of maps $z \mapsto az + b$, for $a \neq 0$.

(b) Show that the automorphisms of the Riemann sphere are given by Möbius maps $PSL(2, \mathbf{C})$.

(c) Use the Schwarz Lemma to identify the stabiliser of 0 in Aut(D) and hence identify Aut(D) and Aut(H).
(This is intended as a review of some fairly standard results which can be found in complex-analysis texts.)

9. Show that a discrete group $\Gamma \subset PSL(2, \mathbf{R})$ acts freely on H if and only if it is torsion-free.

10. Show that in fact Γ_p acts freely on H for all primes p.

11. Let Γ be the set of Möbius maps of the form $z \mapsto z + n$ for $n \in \mathbf{Z}$. Show that the quotient H/Γ is equivalent to the punctured disc. For $\rho < 1$, find a discrete group $\Gamma_\rho \subset PSL(2, \mathbf{R})$ such that the quotient H/Γ_ρ is isomorphic to the annulus $\{z \in \mathbf{C} : \rho < |z| < 1\}$.

4 Maps between Riemann surfaces

4.1 General properties

The foundation for this chapter is provided by two simple lemmas from complex analysis.

Lemma 2. *Let f be a holomorphic function on an open neighbourhood U of 0 in $\mathbf{C}$ with $f(0) = 0$. Suppose that the derivative $f'(0)$ does not vanish. Then there is another open neighbourhood $U' \subset U$ of 0 such that f is a homeomorphism from U' to its image $f(U') \subset \mathbf{C}$ and the inverse map is also holomorphic.*

The proof of this is very similar to that of Theorem 1. (The above lemma is an instance of the general 'inverse function theorem', while Theorem 1 is an instance of the general 'implicit function theorem'.) Thus we choose a small disc D_ϵ about 0 and use the fact that the number of roots of $f(z) = w$ in D_ϵ is given by the contour integral

$$\int_{\partial D_\epsilon} \frac{f'(z)}{f(z) - w} dz,$$

provided there are no roots on the boundary ∂D_ϵ. The argument then runs parallel to that for Theorem 1, and we leave the details to the reader. (The lemma can also be deduced from a slightly more general form of Theorem 1, where we take the function of two variables $P(z, w) = f(z) - w$ and interchange the roles of z and w.)

Lemma 3. *Let f be a holomorphic function on an open neighbourhood U of 0 in $\mathbf{C}$ with $f(0) = 0$, but with f not identically zero. Then there is a unique integer $k \geq 1$ such that on some smaller neighbourhood U' of 0, we can find a holomorphic function g with $g'(0) \neq 0$ and $f(z) = g(z)^k$ on U'.*

To see this, we consider the power series expansion of f about 0 and let k be the order of the first non-zero term:

$$f(z) = a_k z^k + a_{k+1} z^{k+1} + \cdots, \quad a_k \neq 0.$$

Thus

$$f(z) = a_k z^k (1 + b_1 z + b_2 z^2 + \cdots),$$

where $b_i = a_{k+i}/a_k$. If z is sufficiently small, there is a well-defined holomorphic function

$$h(z) = (1 + b_1 z + b_2 z^2 + \cdots)^{1/k}$$

(more precisely, we need $|\sum b_i z^i| < 1$). Then $f(z) = g(z)^k$, where

$$g(z) = a_k^{1/k} z h(z),$$

taking any choice of the root $a_k^{1/k}$. The derivative of g at 0 is $a_k^{1/k}$, and hence non-zero, so we have established the existence asserted in the lemma. The uniqueness of k is also clear. Note that $k = 1$ if and only if $f'(0) \neq 0$, and otherwise $k - 1$ is the multiplicity of the zero of f' at $z = 0$.

These simple lemmas yield a complete local description of holomorphic maps between Riemann surfaces.

Proposition 5. *Let X and Y be connected Riemann surfaces and $F : X \to Y$ a non-constant holomorphic map. For each point x in X, there is a unique integer $k = k_x \geq 1$ such that we can find charts around x in X and $F(x)$ in Y in which F is represented by the map $z \mapsto z^k$.*

To spell out the statement in more detail, we mean that there are a chart $(U, \tilde{U}, \psi)$ about $x \in X$ with $\psi(x) = 0 \in \tilde{U} \subset \mathbf{C}$ and a chart $(V, \tilde{V}, \phi)$ about $F(x) \in Y$ with $\phi(F(x)) = 0 \in \tilde{V} \subset \mathbf{C}$, such that the composite $\phi \circ F \circ \psi^{-1}$ is equal to the map $z \mapsto z^k$ on its domain of definition.

To prove the proposition, we begin by choosing arbitrary charts about x and $F(x)$. In these charts, F is represented by a holomorphic function, which we denote by f. We apply Lemma 3 to write f as g^k. Then the derivative of g at 0 does not vanish, so we can apply Lemma 1 to see that, after restricting the domain of definition, g gives a holomorphic homeomorphism with a holomorphic inverse. Thus we can change the chart about x by composing with g to get a new chart having the desired property. Again the uniqueness of $k = k_x$ is clear.

To get a straightforward global theory, it is natural to impose some conditions on the holomorphic maps we wish to study. A good class to work with is that of *proper* holomorphic maps. Recall that a map $F : S \to T$ between topological spaces S, T is called proper if, for any compact set $K \subset T$, the pre-image $f^{-1}(K)$ is also compact. Note that if S itself is compact then any map F is proper, since $F^{-1}(K)$ is a closed subset of S, and hence compact.

Recall also that a subset Δ of a topological space S is *discrete* if for any point $\delta \in \Delta$ there is a neighbourhood U of δ in S such that $\Delta \cap U = \{\delta\}$.

Proposition 6. *Let $F : X \to Y$ be a non-constant holomorphic map between connected Riemann surfaces.*

1. *Let $R \subset X$ be the set of points x where $k_x > 1$; then R is a discrete subset of X.*
2. *If F is proper, then the image $\Delta = F(R)$ is discrete in Y.*
3. *If F is proper, then for any y in Y the pre-image $f^{-1}(y)$ is a finite subset of X.*

The first item follows from the fact that, in local charts, R is given by the set of zeros of the derivative, using the standard fact that the zeros of a non-constant holomorphic function are discrete. The second item is a consequence of the general fact that a proper map takes discrete sets to discrete sets. The third item uses the fact that $f^{-1}(y)$ is compact and the same standard result from complex analysis.

Terminology We call the points of the set R the *critical points* of f and the points in the image $F(R)$ the *critical values*. (These points are also called 'branch points' or 'ramification points'.) For x in X, we call the integer k_x the *multiplicity* of F at x.

Suppose again that $F : X \to Y$ is a proper, non-constant holomorphic map between Riemann surfaces, with Y, connected. For each $y \in Y$, we define an integer $d(y)$ by

$$d(y) = \sum_{x \in F^{-1}(y)} k_x.$$

The sum runs over a finite set, by item 3 of the previous proposition. Notice that if $y \notin \Delta$ then $d(y)$ is just the number of points in $F^{-1}(y)$, and in general we will refer to $d(y)$ as the number of points in $f^{-1}(y)$, counted with multiplicity.

Proposition 7. *Let $F : X \to Y$ be a proper, non-constant holomorphic map between connected Riemann surfaces. Then the integer $d(y)$ does not depend on $y \in Y$.*

First, observe that this is true in the special case when $X = Y = \mathbf{C}$ and F is the map $F(z) = z^n$ for some $n \geq 1$. The general case can be reduced to this using the local description, in Proposition 5, of holomorphic maps. Fix $y \in Y$. We can find charts $U_x \subset X$ about each point $x \in f^{-1}(y)$ and a corresponding $V_x \subset Y$ about y, with respect to which F is expressed locally as $z \mapsto z^{k_x}$. Let V be the intersection of the V_x; this is an open neighbourhood of y in Y, since there are only finitely many x's. Using the properness of F, we can arrange that $F^{-1}(V)$ is contained in the union of the U_x's. Thus, for another point $y' \in V$, we can study the set $F^{-1}(y')$ using the local models. It follows from the special case we began with that $d(y') = d(y)$. Thus $d(y)$ is locally constant on Y and hence constant, since Y is connected.

The upshot of this is that we have defined an integer invariant, the *degree*, of a proper holomorphic map between connected Riemann surfaces. This is just the integer $d(y)$ for any y in the target space. (In the special case of a constant map, we define the degree to be 0.) While we have defined this in a holomorphic setting, it is in fact essentially a topological invariant: See also Exercise 2 in Chapter 5.

While the proofs above are not difficult, the results give us a striking corollary. First, some terminology. A *meromorphic function* F on a Riemann surface is a holomorphic map to the Riemann sphere which is not identically equal to ∞. In local charts, this agrees with the ordinary notion of a meromorphic function, having a Laurent series

$$\sum_{i=-n}^{\infty} a_i z^i.$$

The *poles* of F are just the points in $F^{-1}(\infty)$, and if x is a pole, its *order* is the integer k_x.

Corollary 1. *Let X be a compact connected Riemann surface. If there is a meromorphic function on X having exactly one pole, and that pole has order 1, then X is equivalent to the Riemann sphere.*

Let $F : X \to S^2$ be the given meromorphic function. It is proper, since X is compact. The hypotheses imply that the degree of F is 1 (computing using $y = \infty$). This means that for any $y \in S^2$ there is exactly one point x in $f^{-1}(y)$ (and that $k_x = 1$). Thus F is a bijection. The inverse map is continuous (since the image under F of a closed set in X is compact in S^2 and hence closed), so F is a homeomorphism. It also follows from Lemma 2 that the inverse map is holomorphic.

4.2 Monodromy and the Riemann Existence Theorem

4.2.1 Digression into algebraic topology

To take our study further, we recall some algebraic topology. Obviously, we do not have the space to give a full account, so much of what follows is intended as a review and summary of material which the reader has either met before or can study in standard texts such as Hatcher (2002).

Let $F : P \to Q$ be a map between topological spaces.

Definition 5. *F is a local homeomorphism if, around each point x in P, there is an open neighbourhood U such that $F|_U$ is a homeomorphism to its image in Q.*

Definition 6. *F is a covering map if, around each point $y \in Q$, there is an open neighbourhood V such that $F^{-1}(V)$ is a disjoint union of open sets U_α in P and $F|_{U_\alpha}$ is a homeomorphism from U_α to V.*

Clearly, a covering map is a local homeomorphism, but the converse is not true in general. (Exercise: give an example). However, we have the following proposition.

Proposition 8. *A proper local homeomorphism is a covering map.*

In fact, a proper local homeomorphism is the same as a *finite* covering map, where the number of points in $f^{-1}(y)$ is finite for each $y \in Q$.

We need to recall the relation between these notions and the *fundamental group*. Let Q be a topological space and let $q_0 \in Q$ be a 'base point'. The fundamental group $\pi_1(Q, q_0)$ consists of homotopy classes of loops based at q_0. One often drops the base point from the notation. Here, we give some examples which will be important for us:

- If Q is $\mathbf{C}$ or the disc D, then $\pi_1(Q)$ is trivial.
- If Q is the punctured plane $\mathbf{C} \setminus \{0\}$, then $\pi_1(Q) = \mathbf{Z}$.
- If Q is a multiply punctured plane $\mathbf{C} \setminus \{z_1, \ldots, z_n\}$, then $\pi_1(Q)$ is the free group on n generators.
- If Q is the torus T^2, then $\pi_1(Q) = \mathbf{Z} \times \mathbf{Z}$.
- If Q is the standard compact surface of genus g, then $\pi_1(Q)$ is a group with $2g$ generators $a_1, b_1, \ldots a_g, b_g$ and a single relation

$$[a_1, b_1][a_2, b_2] \ldots [a_g, b_g] = 1,$$

where $[a, b]$ denotes the commutator $aba^{-1}b^{-1}$.

There is a precise connection between the fundamental group and coverings, at least for spaces with a 'reasonable' local structure. Following Hatcher (2002), we can take 'reasonable' to be 'locally path connected and semi-locally simply connected', but such technicalities are irrelevant for our purposes, since we are concerned with surfaces where the local structure is just that of a disc. So we shall state our result just for surfaces, although it is valid much more generally.

Proposition 9. *Let Q be a connected surface and q_0 a base point in Q. There is a one-to-one correspondence between:*

- *equivalence classes of coverings $F : P \to Q$ where P is connected;*
- *conjugacy classes of subgroups of $\pi_1(Q, q_0)$.*

Here, coverings $F : P \to Q$ and $F' : P' \to Q$ are equivalent if there is a homeomorphism $g : P \to P'$ such that $F = F' \circ g$. The correspondence is realised in the following way. Any map $F : P \to Q$ induces a homomorphism of fundamental groups $F_* : \pi_1(P, p_0) \to \pi_1(Q, F(p_0))$. The subgroup corresponding to a covering $F : P \to Q$ is the image of $F_*(\pi_1(P, p_0)) \to \pi_1(Q, q_0)$ for any choice of $p_0 \in F^{-1}(q_0)$. Different choices of p_0 change the subgroup by conjugation.

To construct the covering corresponding to a subgroup of $\pi_1(Q)$, we can begin with the case of the trivial subgroup. The covering $G : \tilde{Q} \to Q$ which corresponds to this is characterised by the property that $\pi_1(\tilde{Q})$ is trivial. We define $\tilde{Q}$, as a set, to be the pairs (q, A), where q is a point in Q and A is a homotopy class of paths in Q from q_0 to q. Then we define $G(q, A) = q$. We refer to standard textbooks, (e.g. Hatcher 2002) for the details of how to put a topology on $\tilde{Q}$ such that G is a covering map. We also have an action of $\pi_1(Q, q_0)$ on $\tilde{Q}$ given by concatenating a path with a loop based at q_0, and

$$Q = \tilde{Q}/\pi_1(Q, q_0).$$

Granted this, for any subgroup $H \subset \pi_1(Q, q_0)$ we define the associated covering space S_H to be $\tilde{Q}/H$. We then have a factorisation of the universal covering as

$$\tilde{Q} \to S_H = \tilde{Q}/H \to Q = \tilde{Q}/\pi_1(Q, q_0).$$

The basic idea required in the proofs of the assertions above is that of *lifting* of paths, and homotopies of paths. Let $F : P \to Q$ be any map and let $\gamma : [0, 1] \to Q$ be a path. Given a point $p_0 \in P$ with $F(p_0) = \gamma(0)$, a lift of γ starting at p_0 is just a path $\tilde{\gamma} : [0, 1] \to P$ with $F \circ \tilde{\gamma} = \gamma$ and $\tilde{\gamma}(0) = p_0$.

Proposition 10.

1. *If $F : P \to Q$ is a local homeomorphism, then a path lift (with a given initial point) is unique, if it exists. If F is a covering map, then path lifts (with a given initial point) always exist.*
2. *If $F : P \to Q$ is a local homeomorphism and γ_0, γ_1 are paths in Q with the same end points which are homotopic (through maps with fixed end points) through liftable paths, with lifts $\tilde{\gamma}_s$ having the same initial point in P, then $\tilde{\gamma}_s(1) = \tilde{\gamma}_0(1)$ for all $s \in [0, 1]$.*

4.2.2   Monodromy of covering maps

Now let us return to a proper holomorphic map $F : X \to Y$ of connected Riemann surfaces, with degree $d \geq 1$. It follows immediately from Lemma 2 that the restriction of F to $X \setminus R$ is a local homeomorphism. This restriction need not be a proper map, but if we set $R^+ = F^{-1}(\Delta) = F^{-1}(F(R))$ then the restriction of F to $X \setminus R^+$ is proper, as one can easily check from the definition. So we have a covering map

$$F : X \setminus R^+ \to Y \setminus \Delta.$$

This is classified by a subgroup $H \subset \pi_1(Y \setminus \Delta)$ (or, more precisely, a conjugacy class of subgroups). There is another way to think about this algebrotopological data, which we will generally prefer to use. It follows from the definitions that H has finite index in $\pi_1(Y \setminus \Delta)$; indeed, the index is just the number of sheets of the cover, which is the degree d.

In general, the subgroups of index d in a group π correspond to transitive permutation representations. More precisely, suppose we have a set F of d elements and a choice of one element $f_0 \in F$. Then, given an action of π on F (which is the same as a homomorphism from π to the group of permutations $S(F)$), the stabiliser of f_0 is a subgroup of π. If the action is transitive, the stabiliser has index d. Conversely, if we have a subgroup H of index, then π acts transitively on the set of cosets π/H, which has d elements, and H is the stabiliser of the coset of the identity. Changing the choice of preferred point $f_0 \in F$ just changes the stabiliser by conjugation. Of course, we can always take our finite set F to be the standard set $\{1, \ldots, d\}$, so what we are talking about are representations of π in the symmetric group $\mathcal{S}_d$. Returning to our situation, our proper holomorphic map F yields a transitive representation $\rho : \pi_1(Y \setminus \Delta) \to \mathcal{S}_d$, determined up to conjugacy. This is the *monodromy* of the covering, and we can give a more intuitive description of it as follows. Suppose we have a loop γ in $Y \setminus \Delta$ beginning and ending at y_0. We label the points in the set $F^{-1}(y_0)$ by $1, \ldots, d$. Now we move around the loop γ and 'transport' the points, with their labelling, in $F^{-1}(\gamma(t))$ continuously around in X. When we return to y_0, we recover the same set $F^{-1}(y_0)$ but the labelling may have changed. This change is given by a permutation in $\mathcal{S}_d$ which is $\rho([\gamma])$, where $[\gamma]$ denotes the homotopy class of the loop γ.

This point of view is close to that traditionally adopted when introducing Riemann surfaces: that of regarding them as formed from sheets over domains in $\mathbf{C}$ joined along 'cuts'. For a very simple example, consider the Riemann surface X defined by the equation $w^2 = f(z)$, where

$$f(z) = (z - z_1)(z - z_2) \ldots (z - z_{2n}),$$

and where $F : X \to \mathbf{C}$ is the projection to the z factor. Then $\Delta = \{z_1, \ldots, z_{2n}\}$ and $\pi_1(\mathbf{C} \setminus \Delta)$ is generated by $2n$ loops $\gamma_1, \ldots, \gamma_{2n}$, where γ_i is a standard loop going once around z_i. The degree d is 2, and the representation ρ maps each generator γ_i to the non-trivial element of $\mathcal{S}_2$ (a transposition of the two objects). In traditional language, we make cuts along n disjoint paths joining z_{2i-1} to z_{2i} for $i = 1, \ldots, n$. Then we take two copies of the cut plane and form $X \setminus R$ by gluing these along the cuts. More generally, we can express the procedure as saying that we make cuts so that ρ becomes trivial on π_1 of the cut plane, and then ρ is just the combinatorial data required to specify the gluing along the cuts.

We now summarise our work so far. Starting with a proper, non-constant, holomorphic map between connected Riemann surfaces X, Y, we get a degree d, a discrete set $\Delta \subset Y$ and a transitive permutation representation (up to conjugacy) $\rho : \pi_1(Y \setminus \Delta) \to \mathcal{S}_d$. The next result, *Riemann's Existence Theorem*, shows that we can go in the other direction.

Theorem 2. *Let Y be a connected Riemann surface and Δ a discrete subset of Y. Given $d \geq 1$ and a transitive permutation representation $\rho : \pi_1(Y \setminus \Delta) \to \mathcal{S}_d$, there is a connected Riemann surface X and a proper holomorphic map $F : X \to Y$ which realises ρ as its monodromy homomorphism. Moreover, X and F are unique up to equivalence.*

First, the theory of covering spaces recalled above gives us a covering map $F_0 : X_0 \to Y \setminus \Delta$. It is easy to see that there is a unique way to make X_0 into a Riemann surface such that the map is holomorphic. At the end of the proof, the Riemann surface X_0 will correspond, of course, to $X \setminus R^+$, so what we need to see is how to 'fill in' the points of R^+ lying over Δ. Let y_1 be a point of Δ, and choose a small disc D_1 in Y about y_1, not containing any other points of Δ. The boundary of the disc D_1 defines an element of $\pi_1(Y \setminus \Delta)$ (or, more precisely, a conjugacy class). The homomorphism ρ maps this to a permutation σ of $(1, \ldots, d)$. Now σ may not act transitively on $(1, \ldots, d)$. This corresponds to the fact that $F_0^{-1}(D_1 \setminus \{y_1\})$ may not be connected. If we write σ as a product of disjoint cycles, then it is easy to see from the definitions that the cycles naturally correspond to the components of $F_0^{-1}(D_1 \setminus \{y_1\})$. Thus if Z is one such connected component, corresponding to a cycle of length d', the restriction of F_0 to Z gives a connected covering of $D_1 \setminus \{y_1\}$, determined by the homomorphism which maps the generator of $\pi_1(D_1 \setminus \{y_1\})$ to the d'-cycle in $\mathcal{S}_{d'}$. But we know a covering which realises this data: if we identify $D_1 \setminus \{y_1\}$ with the standard punctured unit disc $D^* \subset \mathbf{C}$, it is given by the map $z \mapsto z^{d'}$ from D^* to D^*. So we conclude that Z is equivalent as a Riemann surface to D^* by an isomorphism $\phi : D^* \to Z$, say.

We now define a set X^+ by

$$X^+ = X_0 \cup_\phi D,$$

where D is the unit disc in $\mathbf{C}$ and the notation means that we identify $z \in D^* \subset D$ with $\phi(z) \in Z \subset X_0$. We make X^+ into a Riemann surface as follows. We take an atlas of charts in X^+ to be an atlas for X_0 with one further chart, the inverse of the obvious map from D to X^+ arising from the definition. There is then a unique way to introduce a topology on X^+ making all these charts homomorphisms, but we have to check that this topology is Hausdorff, i.e. that for any two points a, b in X^+ there are disjoint open sets U, V containing a, b, respectively. If a and b lie in the copy of X_0 in X^+, this is clear: we just take corresponding open sets in X_0. So suppose a lies in X_0 and b is the point corresponding to 0 in D. Then $F_0(a)$ is not equal to y_1 in Y, so we can find a small neighbourhood N of $F_0(a)$ in $Y \setminus \Delta$ which is disjoint from a smaller disc $D_2 \subset D_1$ containing y_1. The open sets $F_0^{-1}(N)$ and

$$\{0\} \cup \phi^{-1}(F_0^{-1}(D_2))$$

are disjoint in X^+ and contain b, a, respectively. This shows that X^+ is indeed a Riemann surface. Moreover, the map F_0 obviously extends to a holomorphic map from X^+ to Y. (The point of dealing carefully with the Hausdorff condition here is this. Suppose W is any Riemann surface and $\phi : D^* \to W$ is a holomorphic map. Then we can form the set $W \cup_\phi D^*$ as above and equip it with charts, but in general the resulting space may not be Hausdorff—it will fail to be Hausdorff precisely when ϕ extends to a holomorphic map from D to W—that is, when the 'new' point we are trying to add in was already there!)

We repeat the procedure above for each point of Δ and for each cycle of the corresponding monodromy. This gives us a Riemann surface X with a holomorphic map F to Y, and we check that this map is indeed proper.

4.2.3 Compactifying algebraic curves

This construction has an important application to algebraic curves. Suppose $P(z, w)$ is a polynomial in two complex variables, and consider again the set X of solutions to the equation $P(z, w) = 0$ in $\mathbf{C}^2$ with the projection map $\pi : X \to \mathbf{C}$ onto the z-factor. Suppose that P is an *irreducible* polynomial, i.e. it cannot be written as $P = QR$ for non-constant polynomials Q, R. We also exclude the rather trivial case when P is a linear polynomial $z - z_0$. We use two facts whose proof we postpone until Chapter 11:

- There are only finitely points (z, w) where P and $\partial P / \partial w$ both vanish.
- The set X is connected.

Let S be the set of points in X where both partial derivatives P_z, P_w vanish. Clearly, the first fact above implies that S is finite. The proof in Subsection 3.2.2 shows that the complement $X \setminus S$ is a Riemann surface. Now we let F be the finite subset of $\mathbf{C}$ defined by those values of z for which the term in P of highest degree in w (a polynomial in z) vanishes. We put

$$S^+ = \pi^{-1}(\pi(S) \cup F) \subset X.$$

The set S^+ is again finite, for there are only finitely many points in $\pi(S) \cup F$, and if z_0 is such a point the set $\pi^{-1}(z_0)$ consists of the roots of the polynomial equation $P(z_0, w) = 0$ in the single variable w. This has only finitely many roots unless the polynomial vanishes identically, which would imply that $(z - z_0)$ divides P. Now let E be the discrete subset $\pi(S) \cup F \cup \{\infty\}$ of the Riemann sphere S^2. We get a proper holomorphic map

$$\pi : X \setminus S^+ \to S^2 \setminus E.$$

Applying the theory above, we have a set of critical values $\Delta \subset S^2 \setminus E$ and we can recover $X \setminus S^+$ from the monodromy homomorphism $\rho : \pi_1(S^2 \setminus (\Delta \cup E)) \to \mathcal{S}_d$. On the other hand, this data also defines a *compact* Riemann surface X^*, containing $X \setminus S^+$ as a dense open set mapping holomorphically to S^2. The fact that X is connected implies the same for X^*.

   Now recall that, on the other hand, we have a compact set $\overline{X} \subset \mathbf{CP}^2$ defined by the homogeneous polynomial corresponding to P. This contains X and hence $X \setminus S^+$, again as dense open sets.

Proposition 11. *The inclusion of $X \setminus S^+$ in $\overline{X}$ extends to a holomorphic map from X^* to $\mathbf{CP}^2$, mapping onto $\overline{X}$.*

   Here a holomorphic map from a Riemann surface to $\mathbf{CP}^2$ is defined in the obvious way, as a continuous map which is holomorphic with respect to the three charts $W_i \equiv \mathbf{C}^2$ covering $\mathbf{CP}^2$.

   To prove Proposition 11, it suffices to work in the affine plane $\mathbf{C}^2$. What we need to show is that when we attach discs to $X \setminus S^+$, the inclusion of the punctured disc extends meromorphically over 0. Thus the proposition boils down to the following.

Lemma 4. *Suppose P is an irreducible polynomial in two variables and n is a positive integer. Suppose f is a holomorphic function on the punctured disc $D \setminus \{0\}$ with $P(z^n, f(z)) = 0$ for all $z \in D \setminus \{0\}$. Then f is a meromorphic function.*

The irreducibility of P implies, as above, that there are only finitely many roots $w_1, \ldots w_N$, say, of the equation $P(0, w) = 0$. Thus when $|z|$ is small, $f(z)$ must be close to one of the w_i. We recall a result from complex analysis: if f has an essential singularity at 0, then for all w in $\mathbf{C}$ and $\epsilon, \delta > 0$ there is a z with $0 < |z| < \delta$ and $|f(z) - w| < \epsilon$. This clearly does not hold in our case when w is not one of the w_i and ϵ, δ are sufficiently small. Thus f is meromorphic, as asserted.

Given the above lemma, we know that for suitable m, $z^m f(z)$ is holomorphic and non-vanishing for small z. Then the map $z \mapsto [z^n, f(z), 1]$ from the punctured disc to $\mathbf{CP}^2$ is equal to the map $z \mapsto [z^{n+m}, z^m f(z), z^m]$ and extends to a holomorphic map of the disc to $\mathbf{CP}^2$.

The conclusion is that we have associated a compact connected Riemann surface X^* to any irreducible polynomial. This is called the 'normalisation' of the projective curve $\overline{X}$.

Some examples are given below:

1. Suppose P is the polynomial $w^2 - z^2(1 - z)$. Both partial derivatives vanish at the point $(0, 0)$. To help our intuition, we can sketch the corresponding real curve as in Figure 4.1. The origin is a singular point, where two branches of the curve cross. Following through the constructions, one finds that X^* is equivalent to the Riemann sphere and the map from X^* to $\mathbf{CP}^2$ is given by $\tau \mapsto [1, 1 - \tau^2, \tau - \tau^3]$. (The point at infinity in S^2 maps to $[0, 0, 1]$.) Thus we obtain X^* by separating the two branches of the curve passing through the origin.

2. Let P be the polynomial $w^2 - z^3$. The real curve looks as shown in Figure 4.2. It has a 'cusp' singularity at the origin. Again the normalisation is the Riemann sphere, with the map $\tau \mapsto [1, \tau^2, \tau^3]$.

There are more algebraic constructions of the normalisation. We outline one approach here, using *Puiseaux expansions*. Suppose that $(0, 0)$ is a point in the set S, that is to say that P and both its partial derivatives vanish there. So,

$$P(z, w) = \sum_{m,n} p_{m,n} z^m w^n$$

and $p_{m,n} = 0$ if $m + n \leq 1$. From the topological point of view, which we have adopted, if $\epsilon \in \mathbf{C}$ is small but not zero there are a certain number d_1 of points in $F^{-1}(\epsilon)$ 'close to' the origin, and as ϵ moves around a small circle we get a monodromy which is a permutation of these d_1 points. Constructing the normalisation boils down to constructing holomorphic maps from a small disc into the curve X, mapping 0 to the origin in one-to-one correspondence with cycles in the monodromy. Suppose first that we have a case when the

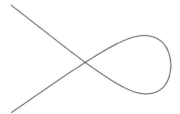

Figure 4.1 *Curve with self-intersection*

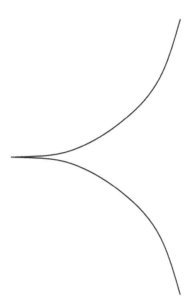

Figure 4.2 *A cusp*

monodromy is trivial. Then the composites of these maps with projection to the z-co-ordinate are injective, and we can suppose the maps are given by $z \mapsto (z, f_i(z))$ for holomorphic functions $f_i(z)$ and $i = 1, \ldots d_1$. Algebraically, what this means is that we can factorise P as

$$P(z, w) = (w - f_1(z)) \ldots (w - f_{d_1}(z)) Q(z, w),$$

and Q does not vanish at $(0, 0)$. More generally, suppose that the monodromy contains a cycle of length a. Then we set $z = \zeta^a$ and we seek a factor of $P(\zeta^a, w)$ of the form $w - f(\zeta)$. For $b < a$, let $f_b(\zeta) = f(\theta^b \zeta)$, where $\theta = e^{2\pi i/a}$ is a root of unity. Then it follows that if $w - f(\zeta)$ is a factor then so are all the $w - f_b(\zeta)$, so we seek to find a factorisation of the form

$$P(\zeta^a, w) = (w - f(\zeta)) \ldots (w - f_1(\zeta)) \ldots (w - f_{a-1}(\zeta)) Q(\zeta, w).$$

We could also write this in a different notation as

$$P(z, w) = (w - f(z^{1/a})) \ldots (w - f_{a-1})(z^{1/a}) Q(z^{1/a}, w).$$

In the general case, the idea then is that we adjoin formal variables $z^{1/\nu}$; for a suitable ν, we arrive at a factorisation

$$P(z, w) = \prod_{j=1}^{d_1} (w - f_j(z^{1/\nu})) Q(z^{1/\nu}, w).$$

For a very simple example of this, suppose $P(z, w) = w^2 - z^3$. The monodromy is a single 2-cycle. Adjoining $z^{1/2}$, we can factor $P(z, w)$ as

$$P(z, w) = (w - z^{3/2})(w + z^{3/2}).$$

The two factors correspond to the same disc, which is $\zeta \mapsto (\zeta^2, \zeta^3)$.

There is a systematic theory, using 'Newton diagrams', for finding the 'branches' of an algebraic function; see Kirwan (1992), Section 7.2.

4.2.4   The Riemann surface of a holomorphic function

Throughout this chapter we have emphasised the case of 'proper' maps, which arise most naturally when one considers algebraic functions. We will say a little more now about the general case. Suppose $F : X \to Y$ is a holomorphic map between connected Riemann surfaces, without branch points, and let $\Psi : X \to \mathbf{C}$ be any holomorphic function. Then F is a local homeomorphism; so, given a point $x_0 \in X$, we can define a holomorphic function $\psi_0 = \Psi \circ F_{x_0}^{-1}$ on a neighbourhood of $y_0 = F(x_0)$ in Y, where $F_{x_0}^{-1}$ denotes a local inverse mapping y_0 to x_0. More generally, given a path $\gamma : [0, 1] \to Y$, a lift $\tilde{\gamma}$ of γ starting at x_0 defines an analytic continuation of ψ_0 along γ (as defined in Chapter 1).

The converse to this construction is expressed in the following proposition.

Proposition 12. *Given a point y_0 in a connected Riemann surface Y and a holomorphic function ψ_0 defined on a neighbourhood of y_0, there are then a Riemann surface X, a holomorphic map $F : X \to Y$ without branch points, a point x_0 in $F^{-1}(y_0)$ and a holomorphic function Ψ on X such that:*

- *ψ_0 can be analytically continued along a path γ in Y if and only if γ has a lift to X starting at x_0.*
- *The analytic continuation of ψ_0 along γ has ψ_1 equal to $\Psi \circ F_{\tilde{\gamma}}^{-1}$ in a neighbourhood of $\gamma(1)$.*

The Riemann surface X is the 'Riemann surface of the function germ ψ_0', and the data about all possible analytic continuations of ψ_0 is encoded in the holomorphic map $F : X \to Y$.

The construction of the Riemann surface X associated to ψ_0 is a variant of the construction of covering spaces. We define an equivalence relation on analytic continuations of ψ_0 as follows. Let γ, γ' be paths starting at y_0. Recall that analytic continuations are given by one-parameter families of holomorphic functions ψ_t, ψ_t', say. We say these are equivalent if $\gamma(1) = \gamma'(1)$ and $\psi_1 = \psi'(1)$ on some neighbourhood of this point in Y. Now we define X to be the set of equivalence classes of analytic continuations of ψ_0, with a map $F : X \to Y$ induced by $(\gamma_t, \psi_t) \mapsto \gamma(1)$. We define $\Psi : X \to \mathbf{C}$ to be the map induced by $(\gamma_t, \psi_t) \mapsto \psi_1(\gamma(1))$. Then we have to check that there is a natural (and in fact unique) way to introduce a Riemann surface structure on X such that F and Ψ are holomorphic and F is a local homeomorphism. We leave it to the interested reader to work out the details or to consult suitable texts.

This construction could be generalised. In one direction, we can replace holomorphic functions by maps to a third Riemann surface Z. This makes essentially no difference. In the other direction, we can sometimes extend X by including suitable ramification points, as we did in the case of proper maps.

4.2.5 Quotients

In Chapter 3, we saw that if Γ is any discrete subgroup of $PSL(2, \mathbf{R})$, the quotient H/Γ has a natural Riemann surface structure, even if the action is not free. We can now relate this phenomenon to our analysis of holomorphic maps. In this situation, the natural map from H to H/Γ is holomorphic but has branch points at those points which have a non-trivial stabiliser. In a similar way, let G be a finite group and Σ a compact Riemann surface. (These hypotheses can be relaxed, but simplify our discussion.) Then we make Σ/G into a Riemann surface as follows. Let p be a point of Σ and let $Z \subset G$ be the stabiliser of p. Then Z acts on the *tangent space* of Σ at p (a notion which, strictly, we do not define until the next chapter but which we take as known here). This gives a homomorphism from Z to a finite cyclic subgroup of S^1, of order r, say. So, for $g \in Z$, we have an $\alpha(g) \in \{0, 1, \ldots, r - 1\}$ such that the derivative of g at z is $\exp(2\pi i \alpha(g)/r)$. Now Z acts on the holomorphic functions defined near p. Let z be any local co-ordinate, and set

$$w = \sum_{g \in Z} \exp(-2\pi i \alpha(g)/r) g(z).$$

Then w is another local co-ordinate and, by construction, $g(w) = \exp(2\pi\alpha(g)/r)w$. In other words, we have 'linearised' the action of Z near to p. If $[p]$ denotes the equivalence class of p in the quotient Σ/G, then a neighbourhood of $[p]$ in Σ/G can be identified with the quotient of a disc by the standard action of the cyclic group of order r. Just as before, the map $w \mapsto w^r$ identifies this quotient with a disc, and we get an atlas of charts making X/G into a Riemann surface. The natural map from X to X/G is a branched covering map.

For a very simple example of this, let G be the cyclic group of order r acting on the Riemann sphere $\mathbf{C} \cup \{\infty\}$ by multiplication. So, the stabiliser of any $z \neq 0, \infty$ is trivial, while the stabiliser of both 0 and ∞ is the full group G. The quotient is again the Riemann sphere. For another kind of example, consider a discrete group $\Gamma \subset PSL(2, \mathbf{R})$ and a normal subgroup $\Gamma_0 \subset \Gamma$. Then the quotient group G acts on $\Sigma = H/\Gamma_0$ and Σ/G is naturally identified with H/Γ, and thus (when Σ is compact and G is finite) we are in the situation considered above.

Exercises

1. Suppose $f : \Sigma_1 \to \Sigma_2$ and $g : \Sigma_2 \to \Sigma_3$ are non-constant holomorphic maps between compact connected Riemann surfaces. Show that the degree of $g \circ f$ is the product $\deg f \cdot \deg g$.
2. Show that a proper local homeomorphism is the same thing as a finite covering map.
3. (For those with some knowledge of differential geometry.) Suppose that $S \subset \mathbf{R}^3$ is a *minimal surface*, i.e. the mean curvature of S is zero. Show that with suitable orientation conventions, the Gauss map, from S to the 2-sphere, is holomorphic.
4. Let $p(z)$ be a polynomial of degree d with distinct roots and let X be the compact Riemann surface associated to the functions $w^2 = p(z)$. Show that if $d = 3$, the natural map from X to $\mathbf{CP}^2$ has an image that is a smooth projective curve, but that this is not true for $d > 3$.
5. Let G be the group of automorphisms of the Riemann sphere generated by the maps $z \mapsto -z$ and $z \mapsto 1/z$. Show that the quotient is isomorphic to the Riemann sphere, and identify the critical points, critical values and multiplicities of the map $\mathbf{CP}^1 \to \mathbf{CP}^1/G$
6. Show that the meromorphic functions on the Riemann sphere have the form $p(z)/q(z)$, where p, q are coprime polynomials.
7. Let $\alpha : \mathbf{CP}^1 \to \mathbf{CP}^1$ be the map $z \mapsto z^{-1}$. Then the group $G = \mathbf{Z}/2 \times \mathbf{Z}/2$ acts on $\mathbf{CP}^1 \times \mathbf{CP}^1$ with generators $(\alpha, 1), (1, \alpha)$. We say that a subset $Z \subset \mathbf{CP}^1 \times \mathbf{CP}^1$ is a smooth algebraic curve if for each $g \in G$, the intersection of gZ with $\mathbf{C} \times \mathbf{C} \subset \mathbf{CP}^1 \times \mathbf{CP}^1$ is a smooth affine curve in the sense considered above. Find the condition on a polynomial $P(z, w)$ for the corresponding affine curve to extend to a smooth algebraic curve in $\mathbf{CP}^1 \times \mathbf{CP}^1$. Give examples to show that this is not the same as the condition of extending to a smooth algebraic curve in $\mathbf{CP}^2$.

5 Calculus on surfaces

In this chapter, we develop the theory of *differential forms* on smooth surfaces and Riemann surfaces. This will be our main technical tool in the proofs of the major structural results in the following sections. Most of the material will be familiar to readers who have taken a standard course on manifold theory, but for those who have not, we attempt to give a self-contained treatment.

5.1 Smooth surfaces

5.1.1 Cotangent spaces and 1-forms

Lemma 5. *Let f be a smooth, real-valued function defined on an open neighbourhood U of 0 in $\mathbf{R}^2$ and let $\gamma_1 : (-\epsilon_1, \epsilon_1) \to U$ and $\gamma_2 : (-\epsilon_2, \epsilon_2) \to U$ be smooth maps (for some $\epsilon_1, \epsilon_2 > 0$), with $\gamma_1(0) = \gamma_2(0) = 0$. Let $\chi : U \to V$ be a diffeomorphism to another open set in $\mathbf{R}^2$ with $\chi(0) = 0$. Set $\tilde{\gamma}_i = \chi \circ \gamma_i$ and $\tilde{f} = f \circ \chi^{-1}$.*

- *If both partial derivatives $\partial f / \partial x_1, \partial f / \partial x_2$ vanish at 0, then the same is true of $\tilde{f}$.*
- *If the derivatives $d\gamma_i / dt$ at $0 \in \mathbf{R}$ are equal, then the same is true of $\tilde{\gamma}_i$.*

The assertions here follow immediately from the chain rule for partial derivatives. Now let S be a smooth surface, p be a point in S, f be a smooth function on S and $\gamma_i : (-\epsilon_i, \epsilon_i) \to S$ $(i = 1, 2)$ be a pair of smooth paths with $\gamma_i(0) = p$. We say that f is *constant to first order* at p if the derivative of the function representing f in a local co-ordinate chart about p vanishes at the point corresponding to p. By the first item of the lemma, this notion is independent of the choice of co-ordinate chart. Similarly, we say that γ_1, γ_2 are equal to the first order at p if the derivatives of the paths representing them in a local chart are equal.

Definition 7.

- *The tangent space TS_p of S at p is the set of equivalence classes of maps $\gamma : (-\epsilon, \epsilon) \to S$ with $\gamma(0) = p$ under the equivalence relation $\gamma_1 \sim \gamma_2$ if γ_1 and γ_2 are equal to first order at p.*
- *The (real) cotangent space T^*S_p is the set of equivalence classes of smooth functions on an open neighbourhood of p in S under the equivalence relation $f_1 \sim f_2$ if $f_1 - f_2$ is constant to first order at p.*

The cotangent space T^*S_p has a natural vector space structure induced from that on the smooth functions on open neighbourhoods of p. From the definition, if U is an open neighbourhood of p, there is a map from $C^\infty(U)$ to T^*S_p, which we denote by

$$f \mapsto [df]_p.$$

Let x_1, x_2 be local co-ordinates about p. They are smooth functions, so we have elements $[dx_1]_p, [dx_2]_p \in T^*S_p$. If f is any smooth function on a neighbourhood of p, we write $f = f(x_1, x_2)$, using the usual notation suppressing explicit dependence on the co-ordinate charts. Then the reader can readily verify, from the definition, that

$$[df]_p = \frac{\partial f}{\partial x_1}[dx_1]_p + \frac{\partial f}{\partial x_2}[dx_2]_p.$$

One sees from this that $[dx_1]_p, [dx_2]_p$ form a basis for the vector space T^*S_p. If $\gamma : (-\epsilon, \epsilon) \to S$ is a smooth path with $\gamma(0) = 0$ and f is a function on a neighbourhood of p, the composite $f \circ \gamma$ is defined, as a real-valued function, on some possibly smaller interval, and one can check that the derivative is independent of the choice of f or γ in the equivalence classes defining the tangent space and cotangent space. Thus this derivative induces a map

$$TS_p \times T^*S_p \to \mathbf{R}.$$

Again we leave it to the reader to check that there is a unique vector space structure on TS_p with respect to which this is a bilinear map. In fact, the map induces a duality between TS, T^*S—there is a canonical isomorphism

$$T^*S_p \cong Hom(TS_p, \mathbf{R}).$$

Now we define the *cotangent bundle* T^*S to be

$$T^*S = \bigcup_{p \in S} T^*S_p.$$

A smooth 1-form α on S is a map $\alpha : S \to T^*S$ with $\alpha(p) \in T^*S_p$ for all $p \in S$ and which varies smoothly with p in the following sense. In local co-ordinates (x_1, x_2) about a point p_0, we can write

$$\alpha = \alpha_1 \, dx_1 + \alpha_2 \, dx_2,$$

where α_1, α_2 are functions of x_1, x_2, and we have dropped the $[\ ]_p$ from our notation. Then we require that α_1, α_2 are smooth functions of the local co-ordinates. Again, one needs to check that this is notion is independent of the co-ordinate system. We do this in detail because it gives us an opportunity to illustrate how to compute with these forms. Suppose y_1, y_2 is another system of local co-ordinates, so x_1, x_2 are smooth functions of y_1, y_2 with partial derivatives $\partial x_i / \partial y_j$. Then, by the chain rule, applying the definition, we have

$$dx_i = \frac{\partial x_i}{\partial y_1} dy_1 + \frac{\partial x_i}{\partial y_2} dy_2.$$

Thus if α is represented locally by $\alpha_1(x_1, x_2) \, dx_1 + \alpha_2(x_1, x_2) \, dx_2$ in the x_1, x_2 co-ordinates, it is represented by

$$\alpha_1(x_1(y_1, y_2), x_2(y_1, y_2)) \left(\frac{\partial x_1}{\partial y_1} dy_1 + \frac{\partial x_1}{\partial y_2} dy_2 \right)$$

$$+ \alpha_2(x_1(y_1, y_2), x_2(y_1, y_2)) \left(\frac{\partial x_2}{\partial y_1} dy_1 + \frac{\partial x_2}{\partial y_2} dy_2 \right) \tag{5.1}$$

in the y_1, y_2 co-ordinates. The coefficients of dy_1, dy_2 are obviously smooth functions of y_1, y_2, as required. An example of a smooth 1-form is furnished by the derivative of a function. If f is a function on S, then we define a 1-form df by

$$df(p) = [df]_p.$$

In local co-ordinates, this is just

$$df = \frac{\partial f}{\partial x_1} dx_1 + \frac{\partial f}{\partial x_2} dx_2.$$

There is another important notion, which leads to the same formula (5.1). Suppose $F : S \to Q$ is a smooth map between surfaces. Then a moment's thought about the definitions will show the reader how to define, for any $p \in S$, natural linear maps

$$dF : TS_p \to TQ_{F(p)},$$

$$dF_p^* : T^* Q_{F(p)} \to T^* S_p,$$

compatible with the dual pairings between tangent and cotangent spaces. Suppose α is a smooth 1-form on Q. Then we define the pull-back form $F^*(\alpha)$ by

$$F^*(\alpha)(p) = dF^*(\alpha(F(p)).$$

Then $F^*(\alpha)$ is a smooth 1-form on S. If, now, x_1, x_2 are local co-ordinates about $F(p)$ in Q and y_1, y_2 are local co-ordinates about p in S, the formula (5.1) gives the local representation of $F^*(\alpha)$, where F is locally represented by the functions $x_i(y_j)$.

We now sum up the work so far. Write Ω_S^0 for the smooth functions on S and Ω_S^1 for the smooth 1-forms. Then we have defined

$$d : \Omega_S^0 \to \Omega_S^1,$$

with the following properties:

- $d(fg) = f\,dg + g\,df$, where fg denotes the pointwise product;
- if $F : S \to Q$ is smooth, then $d(F^* f) = F^*(df)$, where $f \in \Omega_Q^0$ and $F^*(f) = f \circ F$.

A difficulty with 1-forms is, perhaps, that they do not have a very obvious geometric meaning, and it takes time on first encountering the notion to become comfortable working with them. (The dual notion, of a *vector field*, which we will discuss briefly below, is probably more intuitively appealing.) One important property is that 1-forms are the objects which can naturally be integrated over one-dimensional sets. There are two slightly different notions here. One is to work with smooth paths

$$\gamma : [0, 1] \to S.$$

Suppose first that the image of γ lies inside some local co-ordinate system x_1, x_2. If α is a 1-form on S, we define

$$\int_\gamma \alpha = \int_0^1 \alpha_1 \frac{d\gamma_1}{dt} + \alpha_2 \frac{d\gamma_2}{dt}. \tag{5.2}$$

Here $\gamma_1(t), \gamma_2(t)$ are the x_1 and x_2 co-ordinates of the local representation of γ, and α_1, α_2 are the co-efficients of dx_1, dx_2 in the local represenion of α. Thus, in a less compressed notation, we would write $\alpha_i(\gamma_1(t)\gamma_2(t))$ in equation (5.2). One can check that this is independent of the local co-ordinate system. Thus, if the image of γ does not lie in a single chart, one can define the integral by breaking up $[0, 1]$ into subintervals and proceeding in the obvious way. The integral has another invariance property. Suppose $\psi : [0, 1] \to [0, 1]$ maps 0 to 0 and 1 to 1. Then we get another smooth path $\gamma \circ \psi$, and

$$\int_{\gamma \circ \psi} \alpha = \int_\gamma \alpha.$$

Essentially, this expresses the fact that the integral depends only on the image of γ. (A more sophisticated way to formulate these definitions is to introduce the notion of a 1-form on an interval in $\mathbf{R}$—or indeed on any smooth manifold—and the integral of a 1-form over an interval. Then the map γ gives a pull-back form $\gamma^*(\alpha)$, which we integrate over $[0, 1]$.)

Now suppose that C is an *oriented embedded curve* in a surface S. Then we can define the integral of α over C by decomposing C into pieces which can be parametrised by smooth paths as above. All of this is essentially the same as the definition of contour integrals in elementary complex analysis, so we will not dwell on the details.

5.1.2   2-forms and integration

Next we want to define smooth 2-forms on a surface. To motivate the definitions here, we consider the following question. Given a 1-form α on a surface S, when can it be written as $\alpha = df$ for some function f? Consider first the case when S is $\mathbf{R}^2$, so

$$\alpha = \alpha_1 \, dx_1 + \alpha_2 \, dx_2,$$

where α_1, α_2 are arbitrary smooth functions of x_1, x_2. Writing $\alpha = df$ means finding a function f with

$$\frac{\partial f}{\partial x_1} = \alpha_1, \quad \frac{\partial f}{\partial x_2} = \alpha_2.$$

The symmetry of second partial derivatives means that an obvious necessary condition is that the function

$$R = \frac{\partial \alpha_1}{\partial x_2} - \frac{\partial \alpha_2}{\partial x_1}$$

vanishes everywhere. The converse is a classical result (the 'criterion for an exact differential') which the reader has very likely encountered: if $R = 0$, then we can find an f. Let us recall the proof. Given α_1, α_2, we define functions f_1, f_2 by

$$f_2(x_1, x_2) = \int_0^{x_1} \alpha_1(t, 0) \, dt + \int_0^{x_2} \alpha_2(x_1, t) \, dt,$$

$$f_1(x_1, x_2) = \int_0^{x_2} \alpha_2(0, t) \, dt + \int_0^{x_1} \alpha_1(t, x_2) \, dt.$$

By construction, $\partial f_i / \partial x_i = \alpha_i$. But *Stokes' Theorem*, applied to a rectangle V with vertices at $(0, 0), (x_1, 0), (0, x_2), (x_1, x_2)$, shows that

$$f_1(x_1, x_2) - f_2(x_1, x_2) = \int_V R,$$

so our hypothesis that $R = 0$ shows that $f_1 = f_2$, and the proof is complete. What we see from this argument in the plane is that the three notions

- the criterion for an exact differential,
- integration over two-dimensional regions,
- Stokes' Theorem

are tightly bound together, and the definition of a 2-form is framed to allow us to extend these notions to surfaces.

Let E be a real vector space. We define $\Lambda^2 E^*$ to be the set of bilinear maps

$$B : E \times E \to \mathbf{R},$$

which are skew-symmetric, i.e. $B(e, f) = -B(f, e)$. We define a 'wedge product'

$$\wedge : E^* \times E^* \to \Lambda^2 E^*$$

by

$$(\alpha \wedge \beta)(e, f) = \alpha(e)\beta(f) - \beta(e)\alpha(f).$$

So the wedge product is linear in each variable, and $\alpha \wedge \beta = -\beta \wedge \alpha$. Now suppose that E has dimension 2; then (exercise for the reader) $\Lambda^2 E^*$ is a one-dimensional real vector space, and if α_1, α_2 is a basis for E^* the wedge product $\alpha_1 \wedge \alpha_2$ furnishes a basis element in $\Lambda^2 E^*$.

We apply this algebra to the case when $E = TS_p$, so $E^* = T^* S_p$. Thus, for each point p in a surface S, we have a one-dimensional space $\Lambda^2 T^* S_p$. If x_1, x_2 are local coordinates around p, we get a basis element $dx_1 \wedge dx_2$ for $\Lambda^2 T^* S_p$. One often omits the wedge product symbol to write this as $dx_1\, dx_2$. We now proceed as before. We define a smooth 2-form ρ on S to be a map from S to the union

$$\bigcup_{p \in S} \Lambda^2 T^* S_p$$

such that $\rho(p)$ lies in $\Lambda^2 T^* S_p$ and varies smoothly with p in the following sense. In local co-ordinates we can write

$$\rho = R(x_1, x_2)dx_1\, dx_2,$$

and we require that R be a smooth function. Applying the definitions, one finds that in a different system of co-ordinates y_1, y_2 this same 2-form is represented by

$$R(x_1(y_1, y_2), x_2(y_1, y_2))J(y_1, y_2)\, dy_1\, dy_2,$$

where

$$J(y_1, y_2) = \frac{\partial x_1}{\partial y_1}\frac{\partial x_2}{\partial y_2} - \frac{\partial x_1}{\partial y_2}\frac{\partial x_2}{\partial y_1}.$$

The reader will recognise this as the usual *Jacobian*: the determinant of the matrix of partial derivatives $\partial y_i / \partial x_j$. Again, this formula can be read in a different way. If $F : S \rightarrow Q$ is a smooth map and ρ is a 2-form on Q, there is a natural way to define a pulled-back form $F^*(\rho)$ on S, and the formula (5.1) expresses this in local co-ordinates.

We write Ω_S^2 for the set of 2-forms on a surface S.

Now these 2-forms provide a natural 'home' for the expression appearing in the critrion for an exact differential above. We have the following lemma.

Lemma 6. *There is a unique way to define an* **R***-linear map*

$$d : \Omega_S^1 \rightarrow \Omega_S^2$$

such that:

- *If $\alpha_1 = \alpha_2$ on an open set $U \subset S$, then $d\alpha_1 = d\alpha_2$ over U.*
- *If f is a function on S, then $ddf = 0$.*
- *If f is a function on S and α is a 1-form on S, then*

$$d(f\alpha) = df \wedge \alpha + f\, d\alpha.$$

To prove this, we first check the uniqueness. Suppose that we have an operator satisfying the conditions of the lemma. By the first condition, we can calculate $d\alpha$ in local co-ordinates. Then we have

$$d(\alpha_1\, dx_1 + \alpha_2\, dx_2) = d\alpha_1 \wedge dx_1 + d\alpha_2 \wedge dx_2,$$

using the second and third conditions (the second condition gives $ddx_i = 0$). Explicitly,

$$d(\alpha_1\, dx_1 + \alpha_2\, dx_2) = \left(\frac{\partial \alpha_2}{\partial x_1} - \frac{\partial \alpha_1}{\partial x_2} \right) dx_1\, dx_2.$$

On the other hand, if we take this formula as the definition of $d\alpha$, we can check that it is independent of the choice of co-ordinate system.

What we see in the course of the proof is that in local co-ordinates, $d\alpha$ is just $R\, dx_1\, dx_2$, where R is the function discussed above which enters into the criterion for an exact differential. So, we can reformulate that result as saying that for a 1-form α on the surface $S = \mathbf{R}^2$, we can find a function with $\alpha = df$ if and only if $d\alpha = 0$.

We now turn to integration. Suppose S is an *oriented* surface and ρ is a 2-form with compact support and supported in the domain of a co-ordinate chart on S. Write $\rho = R(x_1, x_2)\, dx_1\, dx_2$ in these local co-ordinates. Then we define the integral of ρ on S by the following apparently tautological formula

$$\int_S \rho = \int_{\mathbf{R}^2} R(x_1, x_2)\, dx_1\, dx_2,$$

where, on the right-hand side, we mean the ordinary Lebesgue integral on the compactly supported functions on $\mathbf{R}^2$. If y_1, y_2 is another oriented chart, then the Jacobian J relating the two is positive by definition, and the fact that we get the same value of the integral just expresses the usual transformation law for multiple integrals. To define the integral more generally, we use the following lemma, which we will also need later. (In the statement, recall that the *support* of a function is the closure of the set on which it is non-zero.)

Lemma 7. *Let K be a compact subset of a surface S and let $U_1, \ldots, U_n$ be open sets in S with $K \subset U_1 \cup \ldots U_n$. Then there are smooth non-negative functions $f_1, \ldots, f_n$ on S, each of compact support and with the support of f_i contained in U_i, such that $f_1 + \cdots + f_n = 1$ on K.*

Such a system of functions is called a *partition of unity* subordinate to the cover U_i.

To prove this, we begin with the case when $n = 1$. First, consider the very special case when $n = 1$, $S = U_1$ is the unit disc in $\mathbf{R}^2$ and K is the closed disc of radius $1/2$. Then we take a non-negative function $f(x_1, x_2) = F(\sqrt{x_1^2 + x_2^2})$, where F is a function of one variable with $F(r) = 1$ for $r \le 1/2$ and $F(r) = 0$ for $r \ge 3/4$, say.

Now consider the general case when $n = 1$. For each point $p \in K$, we take a local co-ordinate chart mapping a disc D_p about p to the open unit disc in $\mathbf{R}^2$ and the closure $\overline{D_p}$ to the closed unit disc in $\mathbf{R}^2$. Let $\frac{1}{2}D_p$ be the pre-image of the half-sized open disc. We can suppose (scaling the chart) that the closure of D_p is contained in U_1. The set of open discs $\frac{1}{2}D_p$ as p ranges over K forms an open cover of K, so we can find a finite subcover, corresponding to points $p_1, \ldots, p_N$, say. Then for each $j \le N$ we have a function, g_j say, of compact support on D_{p_j} and equal to 1 on the closure of $\frac{1}{2}D_{p_j}$, using the very special case above. We extend g_j by zero to regard it as a function on S. Now $g = \sum g_j$ has the following properties:

- $g \ge 1$ on K, since each point of K lies in at least one disc $\frac{1}{2}D_{p_j}$ on which $g_j = 1$;
- g has compact support contained in U_1, since the support of g is the (finite) union of the supports of the g_i which are contained in the compact discs $\frac{1}{2}D_{p_j} \subset U_1$.

Now take a smooth non-negative function H of one variable with $H(t) = 1$ if $t \ge 1$ and $H(t) = 0$ if $t \le 1/2$. Then $f_1 = H \circ g$ has the desired property (i.e. $f_1 = 1$ on K, and the support of f_1 is a compact subset of U_1).

Finally, we consider the general case when $K \subset U_1 \cup \cdots \cup U_n$. Proceeding just as before, we get discs $\frac{1}{2}U_{p_i} \subset U_{p_i}$ for $i = 1, \ldots N$, where

- $K \subset \frac{1}{2} U_{p_1} \cup \cdots \cup \frac{1}{2} U_{p_N}$;
- for each j, there is an $i(j)$ such that the closed (compact) disc $\overline{D_{p_j}}$ is contained in $U_{i(j)}$.

Now, for $i = 1, \ldots, n$, let

$$K_i = \bigcup_{i(j)=i} \overline{\frac{1}{2} D_{p_j}}, \quad N_i = \bigcup_{i(j)=i} D_{p_j}$$

and

$$J_i = \bigcup_{i(j)=i} \overline{D_{p_j}}.$$

Thus the K_i and J_i are compact, N_i is open, we have

$$K_i \subset N_i \subset J_i \subset U_i,$$

and $K \subset \bigcup_i K_i$. Applying the discussion above to each $J_i \subset U_i$, we find smooth functions h_i on S with $h_i = 1$ on J_i and with h_i compactly supported in U_i. Thus if $h = \sum_{i=1}^{n} h_i$, we have $h \geq 1$ on $J_1 \cup \cdots \cup J_n$. Let N be the open subset of S

$$N = N_1 \cup \cdots \cup N_n.$$

Applying the previous case again, we can find a function A of compact support in N and with $A = 1$ on K. Thus $h \geq 1$ on the support of A, so the ratio A/h extends to a smooth function on S. Finally, put

$$f_i = \frac{A h_i}{h}.$$

Then f_i has compact support in U_i, and $\sum f_i = 1$ on K since $A = 1$ there.

Given this lemma and any 2-form ρ of compact support on an oriented surface S, we proceed as follows. We cover $K = \mathrm{supp}(\rho)$ by open sets U_i, each the domain of a local coordinate chart. By the compactness of K, we can do this with a finite collection of sets U_i. Then let f_i be a system of functions as in the lemma. For each i, the support of $f_i \rho$ is contained in a co-ordinate chart and we can define the integral of $f_i \rho$ as above. Now we define

$$\int_S \rho = \sum_i \int_S f_i \rho.$$

Of course, this formula *must* hold true if we are to have an integral with the obvious linearity properties, since $\sum f_i = 1$ on $\mathrm{supp}(\rho)$ implies that

$$\rho = \sum f_i \rho.$$

Conversely, one can readily show that the linearity of the Lebesgue integral implies that this value of the integral of ρ is independent of the choice of the functions f_i.

While one must be careful to distinguish between 2-forms and functions, on an oriented surface there is a well-defined notion of a *positive* 2-form, just one which is given in local co-ordinates by $R \, dx_1 \, dx_2$ with $R \geq 0$. By the definition of an oriented surface and the transformation law for 2-forms, this is independent of the choice of co-ordinate system. If ρ is a positive 2-form of compact support, then the integral of ρ is positive. With the usual conventions, we can define the integral of any positive 2-form ρ, not necessarily of compact support, taking values in the extended real numbers $\mathbf{R} \cup \{+\infty\}$, by

$$\int_X \rho = \sup \int_X \chi\rho,$$

where χ runs over the smooth compactly supported functions on X with $0 \leq \chi \leq 1$ everywhere.

Notice that it is not really necessary that the 2-forms we integrate are smooth. We can define the notion of a *continuous* 2-form in an obvious way, and the discussion above applies equally well. If ρ is any 2-form, we can define a continuous, positive 2-form $|\rho|$ by the requirement that at each point $|\rho| = \pm\rho$. Thus, for any 2-form ρ whatsoever on a surface S, we can define the integral

$$\int_S |\rho|,$$

taking values in $[0, \infty]$. This notion will be convenient later. By an *area form* on an oriented surface S we mean a strictly positive 2-form. If we have a fixed area form, then we can identify the 2-forms on S with functions, and the notion above becomes the usual notion of integration of functions with respect to a measure. However, there are certain important reasons (the proof of Rado's Theorem) why we do *not* want to assume that our surfaces admit such area forms (see the discussion in Chapter 10 below).

The final item is the general form of Stokes' Theorem.

Proposition 13. *If ρ is a compactly supported* 1*-form on an oriented surface-with-boundary S, then*

$$\int_{\partial S} \alpha = \int_S d\alpha.$$

To sum up, we now have on any surface:

• Spaces of $0-, 1-$ and 2 forms and the *exterior derivative*

$$\Omega_S^0 \xrightarrow{d} \Omega_S^1 \xrightarrow{d} \Omega_S^2.$$

- The integral $\int_C \alpha$ of a 1-form α over a curve C in a surface.
- The wedge product $\Omega_S^1 \times \Omega_S^1 \to \Omega_S^2$.
- If S is oriented, the integral $\int_S \rho$ of a compactly supported 2-form ρ.
- Stokes' Theorem, as above.

5.2 de Rham cohomology

5.2.1 Definition and examples

Let S be a smooth surface. We define the de Rham cohomology groups $H^i(S)$, for $i = 0, 1, 2$, to be the cohomology of the sequence

$$\Omega_S^0 \to \Omega_S^1 \to \Omega_S^2.$$

That is,

$$H^0(S) = \ker(d : \Omega_S^0 \to \Omega_S^1),$$

$$H^1(S) = \ker(d : \Omega_S^1 \to \Omega_S^2)/\mathrm{Im}\,(d : \Omega_S^0 \to \Omega_S^1),$$

$$H^2(S) = \Omega_S^2/\mathrm{Im}\,(d : \Omega_S^1 \to \Omega_S^2).$$

Clearly $H^0(S) = \mathbf{R}$, the constant functions, if S is connected. The classical criterion for an exact differential discussed above amounts to the statement that if S is diffeomorphic to $\mathbf{R}^2$, then $H^1(S) = 0$. It is also clear that, for such S, the cohomology group $H^2(S)$ also vanishes, since any function $R(x_1, x_2)$ can be written as $\partial\alpha_1/\partial x_2 - \partial\alpha_2/\partial x_1$ for some α_1, α_2. (In fact, we can take $\alpha_2 = 0$ and

$$\alpha_1(x_1, x_2) = \int_0^{x_1} R(t, x_2)\, dt.)$$

(Although we are just developing a theory here for surfaces, all the definitions go over to the general case of a smooth manifold.)

Examples

1. Consider the 2-sphere S^2. We write this as the union of two open sets $S^2 = U \cup V$, where U and V are slightly enlarged upper and lower hemispheres, intersecting in a annulus around the equator. So, U and v are each diffeomorphic to $\mathbf{R}^2$. Let α be a 1-form on S^2 with $d\alpha = 0$. Then, by the previous discussion, we can find functions f_U, f_V on U, V, respectively, such that df_U, df_V are the restrictions of α to U, V. Thus $d(f_U - f_V) = 0$ on $U \cap V$ and, since $U \cap V$ is connected, this means that $f_U - f_V$ is constant on the intersection, say $f_U - f_V = c$. There is no loss in supposing that this constant is zero, since we can change f_U to $f_U - c$ without changing df_U.

But if $f_U = f_V$ on $U \cap V$, they arise as the restrictions of a function f on S^2 to U and V, and $\alpha = df$. Hence $H^1(S^2) = 0$.

2. Consider the torus T and take standard angular co-ordinates $\theta, \phi \in [0, 2\pi)$. Let $\gamma_1, \gamma_2 \subset T$ be the standard embedded circles corresponding to $\theta = 0$, $\phi = 0$, respectively. Then the map

$$\alpha \mapsto \left(\int_{\gamma_1} \alpha, \int_{\gamma_2} \alpha \right)$$

induces a linear map from $H^1(T)$ to $\mathbf{R}^2$, since the integral of df around the γ_i vanishes for any function f on T. The forms $d\theta$ and $d\phi$ show that this map is surjective. We claim that the map is also injective, so $H^1(T) = \mathbf{R}^2$. For, if $\alpha = P\, d\theta + Q\, d\phi$ is a closed 1-form with integral 0 around γ_2, then for any fixed ϕ we have, by Stokes' Theorem,

$$\int_0^{2\pi} P(\theta, \phi)\, d\theta = 0.$$

This means that the indefinite integral

$$f(\theta, \phi) = \int_0^\theta P(u, \phi)\, du$$

defines a smooth function on T with $\partial f / \partial \theta = P$. Thus $\tilde{\alpha} = \alpha - df$ is a closed 1-form of the form $\tilde{Q} d\phi$. But the closed condition implies that Q is constant and, if the integral around γ_1 is zero, this constant must be zero and $\alpha = df$.

3. Consider the cylinder $C = (-1, 1) \times S^1$ and let δ be the circle $\{0\} \times S^1$. As before, δ defines a linear map from $H^1(C)$ to $\mathbf{R}$ which is clearly surjective. To see that this map is also injective, we proceed as in the first example, writing $c = U \cup V$ with open sets U, V diffeomorphic to $\mathbf{R}^2$. This time, $U \cap V$ has two components. If α is a closed 1-form on C, then we obtain functions f_U, f_V just as in the first example, but we cannot deduce that $f_U - f_V$ is constant, although it is constant on each component of $U \cap V$. Pick points p, q on δ in the two components of $U \cap V$. Then a moment's thought shows that

$$(f_U - f_V)(p) - (f_U - f_V)(q) = \pm \int_\delta \alpha.$$

So, if the integral of α around δ is zero, then $f_U - f_V$ is constant and we can proceed as in Example 1.

4. Consider the standard surface Σ of genus 2 formed from the connected sum of two copies T, T' of the torus. Thus, in T, we have a pair of standard loops γ_1, γ_2 as above, and likewise γ_1', γ_2' in T'. We form the connected sum by removing two open discs D, D' from T, T' and adding a copy of the cylinder C. We can suppose that the discs do not meet the circles γ_i, γ_i'.

Then, in an obvious way, we get four circles, which we also denote by γ_i, γ_i' in Σ. Integration around these circles defines a linear map from $H^1(\Sigma)$ to $\mathbf{R}^4$, and we claim that this is an isomorphism.

To see that the map is injective, we argue as follows. Suppose α is a closed 1-form on Σ and the integral of α around the four loops vanishes. Let δ be the loop in the cylinder $C \subset \Sigma$ as in Example 3. Then, by Stokes' Theorem, the integral of α around δ vanishes, since δ is obviously the boundary of a portion of Σ. By Example 3, we can write the restriction of α to C as dg for some function g on C. Let P be a smooth function supported in C and equal to 1 on δ (see the lemma below). Then $\tilde{\alpha} = \alpha - d(Pg)$ is in the same class in $H^1(\Sigma)$ as α, and $\tilde{\alpha}$ vanishes on a neighbourhood of C. This means that $\tilde{\alpha}$ defines 1-forms β, β' on T, T', respectively, in an obvious way, vanishing near the centres of the discs D, D'. The integrals of β, β' around the circles in T, T' are still zero, so we know that $\beta = df, \beta' = df'$ for some functions f, f' on T, T' with f, f' constant near the centres of the discs. Then the same argument as before shows that we can suppose f, f' match up to give a function f on Σ with $df = \tilde{\alpha}$.

To see that the map is surjective, we argue as follows. Given any four real numbers, we can find closed 1-forms β, β' on T, T', respectively, realising these as their integrals around the circles. By applying our result for $\mathbf{R}^2$, we can write the restrictions of β, β' to neighbourhoods of $\overline{D}, \overline{D'}$ as dg, dg' for functions g, g' defined on these neighbourhoods. Arguing as in the previous paragraph, we find $\tilde{\beta}, \tilde{\beta}'$ in the same cohomology classes as β, β' and vanishing over these discs. Then β, β' define a 1-form α on Σ in an obvious way, having the given integrals around the γ_i, γ_i'.

5. It should now be clear how to show that for the standard closed oriented surface Σ_g of genus g—the connected sum of g copies of the torus—we have

$$H^1(\Sigma_g) = \mathbf{R}^{2g},$$

the isomorphism being realised by integration over $2g$ standard circles in Σ_g.

6. We now take a more abstract point of view. Let S be a connected smooth surface and s_0 a base point in S. One can show that elements of $\pi_1(S, s_0)$ can be represented by smooth loops $\gamma : [0, 1] \to S$. For such loops, the integral around γ is defined and induces a map

$$\int_\gamma : H^1(S) \to \mathbf{R}.$$

This integral depends only on the homotopy class of the loop and is additive with respect to the product in π_1, so we get a linear map

$$H^1(S) \to Hom(\pi_1(S, s_0), \mathbf{R}).$$

This map is an isomorphism, at least if S satisfies the technical condition of having a countable atlas ('paracompactness').

5.2.2   Cohomology with compact support, and Poincaré duality

There is a variant of the definition of cohomology in which we consider the forms $\Omega^i_{S,c}$ of *compact support*. We define cohomology groups $H^i_c(S)$ in just the same fashion in this case. Thus, for example, if S is connected but not compact, then $H^0_c(S) = 0$, since the non-zero constants do not have compact support. These compactly supported cohomology groups allow us to formulate various results, which are instances of 'Poincare duality'. If S is oriented, the map from $\Omega^2_{S,c}$ to $\mathbf{R}$ defined by integration over S induces a linear map

$$\int_S : H^2_c(S) \to \mathbf{R},$$

since the integral of $d\alpha$, for $\alpha \in \Omega^{1,c}_S$, vanishes by Stokes' Theorem.

Proposition 14. *If S is a connected, oriented, smooth surface, then the map $\int_S$ is an isomorphism from $H^2_c(S)$ to $\mathbf{R}$.*

The proof uses Lemma 7 and the following lemma.

Lemma 8. *Proposition 14 is true for the case when $S = \mathbf{R}^2$.*

To prove this, suppose $\rho = R(x_1, x_2)\, dx_1\, dx_2$ is a 2-form of compact support on $\mathbf{R}^2$ with integral 0. Choose a function ψ on $\mathbf{R}$ of compact support and with

$$\int_{-\infty}^{\infty} \psi(t)\, dt = 1.$$

Let

$$r(x_1) = \int_{-\infty}^{\infty} R(x_1, t)\, dt.$$

Now write

$$\tilde{R}(x_1, x_2) = R(x_1, x_2) - r(x_1)\psi(x_2).$$

So $\tilde{R}$ also has compact support in $\mathbf{R}^2$. For each x_1, we have

$$\int_{-\infty}^{\infty} \tilde{R}(x_1, t)\, dt = 0.$$

Define

$$P(x_1, x_2) = \int_{-\infty}^{x_2} \tilde{R}(x_1, t)\, dt.$$

Then P has compact support, and

$$\frac{\partial P}{\partial x_2} = \tilde{R}(x_1, x_2).$$

Put

$$Q(x_1, x_2) = \psi(x_2) \int_{-\infty}^{x_1} r(t)\, dt.$$

Then Q has compact support, and

$$\frac{\partial Q}{\partial x_1} = \psi(x_2) r(x_1).$$

Thus

$$R = \frac{\partial P}{\partial x_2} + \frac{\partial Q}{\partial x_1}$$

or, in other words, $\rho = d\alpha$, where α is the compactly supported form

$$\alpha = -P\, dx_1 + Q\, dx_2.$$

We can now dispose of the proof of Proposition 14. It is obvious that the map $\int$ is surjective, so what we need to show is that if ρ is a 2-form of compact support on a connected, oriented surface and the integral of ρ is zero, then $\rho = d\alpha$ for some α of compact support. Clearly, we can choose a compact connected set K containing the support of ρ and we can cover K by a finite number of open sets $U_1, \ldots U_n$, each the image of a disc under a local chart. We use induction on n. If $n = 1$, then we are reduced to the case when the surface is $\mathbf{R}^2$, considered above. So suppose $n > 1$. Write $V = U_2 \cup \cdots \cup U_n$, so K is contained in $U \cup V$. If either $K \cap U$ or $K \cap V$ is empty, then we are done by the inductive hypothesis, so suppose these two sets are not empty. Then, since K is connected, there is a point p in $K \cap U \cap V$ and, in particular, in $U \cap V$. Clearly we can choose a 2-form τ with compact support contained in the open set $U \cap V$. Now we apply the lemma to choose functions f_1, f_2 supported in U, V, respectively, and with $f_1 + f_2 = 1$ on K. Then

$$\rho = f_1\rho + f_2\rho,$$

and $f_1\rho, f_2\rho$ have compact support in U, V, respectively. Let

$$I = \int_S f_1\rho = -\int_S f_2\rho.$$

Then $f_1\rho - I\tau$ and $f_2\rho + I\tau$ are 2-forms with compact support in U and V, respectively, and with integral 0. By the inductive hypothesis, we can find a 1-form α of compact support on U with $d\alpha = f_1\rho - I\tau$ and, likewise, a 1-form β of compact support in V with $d\beta = f_2\rho + I\tau$. Then $\rho = d(\alpha + \beta)$ and the proof is complete.

Now let γ be a loop in an oriented surface S. Integration around γ yields a linear map

$$I_\gamma : H^1(S) \to \mathbf{R}.$$

On the other hand, given any closed 1-form θ of compact support, we get a linear map

$$J_\theta : H^1(S) \to \mathbf{R}$$

defined by

$$J_\theta([\phi]) = \int_S \theta \wedge \phi.$$

(By Stokes' Theorem, the integral on the right-hand side is unchanged if we take a different representative ϕ for the same cohomology class.)

Proposition 15. *For any loop γ, there is a compactly supported form θ such that $J_\theta = I_\gamma$.*

To prove this, we choose a subdivision $0 = t_0 < t_1 < t_2 < \cdots < t_N = 1$ of the unit interval such that for each i with $0 \leq i \leq N-1$, there is a co-ordinate chart U_i (diffeomorphic to a disc) containing $\gamma([t_i, t_{i+1}])$. We can also easily arrange that the $\gamma(t_i)$ are distinct except for $i = 0$ and N. (But see the remark at the end of the proof). Now choose small discs D_i containing $\gamma(t_i)$ such that

$$\overline{D_i} \subset U_{i-1} \cap U_i$$

and with $D_N = D_0$, but with $D_i \cap D_j$ empty unless $\{i, j\} = \{0, N\}$. Let ρ_i be a 2-form supported in the interior of D_i with integral 1, and with $\rho_N = \rho_0$. Then $\rho_1 - \rho_0$ is a compactly supported form of integal zero on U_0, and we can find a compactly supported form θ_0 on U_0 such that

$$\rho_1 - \rho_0 = d\theta_0.$$

Similarly, for $i \leq N-1$ we find θ_i such that

$$\rho_{i+1} - \rho_i = d\theta_i.$$

Now put

$$\theta = \theta_0 + \theta_1 + \cdots + \theta_{N-1}.$$

Then

$$d\theta = (\rho_1 - \rho_0) + (\rho_2 - \rho_1) + \cdots + (\rho_N - \rho_{N-1}),$$

and this vanishes since $\rho_N = \rho_0$. We claim that $J_\theta = I_\gamma$. Let $[\phi]$ be a class in $H^1(S)$; we can choose a representative ϕ that vanishes on the discs D_j. Now, on one of the open sets U_i we can write $\phi = df_i$ for some function f, since U_i is diffeomorphic to a disc and $H^1(U_i) = 0$. Thus

$$\int_S \theta_i \phi = \int_S \theta_i \wedge df_i = \int_S f_i(\rho_{i+1} - \rho_i).$$

Now, since ϕ vanishes on all the discs D_j, the function f_i is constant on D_{i+1} and D_i, which contain the supports of ρ_{i+1}, ρ_i, respectively. Thus

$$\int_S \theta_i \wedge \phi = f_i(\gamma(t_{i+1})) - f_i(\gamma(t_i)).$$

On the other hand, the integral of ϕ over the portion of the path parametrised by the subinterval $[t_{i+1}, t_i]$ is the same, since $\phi = df_i$ over the image of the path. Summing over i finishes the proof. (*Remark*: It is not really necessary that the points $\gamma(t_i)$ are distinct—all we need to do is choose $\rho_i = \rho_j$ for any pair with $\gamma(t_i) = \gamma(t_j)$.)

Now the linear map J_θ depends only on the class of θ in $H^1_c(S)$. In other words, we have a bilinear pairing

$$H^1_c(S) \times H^1(S) \to \mathbf{R}.$$

In particular, if S is compact, so that H^1 and H^1_c are the same thing, we have a bilinear form on $H^1(S)$ which is obviously skew-symmetric. Suppose Φ is any class in $H^1(S)$. If Φ is not zero, we know that we can find a loop γ such that $I_\gamma(\Phi)$ is non-zero. By the Proposition above, we can find a 1-form θ such that $J_\theta(\Phi)$ is non-zero. In other words, this bilinear form is non-degenerate. Since a vector space which supports a non-degenerate skew-symmetric form must be even-dimensional, we get the following corollary.

Corollary 2. *For any compact oriented surface, the de Rham cohomology $H^1(S)$ is even-dimensional.*

5.3 Calculus on Riemann surfaces

5.3.1 Decomposition of the 1-forms

Now let X be a Riemann surface and so, *a fortiori*, a smooth oriented surface. Thus, for each point p in X, we have a tangent space TX_p—a two-dimensional real vector space. We also have a cotangent space

$$T^*X_p = Hom_{\mathbf{R}}(TX, \mathbf{R}),$$

such that the derivative of any real-valued function on X yields an element of T^*X_p. We may just as well consider the complex cotangent space

$$T^*X_p^{\mathbf{C}} = \mathrm{Hom}_{\mathbf{R}}(TX, \mathbf{C}),$$

such that the derivative of any complex-valued function on X yields an element of $T^*X_p^{\mathbf{C}}$.

By a *complex structure* on a real vector space V we mean an **R**-linear map $J : V \to V$ with $J^2 = -1$.

Lemma 9. *There is a unique way to define a complex structure on TX_p such that the derivative of any holomorphic function, defined on a neighbourhood of p in X, is complex linear.*

This follows immediately from the definition of a holomorphic function. Now let V be a real vector space and J be a complex structure on V. We say that an **R**-linear map A from V to $\mathbf{C}$ is complex linear if $A(Jv) = iAv$ for all v, and complex antilinear if $A(Jv) = -iA(v)$ for all v.

Lemma 10. *Any **R***-linear map from V to $\mathbf{C}$ can be written in a unique way as a sum of complex linear and antilinear maps.*

For the existence, we write $A = A' + A''$, where

$$A'(v) = \frac{1}{2}(A(v) - iA(Jv)), \quad A''(v) = \frac{1}{2}(A(v) + iA(Jv)),$$

and check that A' and A'' are complex linear and antilinear, respectively. Uniqueness is similarly easy.

Putting this together, we see that we can write the complex cotangent space as a direct sum

$$T^*X_p^{\mathbf{C}} = T^*X_p' \oplus T^*X_p'',$$

in such a way that if f is a local holomorphic function, then the derivative of f lies in T^*X_p' and the derivative of $\bar{f}$ lies in T^*X_p''. Now we can decompose the complex 1-forms on X into corresponding pieces

$$\Omega_{X,\mathbf{C}}^1 = \Omega_X^{1,0} \oplus \Omega_X^{0,1},$$

where elements of $\Omega^{1,0}$ lie in T^*X_p', for each p, and $\Omega^{0,1}$ is the complex conjugate.

We now decompose the exterior derivative operators according to this decomposition of the forms, so we get a diagram

$$\begin{array}{ccc}
\Omega^{0,1} & \xrightarrow{\ \partial\ } & \Omega^2 \\[4pt]
\bar{\partial}\ \uparrow & & \uparrow\ \bar{\partial} \\[4pt]
\Omega^0 & \xrightarrow[\ \partial\ ]{} & \Omega^{1,0}
\end{array}$$

Let us see this more explicitly, in a complex local co-ordinate $z = x + iy$. Thus x, y are real co-ordinates as considered in the previous section. We have

$$dz = dx + i\,dy, \quad d\bar{z} = dx - i\,dy,$$

and these form basis elements for T^*X', T^*X'', respectively. So a $(1,0)$-form is expressed locally as $\alpha\,dz$ and a $(0,1)$- form as $\beta\bar{z}$ for functions α, β. If f is a complex-valued function, then

$$df = \frac{\partial f}{\partial x}dx + \frac{\partial f}{\partial y}dy.$$

We write

$$dx = \frac{1}{2}(dz + d\bar{z}), \quad dy = \frac{1}{2i}(dz - d\bar{z}),$$

so

$$df = \frac{1}{2}\left(\frac{\partial f}{\partial x} - i\frac{\partial f}{\partial y}\right)dz + \frac{1}{2}\left(\frac{\partial f}{\partial x} + i\frac{\partial f}{\partial y}\right)d\bar{z}.$$

This means that

$$\partial f = \frac{\partial f}{\partial z}dz, \quad \bar{\partial}f = \frac{\partial f}{\partial \bar{z}}d\bar{z},$$

where we define

$$\frac{\partial f}{\partial z} = \frac{1}{2}\left(\frac{\partial f}{\partial x} - i\frac{\partial f}{\partial y}\right), \quad \frac{\partial f}{\partial \bar{z}} = \frac{1}{2}\left(\frac{\partial f}{\partial x} + i\frac{\partial f}{\partial y}\right).$$

The equation $\bar{\partial}f = 0$ is, in this local co-ordinate, just the Cauchy–Riemann equation, as of course it should be, since by definition a function is holomorphic if and only if $\bar{\partial}f = 0$. If f is a holomorphic function, then we have, in local coordinates,

$$df = \partial f = f'(z)\,dz,$$

where f' denotes the usual derivative of complex analysis.

Now consider the operators $\partial, \bar{\partial}$ on $\Omega^{0,1}$ and $\Omega^{1,0}$. By following through the definitions we find that, in our local co-ordinate,

$$\partial(A\,d\bar{z}) = \frac{\partial A}{\partial z}dz\,d\bar{z} = 2i\frac{\partial A}{\partial z}dx\,dy,$$

$$\bar{\partial}(B\,dz) = \frac{\partial B}{\partial\bar{z}}d\bar{z}\,dz = -2i\frac{\partial B}{\partial\bar{z}}dx\,dy.$$

Definition 8. *A* $(1,0)$*-form* β *is a holomorphic* 1*-form if* $\bar{\partial}\beta = 0$.

Thus, in local co-ordinates, a holomorphic 1-form has the shape $B\,dz$, where B is a holomorphic function.

Suppose $S \subset X$ is a compact surface-with-boundary and α is a holomorphic 1-form on a neighbourhood of S. Then α is, in particular, a closed 1-form, so Stokes' Theorem gives

$$\int_{\partial S}\alpha = 0.$$

This is one version of Cauchy's Theorem on a Riemann surface.

We define a *meromorphic* 1-form α on X in the obvious way: a holomorphic 1-form on $X \setminus D$, where D is a discrete subset of X, which can be written locally as $f(z)\,dz$, where f is a meromorphic function. (Of course, one needs to check that this is independent of the choice of local chart.) The points of the minimal such set D are the *poles* of the meromorphic 1-form. Let p be such a pole and let C be a small loop in X encircling p. We define the *residue* of α at p to be

$$\mathrm{Res}_p(\alpha) = \frac{1}{2\pi i}\int_C \alpha.$$

Clearly, this is the same as writing, in a local co-ordinate z centred at p,

$$\alpha = f(z)\,dz,$$

where $f(z) = \sum_{-k}^{\infty} a_j z^j$ is a meromorphic function, and taking the usual residue a_{-1} of f; one can just as well take this as the definition.

Proposition 16. *Suppose* α *is a meromorphic* 1*-form on a compact Riemann surface* X. *Then the sum of the residues of* α, *running over all the poles, is zero.*

To see this, we let S be the complement in X of a union of small discs about the poles and apply Cauchy's or Stokes' Theorem.

We should also say a few words about *vector fields*, although these will feature much less later in the book. As before, we start with the case of a smooth surface. If we have local co-ordinates (x_1, x_2) around a point p, we write $\partial/\partial x_i$ for the basis of the tangent space dual to the basis dx_i of the cotangent space. We form the *tangent bundle*, which is the union over all points of the tangent spaces. A vector field is defined (just like a 1-form) as a map from the surface into its tangent bundle. In local coordinates, a vector field can be written as

$$v_1 \frac{\partial}{\partial x_1} + v_2 \frac{\partial}{\partial x_2},$$

where the v_i are smooth functions of x_1, x_2. Now, if we have a Riemann surface structure, we can decompose complexified vector fields into types. In a local complex co-ordinate $z = x + iy$, we write

$$\frac{\partial}{\partial z} = \frac{1}{2} \left(\frac{\partial}{\partial x} - i \frac{\partial}{\partial y} \right), \quad \frac{\partial}{\partial \bar{z}} = \frac{1}{2} \left(\frac{\partial}{\partial x} + i \frac{\partial}{\partial y} \right).$$

A general complexified vector field has the form $a \, \partial/\partial z + b \, \partial/\partial \bar{z}$ for smooth complex-valued functions a, b. The real vector fields are, in this notation, those with $b = \bar{a}$. Thus we often identify the fields of type $(1,0)$ with real vector fields by mapping $a \, \partial/\partial z$ to $a \, \partial/\partial z + \bar{a} \, \partial/\partial \bar{z}$. If $a = v_1 + iv_2$ with v_i real, this is just $v_1 \, \partial/\partial x + v_2 \, \partial/\partial y$. In particular, we have the case of a *holomorphic vector field* $a(z) \, \partial/\partial z$ when $a(z)$ is a holomorphic function.

5.3.2   The Laplace operator and harmonic functions

On a Riemann surface, we have a natural second-order differential operator. We define

$$\Delta = 2i \, \bar{\partial}\partial : \Omega^0 \to \Omega^2.$$

Then, in local co-ordinates,

$$\Delta f = 2i \frac{1}{4} \left(\frac{\partial}{\partial x} + i \frac{\partial}{\partial y} \right) \left(\frac{\partial}{\partial x} - i \frac{\partial}{\partial y} \right) f(dz \, d\bar{z}) = - \left(\frac{\partial^2 f}{\partial x^2} + \frac{\partial^2 f}{\partial y^2} \right) dx \, dy.$$

Thus if, in a given local co-ordinate system, we identify the 2-forms with functions using the area form $dx \, dy$, the operator Δ becomes the standard Laplace operator (up to a sign convention). A function f satisfying the differential equation $\Delta f = 0$ is called a *harmonic function*. If f is a holomorphic function, then the real and imaginary parts of f are harmonic, since

$$\bar{\partial}\partial(f \pm \bar{f}) = -\partial\bar{\partial} f \pm \bar{\partial}(\bar{\partial} f) = 0 \pm 0 = 0.$$

Conversely, we have the following lemma.

Lemma 11. *If ϕ is a real-valued harmonic function on a neighbourhood N of a point p in a Riemann surface X, then there is an open neighbourhood $U \subset N$ of p and a holomorphic function f on U with $\phi = \Re(f)$.*

Being local, this is not really different from the corresponding result for functions on open sets in $\mathbf{C}$, which the reader has very likely encountered in a standard complex-analysis course. However, it may be helpful to see how the proof works in our notation.

Let A be the real 1-form $i\,\overline{\partial}\phi + \overline{(i\,\overline{\partial}\phi)}$. Then the hypothesis that $\overline{\partial}\partial\phi = 0$ shows that $dA = 0$. Thus if U is a small disc about p (or any open set with $H^1(U) = 0$), we can find a real-valued function ψ with $A = d\psi$. This means that $\partial\psi = -i\,\partial\phi$ and $\overline{\partial}\psi = i\,\overline{\partial}\phi$. Then

$$\overline{\partial}(\phi + i\psi) = \overline{\partial}\phi + i\,\overline{\partial}\psi = 0.$$

So we can take $f = \phi + i\psi$.

We will also want to use the 'maximum principle' occasionally.

Lemma 12. *Suppose ϕ is a non-constant real-valued harmonic function on a connected open set U in a Riemann surface X. Then, for a point x in U, there is a point x' in U with $\phi(x') > \phi(x)$.*

This can be seen by writing ϕ, near to x, as the real part of a holomorphic function and then using the fact that holomorphic functions are open maps.

5.3.3 The Dirichlet norm

Let X be a Riemann surface and let α be a $(1,0)$-form on X. We consider the 2-form $i\alpha \wedge \overline{\alpha}$. In a local complex co-ordinate $z = x + iy$, if $\alpha = p\,dz$ then

$$i\alpha \wedge \overline{\alpha} = i|p|^2\,dz\,d\overline{z} = 2|p|^2\,dx\,dy.$$

So $i\alpha \wedge \overline{\alpha}$ is a positive 2-form. We define

$$\|\alpha\|^2 = \int_X i\alpha \wedge \overline{\alpha},$$

taking values in $[0,\infty]$. Of course, if α has compact support, the integral is finite and defines a norm on the space of compactly supported $(1,0)$-forms. This norm (on the compactly supported forms) is derived from a Hermitian inner product

$$\langle \alpha, \beta \rangle = \int_X i\alpha \wedge \overline{\beta}.$$

If we have an area form ω on X, we can define a pointwise norm, the function characterised by

$$i\alpha \wedge \overline{\alpha} = |\alpha|^2\omega.$$

Then we can write, tautologically,

$$\|\alpha\|^2 = \int_X |\alpha|^2\omega.$$

This is perhaps a more familiar point of view. However, the key point is that the 'L^2-norm' on $(1,0)$-forms is actually independent of the choice of an area form.

We can identify the real 1-forms with the $(1,0)$-forms by mapping a real 1-form A to its $(1,0)$ component $A^{1,0}$. Thus we define

$$\|A\|^2 = 2\|A^{1,0}\|^2.$$

Again this norm is associated to a (real-valued) inner product $\langle\, , \rangle$ on the compactly supported real 1-forms,

$$\langle A, B \rangle = 2i \int_X A^{0,1} \wedge B^{1,0}.$$

In Chapter 10, we will need the following simple result.

Lemma 13. *Let A, B be real 1-forms on a Riemann surface X. Then*

$$\int_X |A \wedge B| \leq \|A\|\|B\|.$$

(This needs to be interpreted in the obvious way: if either $\|A\|$ or $\|B\|$ is infinite, then the statement is vacuous; if both are finite, then the left-hand side is also finite and the stated inequality holds.) This is, at bottom, very elementary. Suppose first that A and B are supported inside some local coordinate chart. Thus we can write $A = P\,dz + \overline{P}\,d\bar{z}$, $B = Q\,dz + \overline{Q}\,d\bar{z}$, for complex-valued functions P, Q. Then

$$A \wedge B = (P\overline{Q} - Q\overline{P})\,dz\,d\bar{z} = \text{Im}(P\overline{Q})\,dx\,dy.$$

Thus

$$\int_X |A \wedge B| = \int_C |\text{Im}(P\overline{Q})|\,dx\,dy.$$

By the Cauchy–Schwarz inequality,

$$\int_X |A \wedge B| \leq \left(\int_C |P|^2\right)^{1/2}\left(\int_C |Q|^2\right)^{1/2} = \|A\|\|B\|.$$

Now suppose that A and B are any forms with compact support, both supported in some compact set $K \subset X$. It is convenient to explain the proof by choosing an arbitrary area form ω over a neighbourhood of K. The proof is then essentially the same. We have, pointwise over K,

$$|A \wedge B| \leq |A||B|\omega$$

and the proof is just the Cauchy–Schwarz inequality (for the functions $|A|, |B|$ with respect to the measure defined by ω). In the general case, by the definition of the integral, it suffices to prove that for any function χ of compact support with $0 \leq \chi \leq 1$ we have

$$\int_X \chi |A \wedge B| \leq \|A\| \|B\|.$$

Since $\chi |A \wedge B| = |(\chi A) \wedge B|$ and (obviously)

$$\|\chi A\| \leq \|A\|,$$

we can reduce the problem to the case when A has compact support (replacing A by χA). Suppose A is supported in a compact set J. Then we can choose a function η, with $0 \leq \eta \leq 1$, equal to 1 on J and with compact support. Replacing B by ηB, which has compact support, does not change the wedge product with A, and (obviously) $\|\eta B\| \leq \|B\|$. Thus we can reduce the problem to the case of forms of compact support, considered above.

Suppose f and g are real-valued functions, at least one of compact support, on X. We define the *Dirichlet inner product* to be

$$\langle f, g \rangle_D = \langle df, dg \rangle.$$

Likewise, we define the *Dirichlet norm* by

$$\| f \|_D = \| df \|,$$

with our usual convention that this could be $+\infty$. (Our language here ignores the fact that this is only a semi-norm, since the constants have derivative zero.) The following will be crucial in Chapters 9 and 10.

Lemma 14. *If at least one of f, g have compact support, then*

$$\langle f, g \rangle_D = \int_X g \, \Delta f = \int_X f \, \Delta g.$$

This really amounts to little more than the following elementary identity, for functions f, g (at least one of compact support) on $\mathbf{C}$,

$$\int_{\mathbf{C}} f \left(\frac{\partial^2 g}{\partial x^2} + \frac{\partial^2 g}{\partial y^2} \right) = -\int_{\mathbf{C}} \left(\frac{\partial f}{\partial x} \frac{\partial g}{\partial x} + \frac{\partial f}{\partial y} \frac{\partial g}{\partial y} \right) dx \, dy,$$

which is derived immediately by integration by parts. In our notation, on a general Riemann surface, the proof becomes

$$\langle f, g \rangle_D = \langle df, dg \rangle = 2i \int_X \partial f \wedge \bar{\partial} g = 2i \int_X \partial (f \, \bar{\partial} g) - f \, \partial \bar{\partial} g = \int_X f \, \Delta g,$$

where of course, we have used Stokes' Theorem for the vanishing of the integral of $\partial(f \, \bar{\partial} g)$.

Exercises

1. Let Δ be a set of d distinct points in the plane. Compute the de Rham cohomology of $\mathbf{R}^2 \setminus \Delta$.

2. Show that a smooth proper map from a surface S to a surface S' induces a map on a compactly supported cohomology. If the surfaces are oriented, use Proposition 12 to define the *degree* of the map, and show that this agrees with the definition in Chapter 4 in the case of a holomorphic map.

3. Show that the Riemann surface defined by the equation $w^2 = \sin z$ in $\mathbf{C}^2$ is not the interior of a compact surface-with-boundary.

4. Suppose that Ω is a compact region with a smooth boundary in a Riemann surface and that ϕ is a real-valued function which is positive on Ω and vanishes on the boundary of Ω. Show that

$$\int_{\partial\Omega} i\, \partial\phi \geq 0.$$

(*Hint:* Consider $\Omega \subset \mathbf{C}$ and see that the integral is, in traditional notation, the flux of the gradient of ϕ through the boundary.)

6 Elliptic functions and integrals

In this chapter, we study Riemann surfaces of genus 1. On the one hand, the constructions will give an important model for the more general theory we develop in Part III. On the other hand, the constructions involve classical topics in mathematics, which relate the abstractions of Riemann surface theory to their origin in concrete calculus problems.

6.1 Elliptic integrals

The problem which gives this subject its name was that of finding the arc length between two points on an ellipse. However we will take as our model problem that of finding the motion of a pendulum under gravity. In suitable units and in an obvious notation, the motion is defined by the energy conservation condition

$$\dot{\theta}^2 - \cos\theta = E,$$

where E is constant and θ is the amplitude angle of the pendulum. Thus

$$\dot{\theta} = \sqrt{E + \cos\theta},$$

and

$$t = \int \frac{d\theta}{\sqrt{E + \cos\theta}}.$$

So the problem reduces to doing this indefinite integral. Writing $u = \cos\theta$, this is transformed into

$$t = \int \frac{du}{\sqrt{(E + u)(1 - u^2)}}.$$

More generally, we may consider

$$\int \frac{du}{\sqrt{f(u)}},$$

where f is any polynomial. In the case when f is quadratic, we know how to perform these integrals in terms of elementary functions; the question is what kinds of functions arise for polynomials of higher degree, particularly for cubic polynomials such as that arising in the pendulum problem.

We now have the language required to interpret this in a better way. Suppose $f(z)$ is a polynomial of degree n, with distinct roots $z_1, \ldots, z_n$. Let $X \subset \mathbf{C}^2$ be the set of solutions of the equation $w^2 = f(z)$. The condition that f has distinct roots means that the partial derivatives of $w^2 - f(z)$ do not both vanish anywhere on X, so X is a Riemann surface. By construction, we have a pair of holomorphic functions z, w on X with derivatives dz, dw. Since $w^2 - f(z)$ vanishes on X, we have an identity

$$2w \, dw = f'(z) \, dz.$$

Now dz/w is a holomorphic 1-form away from the points where $w = 0$. In punctured neighbourhoods of such points, we can write

$$\frac{dz}{w} = 2 \frac{dw}{f'(z)},$$

and $f'(z)$ does not vanish, since f has simple roots. So we conclude that dz/w extends to a *holomorphic* 1-form α on X. Moreover, we see that α does not vanish anywhere on X.

To summarise so far: when we write an expression such as

$$\int_{z_0}^{z_1} \frac{dz}{\sqrt{f(z)}},$$

what we really mean is that we choose a path γ on the Riemann surface X, running from a point with $z = z_0$ to a point with $z = z_1$, and form

$$\int_\gamma \alpha.$$

Now, recall that we have defined a compactification X^* of X. We want to consider the extension of α to X^*. This is a good exercise in the theory developed in Chapters 3, 4 and 5. We consider two cases, depending on whether n is odd or even. To construct X^*, we need to examine the monodromy of the covering around a large circle in $\mathbf{C}$ (corresponding to a small circle around $\infty \in S^2$). The monodromy lies in the group S_2 of permutations of the two sheets. Clearly, if n is even, the monodromy is trivial, and if n is odd, it is the non-trivial element of S_2. Consider the odd case. We form X^* by attaching a single disc to X, adjoining one extra point P, say. If

$$f(z) = z^n + a_1 z^{n-1} + \cdots + a_n,$$

then, in terms of a standard co-ordinate τ on the disc,

$$z = \tau^{-2}, \quad w = \sqrt{f(z)} = \tau^{-n}\sqrt{1 + a_1\tau^2 \cdots + a_n\tau^{2n}},$$

where the square root is well defined for small τ. Then we have

$$\frac{dz}{w} = \left(-2\frac{d\tau}{\tau^3}\right)\left(\frac{\tau^n}{\sqrt{1 + a_1\tau^2 + \ldots}}\right) = \frac{-2}{\sqrt{1 + a_1\tau^2 + \ldots}}\tau^{n-3}\, d\tau.$$

So we conclude that if n is odd, α extends to a meromorphic 1-form on X^* and that:

- if $n = 1$, α has a pole of order 2 at the point P;
- if $n = 3$, then α is holomorphic near P and does not vanish at P;
- if $n > 3$, then α is holomorphic near P and has a zero of order $n - 3$ at P.

The even case is similar. Now we adjoin two extra points, P^+ and P^- say, to form X^*; the reader can check, as above, that α is meromorphic on X^* and that:

- if $n = 2$, then α has simple poles at $P_\pm$;
- if $n = 4$, then α is holomorphic near $P_\pm$ and does not vanish at $P_\pm$;
- if $n > 4$, then α is holomorphic near $P_\pm$ and has zeros of order $(n-4)/2$ at $P_\pm$.

We want to focus attention on the cases when $n = 3$ or 4 so α is holomorphic on X^* and does not vanish anywhere. In fact, there is no real distinction between $n = 3, 4$, since we can transform one case to the other by a Möbius transformation of the Riemann sphere: in either case, X^* is a double cover of the sphere with four critical values, and the distinction is just whether we choose ∞ to be a critical value ($n = 3$) or not ($n = 4$). We change point of view slightly and prove a general classification theorem.

Theorem 3. *Let X be a compact Riemann surface and let α be a holomorphic 1-form on X with no zeros. Then there is a lattice $\Lambda \subset \mathbf{C}$ and an isomorphism $\iota : \mathbf{C}/\Lambda \to X$ such that $\pi^*\iota^*(\alpha) = du$, where u is the identity function on $\mathbf{C}$ and $\pi : \mathbf{C} \to \mathbf{C}/\Lambda$ is the projection map.*

First we sketch the idea of the proof, which is quite simple. We try to define an indefinite integral of the holomorphic 1-form α. We can perform the integral along a path in X, but the value depends on the end points since we have a choice of the homotopy class of paths. This indeterminacy means that the indefinite integral is not defined as a $\mathbf{C}$-valued function, but it is defined as a map to $\mathbf{C}/\Lambda$ for a suitable Λ, and this map will turn out to be the inverse of the desired isomorphism ι.

Now for the detailed proof. Consider the universal cover $p : \tilde{X} \to X$. The lift $p^*(\alpha)$ is a holomorphic 1-form on $\tilde{X}$ and, since $\tilde{X}$ is simply connected, the integral of $p^*(\alpha)$ along paths depends only on the end points. So we get a holomorphic map $F : \tilde{X} \to \mathbf{C}$ with $dF = p^*(\alpha)$. Since α has no zeros, the map F is a local homeomorphism. We claim that F is in fact a covering map. For each point $x \in X$, we can find a radius $r > 0$ and an injective holomorphic map $j_x : D_r \to X$, where D_r, is the r-disc $\{u : |u| < r\}$ in $\mathbf{C}$ such that $j_x(0) = x$ and $j_x^*(\alpha) = du$. That is, j_x is the inverse of a locally defined indefinite integral of α. Since X is compact we can, by a simple argument, find a single r which works for all $x \in X$. Now suppose $\tilde{x}$ is a point in $\tilde{X}$. Since the disc is simply connected, we can lift $j_{p(\tilde{x})}$ to get an injective map

$$\tilde{j}_{\tilde{x}} : D_r \to \tilde{X},$$

with $\tilde{j}_{\tilde{x}}(0) = \tilde{x}$ and $\tilde{j}_{\tilde{x}}^*(p^*(\alpha)) = du$. Let $\Delta_{\tilde{x}}$ be the image under $\tilde{j}_{\tilde{x}}$ of the disc $D_{r/2}$ of radius $r/2$. Then, by construction, $F(\Delta_{\tilde{x}})$ is the $r/2$-disc $D_{F(\tilde{x}),r/2}$ in $\mathbf{C}$ centred on $F(\tilde{x})$. Now we observe that, for $\tilde{x}, \tilde{y} \in \tilde{X}$,

$$\tilde{y} \in \Delta_{\tilde{x}} \Leftrightarrow \tilde{x} \in \Delta_{\tilde{y}}.$$

For if $\tilde{y}$ is in $\Delta_{\tilde{x}}$, so that $\tilde{y} = \tilde{j}_{\tilde{x}}(v)$, say, for some $|v| < r/2$, then the whole set $\Delta_{\tilde{y}}$ can be described as

$$\tilde{j}_{\tilde{x}}(\{w : |v - w| < r/2\}),$$

and this obviously contains $\tilde{j}_{\tilde{x}}(0) = \tilde{x}$. Now let z be any point in $\mathbf{C}$ and consider the disc $D_{r/2,z}$ of radius $r/2$ centred on z. Suppose $\tilde{y}$ is a point of $F^{-1}(D_{r/2,z})$. Then z lies in the $r/2$-disc centred on $F(\tilde{y})$, so there is a point $\tilde{x}$ in $\Delta_{\tilde{y}}$ with $F(\tilde{x}) = z$. Equally, by the remark above, $\tilde{y}$ lies in $\Delta_{\tilde{x}}$, where $F(\tilde{x}) = z$. So we have

$$F^{-1}(D_{r/2,z}) = \bigcup_{\tilde{x} \in F^{-1}(z)} \Delta_{\tilde{x}}.$$

Suppose that $\tilde{x}_1$ and $\tilde{x}_2$ are two points in $F^{-1}(z)$ and that $\Delta_{\tilde{x}_1} \cap \Delta_{\tilde{x}_2}$ is not empty. Then there is a point $\tilde{y}$ in the intersection. Then, by the remark above, $\tilde{x}_1, \tilde{x}_2$ both lie in $\Delta_{\tilde{y}}$, but this is a contradiction to the fact that F is injective on $\Delta_{\tilde{y}}$. So we conclude that the union above is a disjoint union, and hence F is indeed a covering map.

Now, since $\mathbf{C}$ is simply connected, it has no non-trivial connected coverings and we conclude that F is an isomorphism from $\tilde{X}$ to $\mathbf{C}$. But we know that X is the quotient of $\tilde{X}$ by an action of $\pi_1(X)$ on $\tilde{X}$. So we conclude that X is isomorphic to the quotient of $\mathbf{C}$ by a group of holomorphic automorphisms. By the classification of these quotients, we see that the only possibility is that

$X = C/\Lambda$ for some lattice Λ. (See Exercise 1 at the end of this chapter.) The identification of the form α follows from the construction.

To sum up: if f is a cubic polynomial with distinct roots and X^* is the compact Riemann surface associated to the equation $w^2 = f(z)$, then there is a lattice $\Lambda \subset C$ and an isomorphism

$$\iota : C/\Lambda \to X^*.$$

This can also be regarded as a Λ-periodic map from C to X^* which can be written as a pair of meromorphic functions $z(u), w(u)$ on C with

$$w(u)^2 = f(z(u)).$$

The map has the property that it pulls the holomorphic form $dz/w = dz/\sqrt{f(z)}$ back to the constant form du on C or, equivalently,

$$\frac{dz}{du} = w = \sqrt{f(z)}.$$

6.2   The Weierstrass $\wp$ function

We now make a fresh start with a lattice Λ in C. We ask the question: can we find a meromorphic function on C/Λ? Since C/Λ is compact, there are no non-trivial holomorphic functions, so we need to allow poles. Moreover, since C/Λ is not homeomorphic to the Riemann sphere, we must have more than one simple pole (or a multiple pole), by Corollary 1 (chapter 4). We can see this more directly as follows. Let P be a parallelogram forming the standard fundamental domain and let Γ be the boundary of P. A meromorphic function F on C/Λ yields a *doubly periodic* meromorphic function $\tilde{F}$ on C. There is no loss in supposing that no pole of $\tilde{F}$ lies on Γ. Then Cauchy's Theorem implies that

$$\int_\Gamma \tilde{F} \, du$$

is the sum of the residues of the poles in P. But the double periodicity means that the integrals around opposite sides of Γ cancel, so we see that the sum of these residues is zero. In particular, we cannot have a single simple pole.

Following the considerations above, we seek a meromorphic function with one *double* pole, and we obtain this through the famous Weierstrass construction. We define $\wp = \wp_\Lambda$ on C by

$$\wp(u) = \frac{1}{u^2} + \sum_{\lambda \in \Lambda \setminus \{0\}} \left(\frac{1}{(u-\lambda)^2} - \frac{1}{\lambda^2} \right).$$

For any u in $\mathbf{C} \setminus \Lambda$, the sum on the right-hand side of this expression converges. For, when $|\lambda|$ is large (and u is fixed),

$$\frac{1}{(u - \lambda)^2} - \frac{1}{u^2} = O(|\lambda|^{-3}).$$

We can compare the sum over large λ in the lattice with the double integral

$$\int_{|\lambda| > 1} |\lambda|^{-3} \, dp \, dq,$$

where $\lambda = p + iq$, to see that the sum converges absolutely. It follows easily that the formula above defines a Λ-periodic meromorphic function on $\mathbf{C}$ with double poles at the points of Λ and no other poles. This then descends to yield a meromorphic function, which we still call $\wp$, on $\mathbf{C}/\Lambda$ with one double pole. Note from the form of the construction that $\wp$ is an even function: $\wp(-u) = \wp(u)$.

Now $\wp$ has a Laurent expansion about 0,

$$\wp(u) = \frac{1}{u^2} + 0 + au^2 + bu^4 + \dots,$$

where the vanishing of the coefficient of u^0 follows from the shape of the construction. This gives

$$\wp''(u) = \frac{6}{u^4} + 2a + \dots,$$

so

$$\wp'' - 6\wp^2 = -10a + \dots$$

is a *holomorphic* Λ-periodic function, and hence constant. Thus $\wp'' - 6\wp^2 = 10a$. We can rewrite this identity as

$$\frac{d}{du}\left(\wp'^2\right) = \frac{d}{du}\left(4\wp^3 - 20a\wp\right),$$

so $\wp'^2 = 4\wp^3 - 20a\wp + a'$, say, with a' another constant. Adopting conventional notation, $\wp$ satisfies an equation

$$\frac{d\wp}{du}^2 = 4\wp^3 - g_2\wp - g_3,$$

for certain constants g_2, g_3 depending on the lattice.

Now, this just expresses the fact that $\mathbf{C}/\Lambda$ arises as the Riemann surface associated to an equation $w^2 = f(z)$ for cubic f, so our conclusion is that the classes of Riemann surfaces obtained in these two ways are identical. Returning finally to our starting point, we see that the solution of the pendulum equation can be written as

$$\theta = \cos^{-1}(\wp_\Lambda(t + t_0)),$$

for a suitable lattice Λ.

6.3   Further topics

6.3.1   Theta functions

The description of meromorphic functions on the torus that we have achieved is not very useful for explicit numerical calculations, because the series converge very slowly. A much better method is to use the theory of θ-functions, which we will now discuss briefly. Of course, the importance of these goes far beyond numerical computation; in particular, the extension of the ideas to higher-dimensional complex tori, and so to the Jacobians of Riemann surfaces, is a very important part of Riemann surface theory, but one we have decided to omit in this book, beyond this short section.

To keep the notation simple, let us begin with the rectangular lattice Λ in $\mathbf{C}$ generated by 2π and i. Thus a Λ-periodic function is one that satisfies $f(z + 2\pi) = f(z)$ and $f(z + i) = f(z)$. The basic idea is to write a meromorphic function as the ratio of two functions which satisfy a different transformation property. Specifically, fix a positive integer k and consider the conditions on a holomorphic function $\phi(z)$ on $\mathbf{C}$

$$\phi(z + 2\pi) = \phi(z), \quad \phi(z + i) = \phi(z)e^{k(-iz+1/2)}. \quad (*)_k$$

If ϕ_1, ϕ_2 both satisfy these conditions, then clearly ϕ_1/ϕ_2 will be a Λ-periodic meromorphic function, that is, a meromorphic function on the torus $\mathbf{C}/\Lambda$. (The factor in $(*)_k$ may look unnatural and complicated but will be convenient later.)

Notice that if ϕ satisfies $(*)_k$, then a power ϕ^m will satisfy $(*)_{mk}$.

Let P be the rectangle $\{z : 0 \leq \text{Re}(z) \leq 2\pi, 0 \leq \text{Im}(z) \leq 1\}$, the standard fundamental domain for Λ.

Lemma 15. *If ϕ satisfies $(*)_k$ and does not vanish on the boundary of P, then there are are exactly k zeros of ϕ in P, counting multiplicity.*

This follows from Rouché's Theorem. Write $D(z) = \phi'/\phi$. Rouché's Theorem states that the number of zeros in P is equal to $(2\pi i)^{-1}$ times the integral of $D(z)\,dz$ around its boundary The modified periodicity condition $(*)_k$ implies that $D(z + 2\pi) = D(z), D(z + i) = D(z) - ik$. This means that the contributions from the vertical sides cancel and the contributions from the horizontal sides combine to give $2\pi k$, and hence the result.

Notice if we did have a zero on the boundary of P we could argue similarly with a suitable translate of P, which also gives a fundamental domain.

We can easily analyse the solutions to $(*)_k$. The first condition means that we can write $\phi(z) = \sum_{p=-\infty}^{\infty} a_p e^{ipz}$. Then

$$\phi(z + i) = \sum (a_p e^{-p}) e^{ipz},$$

while

$$e^{k(-iz+i/1)}\phi(z) = \sum a_p e^{k/2} e^{i(p-k)z} = \sum a_{p+k} e^{k/2} e^{ipz}$$

(replacing $p - k$ by p in the sum). So we get a system of equations for the coefficients

$$a_{p+k} = e^{p-k/2} a_p. \tag{6.1}$$

Clearly, we can specify the co-efficients $a_0, \ldots a_{k-1}$ arbitrarily, and then there is a unique solution to equation (6.1). We can write this explicitly, since $a_{p+Nk} = a_p e^S$, where

$$S = (p - k/2) + (p - k/2 - k) + \cdots + (p - k/2 - (N-1)k) = (pN - kN^2/2)$$

(and the same formula works for negative N). We see then that the size of the a_{p+Nk} decreases rapidly as $N \to \pm\infty$, so there is no problem with convergence.

We conclude then that there is a k-dimensional space of solutions of $(*)_k$. The ratio of any two of these gives a meromorphic function on the torus with k poles and zeros, counted with multiplicity.

Now we fix our attention on the case $k = 1$, when the solution is essentially unique and can be given by the formula

$$\theta(z) = \sum e^{-p^2/2} e^{ipz}. \tag{6.2}$$

Notice that $\theta(z) = \theta(-z)$. We know that there is a single zero of θ on the torus, and this must correspond to a point z where $z = -z \mod \Lambda$. So there is a zero at exactly one of the four points $z = 0, z = \pi, z = i/2, z = \pi + i/2$. On inspection, one finds that the zero is at $c = \pi + i/2$. (Replace p by $-1 - p$ to see that the sum vanishes.) The other values are

$$\theta(0) = 1 + 2 \sum_{p=1}^{\infty} e^{-p^2/2},$$

$$\alpha = \theta(\pi) = 1 - 2(e^{-1/2} - e^{-2} + e^{-9/2} - \ldots) \tag{6.3}$$

and

$$\beta = \theta(i/2)$$

$$= 1 + 4(\cosh(1/2) + \cosh(1)e^{-1/2} + \cosh(3/2)e^{-2} + \cosh(2)e^{-9/2} + \dots. \quad (6.4)$$

Now we go on to the case $k = 2$. There is a two-dimensional space of solutions, so this will lead to essentially one meromorphic function, up to composition with Möbius maps. There are many choices of basis which we can use to write this down explicitly. As one basic element we take $\theta(z)^2$, with θ as defined above. If ϕ is any other solution of $(*)_2$ (not a multiple of θ^2), then $\phi/(\theta^2)$ is a meromorphic function with a double pole at the point $c = \pi + i/2$. This must be of the form $a\wp(z - c) + b$ for constants a, b, and the choice of ϕ is exactly the choice of a, b. However, usually it is not these constants a, b which are most relevant. Whatever ϕ we take, we get a branched covering of the Riemann sphere branched over ∞ and three points in **C**, and what we want to do is to identify these three branch points, for a convenient choice of ϕ. To do this, we arrange that one of the branch points is at 0. Thus we want ϕ to have a double zero at 0. We can do this by 'translating' the known solution θ^2, but the process is less obvious than one might think, and depends crucially on the fact that $2c$ lies in the lattice Λ. We start with the equation $(*)_2$ for a function $\phi(z)$ and write $\phi(z) = \psi(z - c)$. Then $(*)_2$ transforms into

$$\psi(z + 2\pi) = \psi(z), \quad \psi(z + i) = e^{(1-2ic)-2iz}\psi(z). \quad (6.5)$$

This is a *different* equation, and for a general c there would be no direct relation between the solutions of this and $(*)_2$. But for the particular value in question, we proceed as follows. We write $\psi(z) = e^{inz}\chi(z)$ for an integer n. Then the first equation in (6.5) transforms into $\chi(z + 2\pi) = \chi(z)$ (since e^{inz} is 2π-periodic), and the second equation transforms into

$$\chi(z + i) = e^{(1-2ic+n)-2iz}\chi(z). \quad (6.6)$$

Now $e^{-2ic} = e$. If we choose $n = -1$, then equation (6.6) is exactly the original $(*)_2$. So we have a solution $\chi = \theta^2$ and, going backwards, this gives us a solution to the original problem

$$\phi(z) = e^{-i(z-i/2-\pi)}\theta^2(z - i/2 - \pi), \quad (6.7)$$

which vanishes at $z = 0$ as required. Now the other branch points correspond to $i/2, \pi$, and we have

$$\phi(i/2) = e^{-i\pi}\theta^2(-\pi) = -\theta^2(\pi), \quad \phi(\pi) = e^{-1/2}\theta^2(-i/2) = e^{-1/2}\theta^2(\pi).$$

So we conclude that we only need to compute the two values $\alpha = \theta(\pi)$, $\beta = \theta(i/2)$, given by equations (6.3), (6.4) above, and the branch points are at $\lambda_1 = -\alpha^2/\beta^2, \lambda_2 = e^{-1/2}\beta^2/\alpha^2$ in **C**. Thus we get the solution of the equation

$$\frac{df^2}{dz} = f(f - \lambda_1)(f - \lambda_2),$$

in the form $f(z) = \phi(z)/\theta^2(z)$, where ϕ, θ are given by the formulae (6.2), (6.7). For example, by making a simple change of variable, we can use this to solve the pendulum problem for certain special values of the energy E. The advantage of this description is that the series defining the θ function converge extremely quickly, because $e^{-p^2/2}$ quickly becomes very small, and thus the theta function can effectively be computed by hand.

We have confined our attention to one particular lattice so far, in order to simplify the formulae as much as possible. But this restriction is clearly unnatural—for example, in the pendulum problem we want solutions not just for a particular value of the energy. On the other hand, it is a simple matter to extend the discussion to a general lattice, spanned by 2π, τ with Im $\tau > 0$. (The author finds it easiest to think first of rectangular lattices with periods 2π and it, and then go on to the general case.) The starting point is the theta function, which we now write as a function of two variables

$$\theta(z, \tau) = \sum_p e^{i\tau p^2/2} e^{ipz}.$$

This satisfies the periodicity conditions

$$\theta(z, \tau)(z + 2\pi) = \theta(z, \tau)(z), \quad \theta(z + \tau, \tau) = e^{-i\tau/2}e^{-iz}\theta(z, \tau),$$

and vanishes at the point $\pi + \tau/2$. We obtain a description of the Riemann surface as a branched cover of the sphere with branch points $0, 1, \lambda_1(\tau), \lambda_2(\tau)$, where

$$\lambda_1 = -\alpha(\tau)^2/\beta(\tau)^2, \quad \lambda_2 = e^{-i\tau/2}\beta(\tau)^2/\alpha(\tau)^2,$$

with

$$\alpha(\tau) = \theta(\pi, \tau) = \sum(-1)^p e^{i\tau p^2} \qquad (6.8)$$

and

$$\beta(\tau) = \theta(\tau/2, \tau) = \sum e^{(p+p^2)i\tau/2}. \qquad (6.9)$$

6.3.2 Classification

It is a fact that a compact Riemann surface has a nowhere-vanishing holomorphic 1-form if and only if it has genus 1, as a topological surface. Hence such a Riemann surface is a double cover of the Riemann sphere, in an essentially unique way. We will prove this basic fact later (in Part III), but meanwhile, assuming it, we now go on to describe all the *isomorphism classes* of compact Riemann surfaces of genus 1. As we will see, this set $\mathcal{M}$ of isomorphism classes is *itself* a Riemann surface, in a natural way. In fact, $\mathcal{M} = \mathbf{C}$. This is the same as saying that we can assign a complex number $\langle \Sigma \rangle$ to each such Riemann surface, in such a way that $\langle \Sigma \rangle = \langle \Sigma' \rangle$ if and only if Σ, Σ' are isomorphic, and that every complex number occurs as a $\langle \Sigma \rangle$. We get two different points of view on this, using the ideas developed above. In each case, we start with a larger space and obtain $\mathcal{M}$ by taking the quotient under a group action.

In one approach, we start with fact that any of these Riemann surfaces can be realised in an essentially unique way as a double branched cover of the Riemann sphere, with four branch points. Our larger space $\mathcal{J}$ can be defined as the set of equivalence classes of a Riemann surface of genus 1 and an *ordering* of the branch points. This is just the set of quadruples (z_1, z_2, z_3, z_4) of distinct points in the Riemann sphere modulo the action of Mobius maps. There is a unique Möbius map which takes z_1, z_2, z_3 to $0, 1, \infty$, respectively, so we have a unique representative of the form $(0, 1, \infty, \lambda)$ in each orbit under the Mobius maps, and $\mathcal{J}$ can be identified with $\mathbf{C} \setminus \{0, 1\}$. The *cross-ratio* $X(z_1, z_2, z_3, z_4)$ of four distinct points z_1, z_2, z_3, z_4 in the Riemann sphere is defined to be the image of z_4 under the Mobius map μ which takes z_1, z_2, z_3 to $0, 1, \infty$. So, in other words, we are saying that $\mathcal{J}$ is identified with $\mathbf{C} \setminus \{0, 1\}$ by cross-ratio the of the branch points, taken with the given ordering. Explicitly,

$$\mu(z) = \left(\frac{z_2 - z_3}{z_2 - z_1} \right) \frac{z - z_1}{z - z_3}$$

and so

$$X(z_1, z_2, z_3, z_4) = \frac{(z_4 - z_1)(z_2 - z_3)}{(z_1 - z_2)(z_3 - z_4)}.$$

Now, from the definition, the permutation group S_4 acts on $\mathcal{J}$, and $\mathcal{M}$ is the quotient $\mathcal{J}/S_4$. So we have to work out how the cross-ratio changes under permutation of the points. There are three ways of partitioning four objects into two pairs. This gives a homomorphism from S_4 onto S_3 with a kernel that is the four-element subgroup V consisting of the identity and the three permutations

$$(12)(34), \quad (13)(24), \quad (14)(23).$$

By easy inspection, this subgroup acts trivially on the cross-ratio. So the action of S_4 on $\mathcal{J}$ is really given by an action of the quotient group $S_4/V = S_3$. Further, we can naturally identify the quotient group with the permutations of $1, 2, 3$, since V acts simply transitively on $\{1, 2, 3, 4\}$. So our action of S_3 can be written as follows. Each permutation σ of $0, 1, \infty$ has a unique extension to a Möbius map f_σ, and this gives an action on the Riemann sphere. Of course, there is nothing special about the choice of three points $0, 1, \infty$, so, to make things clearer, we work temporarily with the points $1, \rho, \rho^2$, where $\rho = e^{2\pi i/3}$ is a cube root of unity. Then the permutations are generated by $z \mapsto \rho z$ and $z \mapsto z^{-1}$. Clearly, the function $h(z) = z^3 + z^{-3}$ is a complete invariant of this action, i.e. points z_1, z_2 are in the same orbit if and only if $h(z_1) = h(z_2)$. But it is more convenient to take $G(z) = 3^{-3}(h(z) - 2)^{-1}$. Then, as z ranges over the Riemann sphere minus $\{1, \rho, \rho^2\}$, the function $G(z)$ takes every value in $\mathbf{C}$, and $G(z_1) = G(z_2)$ if and only if z_1, z_2 are in the same orbit. The Möbius map $z \mapsto (\rho z + \rho^2)/(z + \rho^2)$ takes $0, 1, \infty$ to $1, \rho^2, \rho$, so points λ_1, λ_2 are in the same orbit under the S_3 action on $\mathbf{C} \setminus \{0, 1\}$ permuting $0, 1, \infty$ if and only if $F(\lambda_1) = F(\lambda_2)$, where

$$F(\lambda) = G\left(\frac{\rho\lambda + \rho^2}{\lambda + \rho^2}\right).$$

This simplifies to

$$F(\lambda) = \frac{(\lambda^2 - \lambda + 1)^3}{\lambda^2(1 - \lambda)^2}.$$

We conclude that $\mathcal{M}$ can be identified with $\mathbf{C}$. That is, we define $\langle\Sigma\rangle$ by the following recipe. Write Σ as a branched cover of the Riemann sphere with branch points z_1, z_2, z_3, z_4. Let λ be the cross-ratio $X(z_1, z_2, z_3, z_4)$ and set

$$\langle\Sigma\rangle = F(\lambda) = \frac{(\lambda^2 - \lambda + 1)^3}{\lambda^2(1 - \lambda)^2}.$$

Notice that the action of S_3 on the Riemann sphere which we are considering above has the following description in terms of three-dimensional Euclidean geometry. Take a regular tetrahedron inscribed in the unit sphere. The oriented symmetry group of the tetrahedron is of order 6 and is isomorphic to S_3, and these symmetries act on the unit sphere.

In the second approach, we start with a Riemann surface described as $\mathbf{C}/\Lambda$, where Λ is a lattice in $\mathbf{C}$. For our larger space $\mathcal{H}$, we take the isomorphism classes of such Riemann surfaces together with a choice of integral basis τ_1, τ_0 for the lattice, chosen so that τ_1/τ_0 has positive imaginary part. Now our set of equivalence classes $\mathcal{M}$ is obtained by 'forgetting' the basis. To say this more precisely, the group $SL(2, \mathbf{Z})$ acts on the integral basis with a matrix $\begin{pmatrix} a & b \\ c & d \end{pmatrix}$ mapping (τ_1, τ_0) to

$$(a\tau_1 + b\tau_0, c\tau_1 + d\tau_0).$$

So we have an action of $SL(2, \mathbf{Z})$ on $\mathcal{H}$, and $\mathcal{M}$ is the quotient $\mathcal{H}/SL(2, \mathbf{Z})$.

Now we can make this more explicit. For any non-zero complex number α and lattice Λ, the Riemann surfaces $\mathbf{C}/\Lambda, \mathbf{C}/(\alpha\Lambda)$ are obviously isomorphic. So there is no loss in restricting ourselves to the case when $\tau_0 = 1$. Then τ_1 is any element of the upper half-plane H, so we see that $\mathcal{H}$ can be identified with

H. The action of $SL(2,\mathbf{Z})$ on $\mathcal{H}$ becomes the usual action on H via Möbius maps, factoring through $PSL(2,\mathbf{Z})$. So, from this point of view, we see that $\mathcal{M} = H/PSL(2,\mathbf{Z})$. Hence we arrive at a fundamental example of a Riemann surface obtained as H/Γ for a discrete subgroup $\Gamma \subset PSL(2,\mathbf{R})$.

To 'see' $\mathcal{M}$ more explicitly in this approach, we recall the notion of a *fundamental domain* for the action of a group G on a space X. By definition, this means an open set $\Omega \subset X$ such that every G-orbit meets $\overline{\Omega}$ and no G-orbit meets Ω more than once. Now we have the following theorem.

Theorem 4. *The set* $\Omega = \{z \in H : |z| > 1, -1/2 < \mathrm{Re}(z) < 1/2\}$ *is a fundamental domain for the action of* $PSL(2,\mathbf{Z})$ *on* H.

In the proof, we will refer briefly to some ideas—the hyperbolic metric on H—which are developed in Chapter 11. This is the most natural language to use, although it would not be hard to bypass it for the specific application we need.

Call Im z the 'height' of a point z in H. The idea is to choose a representative in an orbit with maximal height. Given an orbit $SL(2,\mathbf{Z})z_0$, let k be the supremum of the heights of all points in the orbit. This could, *a priori*, be $+\infty$. Choose a sequence $g_n \in PSL(2,\mathbf{Z})$ such that the heights of $z_n = g_n(z_0)$ tend to k. Let T be the map $T(z) = z + 1$ in $PSL(2,\mathbf{Z})$. Since this preserves the height, we can suppose that for every n the point z_n lies in the strip $\{z : -1/2 \le \mathrm{Re}(z) \le 1/2\}$. Now comes the input from hyperbolic geometry. If k were equal to $+\infty$, the hyperbolic distance between z_n and $z_n + 1$ would tend to 0 as n tends to ∞. This means that the hyperbolic distance between z_0 and $g_n^{-1}(z_n + 1)$ tends to zero. But $g_n^{-1}(z_n + 1) = \gamma_n(z_0)$, where $\gamma_n = g_n^{-1}Tg_n$ lies in $PSL(2,\mathbf{Z})$. So the sequence $\gamma_n(z_0)$ converges to z_0, and this contradicts the fact that $PSL(2,\mathbf{Z})$ is a *discrete* subgroup of $PSL(2,\mathbf{R})$. By the same argument (but without the need to refer to hyperbolic geometry), the supremum k is actually *attained*, so there is a point z_* in the orbit which maximises the height, and we can suppose that z_* lies in $\{z : -1/2 \le \mathrm{Re}(z) \le 1/2\}$.

Now consider the transformation $S(z) = -1/z$ in $PSL(2,\mathbf{Z})$. We have $\mathrm{Im}(S(z) = |z|^{-2}\mathrm{Im}(z)$. So if $|z| < 1$, then $\mathrm{Im}(S(z)) > \mathrm{Im}(z)$. Thus our representative z_*, which maximises the height, must have $|z_*| \ge 1$ and we have proved that any orbit meets the closure of Ω. It remains to show that no orbit meets Ω more than once. We shall prove a slightly stronger statement: if $g \in PSL(2,\mathbf{Z})$ is not the identity, then Ω and $g(\Omega)$ are disjoint. To see this, notice first that it is clearly true when g is a power T^k, and these are precisely the elements $\begin{pmatrix} a & b \\ c & d \end{pmatrix}$ of $PSL(2,\mathbf{Z})$ with $c = 0$. So it suffices to consider the case when $c \ne 0$. Now, for any such element g of $PSL(2,\mathbf{R})$, write

$$g(z) = \frac{az+b}{cz+d} = \frac{a}{c} - \frac{c^{-2}}{z - d/c}$$

(using the fact that $ad - bc = 1$.) Let A_g be the set of points in H with $|z - d/c| \geq |c|^{-1}$. The map takes the region A_g to the region $B_g = \{z : |z - a/c| \leq |c|^{-1}\}$. If $|c| \geq 2$, it is clear that $\overline{\Omega}$ lies in A_g but does *not* intersect B_g, whatever the values of a, d. It follows that in this case there can be no point in $\overline{\Omega} \cap g(\overline{\Omega})$. Now the elements g we have to consider are in $PSL(2, \mathbf{Z})$, so if $|c| \leq 2$, we must have $|c| = 1$. Then a few moments thought shows that the only time $g(\overline{\Omega})$ and $\overline{\Omega}$ can intersect is when $|a|, |d| \leq 1$, and in these cases the interiors $\Omega, g(\Omega)$ are still disjoint.

One now sees that $\mathcal{M}$ can be obtained from $\overline{\Omega}$ by making certain identifications on the boundary (Figure 6.1). The identifications in question are by T, which glues the edge $\mathrm{Re}(z) = -1/2$ to the edge $\mathrm{Re}(z) = 1/2$, and by S, which glues one half of the arc $|z| = 1$ in $\overline{\Omega}$ to the other. Then it is clear that the quotient space $H/PSL(2, \mathbf{Z})$ is homeomorphic to $\mathbf{C}$, matching up with the other approach.

If we put these two approaches together, we obtain a complete invariant for the action of $PSL(2, \mathbf{Z})$ on the upper half-plane H. That is, we obtain a function $J : H \to \mathbf{C}$ such that $J(\tau) = J(\tau')$ if and only if τ and τ' are in the same $PSL(2, \mathbf{Z})$ orbit. We start with a point $\tau \in H$, and from this construct a lattice $\mathbf{Z} \oplus \mathbf{Z}\tau$ and hence a Riemann surface. We have seen that the Weierstrass function represents this Riemann surface as a double cover branched over ∞ and the three roots of $4z^3 - g_2z - g_3$. The quantities g_2, g_3 are given by explicit sums over the lattice: see Exercise 2 below. We also have a choice of ordering of

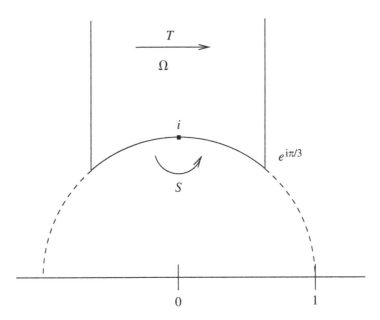

Figure 6.1 *The fundamental domain*

the roots as, say, $\wp(1/2), \wp(\tau/2), -(\wp(1/2) + \wp(\tau/2))$. Now we have to take the cross-ratio $\lambda = X(\wp(1/2), \wp(\tau/2), -(\wp(1/2) + \wp(\tau/2)), \infty)$, and finally define

$$J(\tau) = F(\lambda) = \frac{(\lambda^2 - \lambda + 1)^3}{\lambda^2(\lambda - 1)^2}.$$

A little algebraic manipulation shows that this can be expressed in terms of g_2, g_3 as

$$J = \frac{27}{4} \frac{g_2^3}{g_2^3 - 27g_3^2}.$$

Alternatively, we can use θ functions. Now we get a representation with branch points $0, \lambda_1, \lambda_2, \infty$. The cross-ratio is $\lambda = \lambda_1/\lambda_2$, which is $-e^{-i\tau/2}(\alpha(\tau)/\beta(\tau))^4$, where $\alpha(\tau), \beta(\tau)$ are given by equations (6.8), (6.9), and then $J(\tau) = F(\lambda)$.

Exercises

1. Show that the only Riemann surfaces which arise as quotients of **C** are the cylinder **C** \ {0} and the tori **C**/Λ.

2. Show that

$$g_2 = 60 \sum_{\lambda \in \Lambda'} \lambda^{-4},$$

$$g_3 = 140 \sum_{\lambda \in \Lambda'} \lambda^{-6},$$

   where Λ' denotes $\Lambda \setminus \{0\}$.

3. Show that any smooth curve of degree three in **CP**^{2} is equivalent to a Riemann surface of a function $\sqrt{f(z)}$, for cubic f. (*Hint:* Consider projection from a point of the curve to a line.)

4. Show that the lattice Λ associated to the pendulum equation is *rectangular*, generated by 1 and iq for some $q > 0$, and that for physically meaningful solutions the imaginary part of the constant of integration t_0 must be $q/2$.

5. (From Petr Pushkar.) Let $V(x)$ be a real polynomial of degree four with two local minima. Consider the motion of a particle in the corresponding potential, that is to say, the differential equation $\ddot{x} = -V'(x)$. The energy $E = \dot{x}^2 + V$ is a constant of the motion. For E in a suitable range, there are two distinct periodic solutions with energy E, oscillating about the local minima. Show that the periods of these two oscillations are equal.

6. For $p > 3$, investigate the sums $G_p = \sum_{\lambda \in \Lambda'} \lambda^{-2p}$. Can you express these in terms of g_2, g_3?

7. Use the Poisson summation formula to find a relation between $\theta(z, \tau)$ and $\theta(z, -1/\tau)$.

7 Applications of the Euler characteristic

We have seen that the genus of a compact oriented smooth surface S can be defined as one-half the dimension of the de Rham cohomology group $H^1(S)$. There are many other possibilities. In particular, another way of defining the genus, in some respects more elementary, uses the *Euler characteristic*. This can be done via triangulations of the surface. We do not want to take the time to develop the theory of triangulations in detail, but we will describe this approach slightly informally here and then develop some applications. The reader with a suitable background in topology will know how to make the discussion more rigorous, and in any case we will be able to derive the corollaries as simple consequences of the more advanced theory (the Riemann–Roch formula) later. However, we give this more elementary discussion now, since we do not want the reader to gain the impression that these essentially topological results depend essentially on the rather deeper analysis in Part III.

7.1 The Euler characteristic and meromorphic forms

7.1.1 Topology

Suppose we have a surface, possibly with a boundary, which is *triangulated* in the manner indicated in Figure 7.1. Then the Euler characteristic of the triangulation is defined to be

$$\chi = V - E + F,$$

where V, E, and F are the numbers of vertices, edges and faces, respectively. The first basic fact is that this number is independent of the choice of triangulation, and hence defines an integer $\chi(S)$, the Euler characteristic of S. It is not hard to check, for example by defining explicit triangulations, that for the model surfaces

$$\chi(\Sigma_{g,r}) = 2 - 2g - r, \quad \chi(\Xi_{h,r}) = 2 - h - r.$$

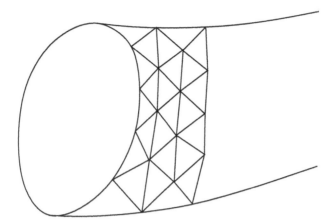

Figure 7.1   *A triangulation*

We can define the genus of a closed oriented surface by $g = 1 - \chi(S)/2$ and, of course, the next thing we need to know is that this coincides with our previous definition. If we are willing to accept the classification of surfaces, of course, there is no need to have any theory here, since we just need to check for the model surfaces. But it is preferable from many points of view (for example, extensions to higher dimensions) to understand the result independently of the classification of surfaces. In any case, let us assume it from now on.

Now suppose S is a compact oriented surface and that α is a real 1-form on S. Suppose that the set $\Delta \subset S$ where α vanishes is discrete. Given any point p of Δ, we choose local co-ordinates centred on p, and represent α locally as

$$\alpha = \alpha_1 \, dx_1 + \alpha_2 \, dx_2.$$

Our hypothesis asserts that for small r, the only zero of the vector-valued function (α_1, α_2) on the closed r-disc about the origin is at the origin itself. Thus the restriction of this function to a circle of radius r gives a map from the circle to $\mathbf{R}^2 \setminus \{0\}$ which has an integer winding number. It is not hard to check that this is independent of the choice of r and the local co-ordinate system. We define the multiplicity $m_p(\alpha)$ of the zero p of α to be this winding number.

Proposition 17. *In the situation above,*

$$\sum_{p \in \Delta} m_p(\alpha) = -\chi(S).$$

We sometimes call the sum on the left-hand side of this formula the 'number of zeros of α (counted with multiplicity)'. Of course, there are many different ways of building up this theory. For example, if one shows that the sum of zeros (counted with multiplicity) is independent of the choice of α, then one can use it to *define* the Euler characteristic, and hence the genus. To relate this

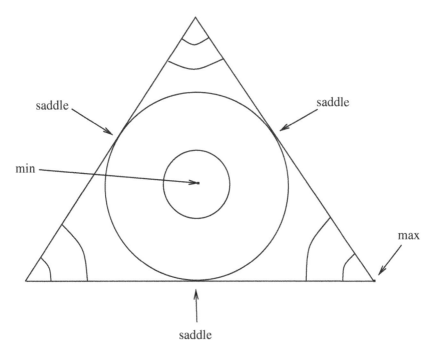

Figure 7.2 *Standard function on a triangle*

to the count in the definition by triangulations, one can consider a standard 1-form on a triangle given by df, where f is the function indicated by the following picture in Figure 7.2.

If we have a triangulation of S, then we can define a 1-form on S which restricts to this model on each triangle. There is then one zero for each vertex, one for each edge and one for each face: the multiplicities are all +1 for the first and third cases and −1 for the second. So the 'count' of zeros gives precisely the count of vertices, edges and faces, with the right signs. In any case, what we will assume to be known is that Proposition 17 holds true for any smooth 1-form α with a discrete zero set and where $\chi(S) = 2 - 2g$, with g the genus defined in Chapter 5.

7.1.2   Meromorphic forms

Now suppose that X is a compact Riemann surface and that α is a holomorphic 1-form on X, not identically zero. We associate to this the real 1-form $A = \alpha + \overline{\alpha}$. In a local co-ordinate z, we write $\alpha = f(z)\,dz$; the zeros of A are the zeros of f and hence discrete. Moreover, the multiplicity of a zero is a positive integer, equal to the multiplicity of the zero of f in the usual sense. Hence in this case Proposition 17 says that the total number of zeros, counted with these positive multiplicities, is $2g - 2$. In particular, if $g = 0$, there can be no such α and if $g = 1$, the situation considered in the previous chapter, a non-trivial holomorphic form is nowhere vanishing.

We can extend this discussion to *meromorphic* 1-forms. To do this, we fix an area form ω on X. This means that we can define a Hermitian metric on T^*X:

$$\xi \wedge \bar{\xi} = |\xi|^2 \omega.$$

Suppose α is a meromorphic 1-form on X. Choose a real-valued function p on $\mathbf{R}$ with $p(t) = 1$ for small t and $p(t) = t^{-1}$ for large t. Now define

$$\tilde{\alpha} = p(|\alpha|^2)\alpha$$

away from the poles of α, and $\tilde{\alpha} = 0$ at the poles of α. Locally, around a pole of α, we have

$$\tilde{\alpha} = \frac{1}{|f(z)|^2} f(z) R \, dz = \frac{1}{\overline{f(z)}} R \, dz,$$

where R is a smooth strictly positive function, determined by the area form $\omega = R \, dx \wedge dy$. Thus $\tilde{\alpha}$ is smooth and its zero set is the union of the zeros and poles of α. It is clear that the zeros of $\tilde{\alpha}$ corresponding to the poles of α have a multiplicity equal to minus the order of the pole. Thus we have the following proposition.

Proposition 18. *If α is a non-trivial meromorphic 1-form on a compact Riemann surface X, then the number of zeros of α minus the number of poles of α, counted with multiplicity, is equal to $2g - 2$.*

Remark Choosing an area form, as we have done above, is the same as choosing a compatible *Riemannian metric*, an aspect which we will discuss further in Chapter 11. Given a metric, we can identify the 1-forms, which we have used here, with *vector fields*.

7.2 Applications

7.2.1 The Riemann–Hurwitz formula

Suppose that $f : X \to Y$ is a non-constant holomorphic map between connected compact Riemann surfaces. In Chapter 4, we associated a multiplicity k_x to each point of X, equal to 1 except for a finite set of critical points. We define the total ramification index to be

$$R_f = \sum_{x \in X} k_x.$$

So this is really a finite sum. We have also defined the degree $d \geq 1$ of the map f. The following result, the *Riemann–Hurwitz* formula, is very useful for calculations.

Proposition 19. *The genus g_X of X and the genus g_Y of Y are related by*

$$2 - 2g_Y = d(2 - 2g_X) - R_f.$$

One way of proving this, which we sketch, is to show (or assume, depending on taste) that there is a triangulation of Y such that each branch point of f is a vertex. Then the triangulation can be 'lifted' to a triangulation of X. Each face or edge of the triangulation of Y gives rise to d faces or edges of the triangulation of X. Likewise, each vertex of the triangulation of X which is *not* a branch point gives rise to d vertices in the triangulation of X. On the other hand, a branch point y in Y gives rise to only

$$d - \sum_{x \in f^{-1}(y)} (k_x - 1)$$

vertices in the triangulation of X, and so the formula follows from the counting formulae for the Euler characteristics.

We can give another proof if we suppose that there is a meromorphic 1-form β on Y. (In Part III, we shall see that these always exist.) Then $f^*(\beta)$ is a meromorphic 1-form on X. A pole or zero of β which is not a branch point gives rise to d poles or zeros of $f^*(\beta)$ on X with the same multiplicity. On the other hand, suppose that x is a ramification point so, in local co-ordinates, the map f can be represented as $z \mapsto w = z^k$, where $k = k_x > 1$. If β is then given in the w co-ordinate by $g(w) \, dw$ for some meromorphic g, the pull-back $f^*(\beta)$ is

$$kz^{k-1}g(z^k) \, dz.$$

If g has a zero of order $l \in \mathbf{Z}$ (where a negative value of l indicates a pole in the obvious way), then $z^{k-1}g(z^k)$ has a zero of order $kl + k - 1$. So the contribution to the count of zeros/poles of $f^*(\beta)$ from the points x in $f^{-1}(y)$ is

$$\sum_{x \in f^{-1}(y)} (k_x l + k_x - 1) = dl + \sum_{x \in f^{-1}(y)} (k_x - 1),$$

since we know that

$$d = \sum_{x \in f^{-1}(y)} (k_x - 1).$$

Then, applying Proposition 18 to β and $f^*(\beta)$, we obtain the Riemann–Hurwitz formula.

7.2.2   The degree–genus formula

Now suppose that X is a smooth complex curve of degree d in $\mathbf{CP}^2$. Recall that this means that X is defined by a homogeneous polynomial $p(Z_0, Z_1, Z_2)$ and that not all of the partial derivatives $\partial p/\partial Z_i$ vanish at any point of X.

Proposition 20. *The genus of X is given by*

$$g_X = \frac{1}{2}(d-1)(d-2).$$

For example, we have already seen that when $d = 1$ or 2, the Riemann surface X is equivalent to the Riemann sphere (genus 0), and when $d = 3$, to a complex torus $\mathbf{C}/\Lambda$ (genus 1). Notice that this shows that some compact Riemann surfaces *cannot* be realised as smooth curves in $\mathbf{CP}^2$, since not all integers can be expressed as $(d-1)(d-2)/2$.

There are various ways of obtaining the formula in the proposition. We will establish the result by constructing a meromorphic form on X and counting the poles and zeros. Suppose, for simplicity, that the curve X meets the line at infinity in d distinct points. (It is easy to show that this can be arranged by a suitable linear transformation of the Z_i.) Let $P(z, w)$ be the polynomial in two variables defining the corresponding affine curve X_0. We follow the same construction that we used, in a special case, in Chapter 6. Thus dz, dw represent holomorphic 1-forms on X_0, and the identity $P(z, w) = 0$ on X_0 yields

$$P_z\, dz + P_w\, dw = 0.$$

At points where P_z, P_w are both non-zero, we have

$$\frac{dz}{P_w} = -\frac{dw}{P_z}.$$

Since, by hypothesis, there are no points on X_0 where P_z, P_w both vanish, we obtain a non-vanishing holomorphic 1-form θ on X_0 equal to dz/P_w or $-dw/P_z$ at the points where these are defined. We have to check that θ is a meromorphic 1-form on X, i.e. that it has at worst poles at the d points of intersection with the line at infinity, and then count the zeros or poles at these points.

Now we switch to homogeneous co-ordinates $[Z_0, Z_1, Z_2]$, so we have a homogeneous polynomial p of degree d and $P(z, w) = p(1, Z, W)$. There is no loss of generality in supposing that $[0, 1, 0]$ is one of the intersection points of the projective curve with the line at infinity. To study the situation around this point, we use a different affine chart, consisting of points $(u, 1, v)$ so that $u = 1/z, v = w/z$ on the intersection of the two charts. Let q be the partial derivative of p with respect to Z_2, so q is a homogeneous polynomial of degree $d-1$ and $P_w(z, w) = q(1, Z, W)$. The hypothesis that the line at infinity meets the curve in d points implies that q does not vanish at $(0, 1, 0)$. Further, we can take u as a local co-ordinate on the curve around this point. Then $dz = -u^{-2} \, du$ and $P_w(z, w) = q(1, z, w) = z^{d-1} q(u, 1, v)$, by the homogeneity of q. Thus our 1-form is

$$-\frac{u^{-2} u^{d-1}}{q(u, 1, v)} du.$$

If $d \geq 3$, this extends as a holomorphic 1-form across the intersection point with the line at infinity, with a zero of order $d - 3$. If $d < 3$, it extends as a meromorphic 1-form. In any case, the contribution to the count of zeros is $d - 3$. Now we apply this to each of the d intersection points to see that the total 'number of zeros' is $d(d - 3)$. So,

$$g = \frac{1}{2}(d(d - 3) + 2) = \frac{1}{2}(d - 1)(d - 2).$$

7.2.3   Real structures and Harnack's bound

In Chapter 2, we discussed non-orientable surfaces; indeed, this was the most interesting case from the point of view of the classification theorem, but since then they have dropped out of the picture, mainly because any Riemann surface is oriented. However, non-orientable surfaces do arise naturally in certain questions, as we will now illustrate.

In Chapter 3, we defined the notion of *holomorphic maps* between Riemann surfaces. One can just as well define *antiholomorphic maps*, given in local complex coordinates by antiholomorphic functions (a function f on an open set in $\mathbf{C}$ is antiholomorphic if $\bar{f}$ is holomorphic). A composite of antiholomorphic maps is holomorphic, and the composite of a holomorphic map and an antiholomorphic map is antiholomorphic. We say that a *real structure* on a Riemann surface X is an antiholomorphic map $\sigma : X \to X$ with $\sigma \circ \sigma$ equal to the identity. The *real points* $X_{\mathbf{R}}$ of such a pair (X, σ) are defined to be the fixed points of σ. For example, the maps $\sigma_0, \sigma_1 : \mathbf{C} \setminus \{0\} \to \mathbf{C} \setminus \{0\}$ given by

$$\sigma_0(z) = \bar{z}, \quad \sigma_1(z) = -1/\bar{z}$$

are two different real structures on $\mathbf{C} \setminus \{0\}$. The real points are the real axis in one case and the empty set in the other. Each of these extends to a real structure on the Riemann sphere, and $S^2_{\mathbf{R}}$ is a copy of a circle in one case and empty in the other.

Lemma 16. *Let (X, σ) be a Riemann surface with a real structure and let x be a point of $X_{\mathbf{R}} \subset X$. Then there is a local holomorphic co-ordinate z around x in which σ is given by the map $z \mapsto \bar{z}$.*

We leave the proof as an exercise for the reader.

If (X, σ) is a surface with a real structure, we can form the quotient space X/σ. Using the proposition above, it is not hard to show the following proposition.

Proposition 21. *The space X/σ is a surface-with-boundary, where the boundary of X/σ can be identified with $X_{\mathbf{R}}$.*

(More precisely, we should say that X/σ can be endowed with the structure of a surface-with-boundary.) The quotient X/σ may be orientable or non-orientable and the boundary may or may not be empty. For example, with the two real structures σ_0, σ_1 on S^2 above, S^2/σ_0 is a disc and S^2/σ_1 is the real projective plane $\mathbf{RP}^2$.

Now suppose that X is a *compact* connected Riemann surface with a real structure σ. Then X/σ is a compact connected surface-with-boundary. We have the following proposition.

Proposition 22. *The Euler characteristics of X and X/σ satisfy*

$$\chi(X) = 2\chi(X/\sigma).$$

Just as with the Riemann–Hurwitz formula, various proofs are possible. In terms of triangulations, if we choose a triangulation of X/σ we can lift to a triangulation of X. Then to each face of X/σ correspond two faces of X, and likewise for each vertex and edge which do *not* lie in the boundary. On the other hand, each edge and vertex in the boundary of X/σ correspond to just one edge or vertex in X. But each component of the boundary of X/σ is a circle, so clearly the numbers of edges and vertices in the boundary are equal. Then the result follows from simple counting.

Now, for any surface-with-boundary Y with r boundary components, we have

$$\chi(Y) \le 2 - r,$$

with equality in the case when Y is a sphere with r discs removed. If we accept the classification of surfaces, we can read this assertion off from that: an independent proof is also possible, of course. In any case, we obtain the following conclusion.

Proposition 23. *Let X be a compact connected Riemann surface of genus g. If σ is a real structure on X, then the number of components of $X_{\mathbf{R}}$ is at most $g + 1$.*

This follows immediately from the discussion above, since if r is the number of components of $X_{\mathbf{R}}$ we have

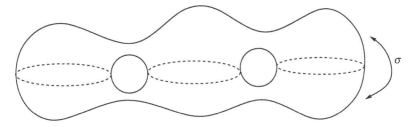

Figure 7.3 *Riemann surface of genus 2 with real structure*

$$1 - g = \frac{1}{2}\chi(X) = \chi(X/\sigma) \leq 2 - r,$$

since r is also the number of boundary components of X/σ.

Equality holds in the case when σ is the reflection map acting on the standard picture of a surface of genus g, and the quotient is a sphere with $g + 1$ discs removed (Figure 7.3).

To give a concrete application of this, suppose that P is a polynomial in two variables with *real coefficients*. Then, regarded as a complex polynomial, P defines an affine curve in $\mathbf{C}^2$ and a projective curve X in $\mathbf{CP}^2$. Suppose, for simplicity, that these are smooth. Complex conjugation of the co-ordinates induces a real structure on X, and the set $X_\mathbf{R}$ just corresponds to the points of the corresponding real projective curve in $\mathbf{RP}^2$. So we have *Harnack's bound*, as stated in the following proposition.

Proposition 24. *Let p be a homogeneous polynomial with real co-efficients and degree d in three variables with the property that the corresponding complex projective curve is smooth. Let $\Gamma \subset \mathbf{RP}^2$ be the real projective curve defined by p. Then the number of components of Γ is at most $\frac{1}{2}(d-1)(d-2) + 1$.*

(It is not hard to relax the hypothesis on the smoothness of the complex projective curve.)

7.2.4   Modular curves

In Chapter 3, we introduced the modular curves X_p, the quotients of the upper half-plane by the group Γ_p. From the definition, it was not clear how to say much about these. We will now see how to understand the topology of these Riemann surfaces.

Proposition 25. *If p is a prime number not equal to 2, then there is a compact Riemann surface $\overline{X}_p$ of genus*

$$g = 1 + \left(\frac{(p-6)(p^2-1)}{24} \right),$$

with a subset $\Delta \subset \overline{X}_a$ containing $(p^2 - 1)/2$ points, such that X_p is equivalent to the complement $\overline{X}_p \setminus \Delta$.

Recall that Γ_p is the quotient by ± 1 of the integer matrices with determinant 1 and equal to $\pm$ the identity modulo p. Thus Γ_p is a normal subgroup of $PSL(2, \mathbf{Z})$ and the quotient group is $PSL(2, \mathbf{Z}/p)$. The number of elements in $GL(2, \mathbf{Z}/p)$ is equal to the number of bases in $(\mathbf{Z}/p)^2$. This is $(p^2 - 1)(p^2 - p)$, since we have $p^2 - 1$ ways to choose the first basis vector and the second can be any vector not on the line generated by the first, which gives $p^2 - p$ ways to choose the second basis vector. The subgroup $SL(2, \mathbf{Z}/p)$ is the kernel of a homomorphism onto the multiplicative group $\mathbf{Z}/p \setminus \{0\}$ of order $p - 1$, so $SL(2, \mathbf{Z}/p)$ has order $p(p^2 - p)$ and $PSL(2, \mathbf{Z}/p)$ has order $N = p(p^2 - p)/2$.

In Chapter 4 we explained that, even though $PSL(2, \mathbf{Z})$ does not act freely on the upper half-plane, the quotient is still a Riemann surface, and in Chapter 6 we identified this explicitly with **C**. Now, $PSL(2, \mathbf{Z}/p)$ acts on $X_p = H/\Gamma_p$ with quotient $H/PSL(2, \mathbf{Z}) = \mathbf{C}$. We claim that the map π from X_p to **C** is proper. Recall that in Chapter 6 we constructed a fundamental domain Ω for the action of $PSL(2, \mathbf{Z})$ on H. Choose any set of $g_1, \ldots, g_N \in PSL(2, \mathbf{Z})$ representing the cosets of Γ_p. Then we have N translates $g_i(\Omega)$ in H and the closure of these maps onto X_p. It is clear that if K is a compact subset of **C**, the intersection of $\pi^{-1}(K)$ with any $g_i(\overline{\Omega})$ is compact, so $\pi^{-1}(K)$ is covered by the union of a finite number of compact sets and hence is itself compact. So we have a proper holomorphic map $\pi : X_p \to \mathbf{C}$, and we can apply the theory of Chapter 4 to construct a compactification $\overline{X}_p$ mapping to the Riemann sphere, and use the same symbol to write $\pi : \overline{X}_p \to \mathbf{CP}^1$.

It is clear that π has degree N. The remaining task is to compute the order of branching of π. Recall that the stabiliser of the special point i in $PSL(2, \mathbf{Z})$ has order 2 and the stabiliser of $\rho = e^{\pi i/3}$ has order 3. Apart from the orbits of these points, all other stabilisers are trivial. We can suppose that i maps to 0 in $H/PSL(2, \mathbf{Z}) = \mathbf{C}$ and ρ maps to 1. Then, since Γ_p acts freely on H, the set $\pi^{-1}(0) \subset X_p$ has $N/2$ elements. The branching of π at each of these points is the same, since they are permuted transitively by $PSL(2, \mathbf{Z}/p)$. So they are simple branch points with multiplicity 2. Similarly, we get $N/3$ points lying over 1, with a multiplicity of order 3. Finally, we have to compute the branching over ∞ in **C**. Let $A = \{z \in \mathbf{C} : |z| > 2\}$. According to our general theory, the pre-image $\pi^{-1}(A)$ is a disjoint union of m copies of the punctured disc for some number m. The group action implies that m divides N and the branching order at each point over ∞ is N/m. Now A corresponds to some subset A' of $\overline{\Omega}$, and $\pi^{-1}(A)$ is covered by N translates of the closure of A'. Take p of our coset representatives to be the maps $z \mapsto z + i$ for $0 \le i \le p - 1$. The corresponding translates of the closure of A' form a connected set in H, and the map $z \mapsto z + p$, which lies in Γ_p, identifies the two vertical boundaries. So it is clear that these cover a single punctured disc in X_p. Thus the number m is N/p and the branching order is p.

To sum up, the branched cover $\pi : \overline{X}_p \to \mathbf{CP}^1$ has

- $N/2 = p(p^2 - 1)/4$ points of branching order 2 over 0;
- $N/3 = p(p^2 - 1))/6$ points of branching order 3 over 1;
- $N/p = (p^2 - 1)/2$ points of branching order p over ∞.

The Riemann–Hurwitz formula for the genus g of $\overline{X}_p$ gives

$$2g - 2 = N\left(\frac{2}{3} + \frac{1}{2} + \frac{p-1}{p} - 2\right),$$

which rearranges to $g - 1 = (p - 6)(p^2 - 1)/24$.

Exercises

1. Let Z be a smooth algebraic curve in $\mathbf{CP}^1 \times \mathbf{CP}^1$ (see Exercise 7 in Chapter 4). Let d_1, d_2 be the degrees of the projection maps from Z to the two factors. By considering a suitable meromorphic form, show that the genus of Z is $(d_1 - 1)(d_2 - 1)$.
2. Let X be the compact Riemann surface associated to the equation $z^{2a} - 2w^b z^a + 1 = 0$, for fixed positive integers a, b. Identify the branch points of the covering of the Riemann sphere defined by the z coordinate and hence show that the genus of X is $ab - a$.
3. Show that a smooth, non-empty, real cubic curve in the projective plane has exactly two components. Investigate the number of components for higher-degree curves. (This is related to a famous problem of Hilbert.)
4. If we put $p = 2$ in the general formula for the genus of the modular curve X_p, we get $g = 1/2$. Explain what needs to be modified, and show that in fact $g = 0$ in this case.

Part III

Deeper theory

8 Meromorphic functions and the Main Theorem for compact Riemann surfaces

In this chapter, we explain our strategy for proving the fundamental structural results about Riemann surfaces. For motivation, we can review the discussion of surfaces of genus 1 in Chapter 6. Suppose we have a general compact Riemann surface X of genus 1. We can show that X is equivalent to one of the families of surfaces studied in the previous chapter if we can show *either*:

- That there is a meromorphic function with a double pole (or two single poles) on X. This then represents X as a two-sheeted cover of the Riemann sphere with four branch points.

Or:

- That there is a nowhere-vanishing holomorphic 1-form on X. Then we can apply Theorem 3 to see that X is a complex torus $\mathbf{C}/\Lambda$.

This motivates our task, which is to get a good understanding of the existence of meromorphic functions, holomorphic 1-forms and the relationship between these for general compact Riemann surfaces.

Recall that on any Riemann surface X we have the 'square' of differential operators $\partial, \overline{\partial}$. We define complex vector spaces

$$H_X^{0,0} = \ker \overline{\partial} : \Omega^0 \to \Omega^{0,1},$$
$$H_X^{1,0} = \ker \overline{\partial} : \Omega^{1,0} \to \Omega^2,$$
$$H_X^{0,1} = \operatorname{coker} \overline{\partial} : \Omega^0 \to \Omega^{0,1},$$
$$H_X^{1,1} = \operatorname{coker} \overline{\partial} : \Omega^{1,0} \to \Omega^2.$$

Thus $H_X^{0,0}$ and $H_X^{1,0}$ are the spaces of holomorphic functions and holomorphic 1-forms, respectively. The significance of the other groups is not so clear. What we want to do now is to explain how $H_X^{0,1}$ arises naturally when one attempts to construct meromorphic functions. (The more general context for these ideas, extending also to higher-dimensional complex manifolds, is the theory of 'Dolbeault cohomology'.)

Let p be a point in X. We ask the question: is there a meromorphic function on X with a simple pole at p and no other poles? (Of course, in a sense, we have an answer: this happens if and only if X is equivalent to the Riemann sphere. But if we are presented with X, how can we tell if it is equivalent to the sphere?) Let z be a local co-ordinate centred on p. Thus $1/z$ can be regarded as a meromorphic function on some open neighbourhood U of p. We introduce a cut-off function β: a smooth function supported in U and equal to 1 near p. Then $\beta\, 1/z$ can be thought of as a function on $X \setminus p$, extending by zero outside U. Finding a meromorphic function with a pole at p is equivalent to finding a smooth function g on X such that $g + \beta\, 1/z$ is holomorphic on $X \setminus p$. Now

$$A = \bar{\partial} \left(\beta \frac{1}{z} \right) = (\bar{\partial}\beta)\frac{1}{z}$$

has compact support in $X \setminus \{p\}$, since β equals 1 near p. So we can regard A as a $(0,1)$-form on X, extending by zero over p. Thus our problem is equivalent to solving the equation

$$\bar{\partial} g = -A$$

for the given element A of $\Omega_X^{0,1}$ and the unknown $g \in \Omega_X^0$. By definition, a solution exists if and only if the class $[A]$ in the quotient $H_X^{0,1} = \operatorname{coker} \bar{\partial} = \Omega^{0,1}/\operatorname{Im} \bar{\partial}$ is zero. In particular, a solution will exist if $H_X^{0,1} = 0$. Even if a solution does not exist, the class $[A]$ is, up to multiplication by a non-zero scalar, a well-defined element of $H_X^{0,1}$ associated to the point p in X. For if ϕ is any smooth function on $X \setminus \{p\}$ which restricts to a meromorphic function with a pole at p on some neighbourhood of p, then, for a suitable choice of $\lambda \in \mathbb{C}$, the difference $\phi - \lambda\beta\, 1/z$ extends to a smooth function on X (holomorphic near p), so

$$[\bar{\partial}\phi] = \lambda[A] \in H_X^{0,1}.$$

Now suppose that we have d distinct points $p_1, \ldots p_d$ in X. We ask if we can find a meromorphic, but not holomorphic, function on X with poles at some or all of the p_i and no other poles. We follow through the same procedure as before, choosing local co-ordinates around the p_i to get $(0,1)$-forms A_i, which we can take to be supported in small disjoint annuli around the p_i if we like. The same argument as before shows that we can find the desired meromorphic function if there are scalars λ_i, not all zero, such that

$$\lambda_1[A_1] + \cdots + \lambda_d[A_d] = 0 \in H_X^{0,1}.$$

Given such a linear relation, we get a meromorphic function with poles at the points p_j for which $\lambda_j \neq 0$. In particular, we have the following result.

Proposition 26. *Suppose* $H_X^{0,1}$ *has finite dimension* h. *Then, given any* $h + 1$ *points* $p_1, \ldots, p_{h+1}$ *on* X, *there is a non-holomorphic meromorphic function on* X *with simple poles at some subset of the* $p_1, \ldots, p_{h+1}$.

This is just because there must be a non-trivial linear relation between any $h + 1$ elements of $H_X^{0,1}$.

The discussion above shows how we can cast our problem in terms of the spaces $H_X^{0,1}$ but it does not, by itself, get us very far. To go further, we need some much deeper input, and we will formulate this in terms of the following 'Main Theorem' for compact Riemann surfaces (this is not standard terminology).

Theorem 5. *Let* X *be a compact connected Riemann surface and let* ρ *be a 2-form on* X. *There is a solution* f *to the equation* $\Delta f = \rho$ *if and only if the integral of* ρ *over* X *is zero, and the solution is unique up to the addition of a constant.*

We will give a proof of this theorem in the next section, but let us see some consequences first.

8.1   Consequences of the Main Theorem

The relation between the 'Dolbeault cohomology' $H^{i,j}$ and the de Rham cohomology H^i can be summarised as follows. We have the following natural maps:

- A map $\sigma : H^{1,0} \to \overline{H^{0,1}}$ induced by $\alpha \mapsto \bar{\alpha}$.
- A bilinear map

$$B : H^{1,0} \times H^{0,1} \to \mathbf{C},$$

defined by

$$B(\alpha, [\theta]) = \int_X \alpha \wedge \theta.$$

(This is well defined, since changing the representative θ to $\theta + \bar{\partial} f$ changes the integral by

$$\int_X \alpha \wedge \bar{\partial} f = -\int_X \bar{\partial}(f\alpha),$$

which vanishes by Stokes' Theorem.)

- A map $i : H^{1,0} \to H^1$, defined by mapping a holomorphic (and hence closed) 1-form to its cohomology class.

- A map $v : H^{1,1} \to H^2$, defined to be the natural map induced from the inclusion Im : $\bar{\partial} : \Omega^{1,0} \to \Omega^2 \subset$ Im : $d : \Omega^1 \to \Omega^2$.

Theorem 6. *Let X be a compact connected Riemann surface.*

1. *The map σ induces an isomorphism from $H^{1,0}$ to $\overline{H^{0,1}}$.*
2. *The pairing B induces an isomorphism $H^{0,1} \cong (H^{1,0})^*$.*
3. *The map $H^{1,0} \oplus H^{0,1} \to H^1$ defined by*

$$(\alpha, \theta) \mapsto i(\alpha) + \overline{i(\sigma^{-1}(\theta))}$$

 is an isomorphism.
4. *The map $v : H^{1,1} \to H^2$ is an isomorphism.*

The proofs are entirely straightforward applications of the Main Theorem. To show that σ is surjective, we start with any class $[\theta]$ in $H^{0,1}$. We want to find a representative $\theta' = \theta + \bar{\partial} f$ such that $\partial \theta' = 0$, for this means that $\alpha = \overline{\theta'}$ is a holomorphic 1-form and $[\theta] = -\sigma(\alpha)$. Thus we want to solve the equation

$$\partial \bar{\partial} f = -\partial \theta.$$

Since $\partial \bar{\partial} = \frac{1}{2} i \Delta$, the Main Theorem tells us that we can solve this equation provided the integral of $\partial \theta$ vanishes, but this is so by Stokes' Theorem.

Now the composite of the map σ with the bilinear pairing B is, up to a factor, the Hermitian form

$$\langle \alpha, \beta \rangle = \int_X \alpha \wedge \bar{\beta},$$

which we know is positive definite. It follows that the map σ must be injective and, in turn, that B is a dual pairing. We leave the other parts as an exercise for the reader.

We see in particular from Theorem 6 that both $H^{1,0}$ and $H^{0,1}$ are complex vector spaces of dimension g. Thus the genus, which was initially a topological invariant, appears also as the crucial numerical invariant of the complex geometry of a Riemann surface.

We can give some simple consequences of the above.

Corollary 3. *Any compact Riemann surface of genus 0 is equivalent to the Riemann sphere.*

Corollary 4. *Any compact Riemann surface of genus 1 is equivalent to a torus $\mathbb{C}/\Lambda$.*

Corollary 5. *Let X be a compact Riemann surface of genus g and let $p_1, \ldots, p_{g+1}$ be distinct points on X. Then there is a non-constant meromorphic function on X with poles at some subset of the p_i.*

8.2 The Riemann–Roch formula

A more careful analysis of the argument leading to Corollary 5 gives the famous *Riemann–Roch* formula, the fundamental tool in the theory of compact Riemann surfaces. In this subsection, we will derive a restricted version of this, leaving the most general form to Chapter 10.

Consider, as before, distinct points $p_1, \ldots, p_d$ in a compact Riemann surface X and denote the set of these points by D. We write $H^0(D)$ for the vector space consisting of meromorphic functions having at worst simple poles at the points p_i. We also write $H^0(K - D)$ for the space of holomorphic 1-forms on X which *vanish* at the points p_i. (This choice of notation will become clearer in Chapter 10.) We write $h^0(D), h^0(K - D)$ for the dimensions of these spaces.

Theorem 7. *We have*

$$h^0(D) - h^0(K - D) = d - g + 1.$$

To see this, we go back to consider the invariant meaning of the 'residue' of a meromorphic function with a simple pole at a point p. Working in a local co-ordinate z centred at p, this of course just means the co-efficient a_{-1} in the Laurent series

$$f(z) = a_{-1}z^{-1} + a_0 + a_1 z + \cdots.$$

If we have another local co-ordinate $\tilde{z} = c_1 z + c_2 z^2 + \cdots$, then the residue $\tilde{a}_{-1}$ in this co-ordinate will be

$$\tilde{a}_{-1} = c_1^{-1} a_{-1}.$$

What this means is that the expression $a_{-1}\, \partial/\partial z$ defines a tangent vector at p, independent of the choice of co-ordinate. So the invariant meaning of the residue is an element of TX_p. (Another way of expressing this is that the residue of a meromorphic 1-form is well defined as a complex number, and the residue of a meromorphic function f paired with a cotangent vector α is the residue of $f\alpha$ for any local extension of α.) These residues give us a map

$$R : H^0(D) \to \bigoplus_i TX_{p_i},$$

which takes a meromorphic function to its residues. The kernel of this map consists of the holomorphic functions, which are just the constants since X is compact.

In this more invariant setting the elements $[A_i] \in H^{0,1}$ considered above, which depended on a choice of local co-ordinates, should be replaced by linear maps $A_i : TX_{p_i} \to H^{0,1}$. Thus we have a map

$$\underline{A} : \bigoplus TX_{p_i} \to H^{0,1}.$$

The image of R and the kernel of $\underline{A}$ are both vector subspaces of $\bigoplus TX_{p_i}$, and our previous argument shows that these are identical. In other words, we have an *exact sequence*

$$0 \to \mathbf{C} \to H^0(D) \to \bigoplus TX_{p_i} \to H^{0,1}. \tag{8.1}$$

Now we have the elementary linear-algebra relation

$$\dim \ker \underline{A} = d - \dim H^{0,1} + \dim \ker \underline{A}^T, \tag{8.2}$$

where $\underline{A}^T$ is the transpose acting on the dual vector spaces,

$$\underline{A}^T : (H^{0,1})^* \to \bigoplus (TX_{p_i})^*.$$

By item 2 of Theorem 6, we can identify the dual space $(H^{0,1})^*$ with $H^{1,0}$ and we can obviously identify the dual of $\bigoplus TX_{p_i}$ with $\bigoplus T^*X_{p_i}$. On the other hand, we have an evaluation map

$$ev : H^{1,0} \to \bigoplus T^*X_{p_i},$$

which simply takes a holomorphic 1-form to its values at the points p_i. Now we reach the crucial point.

Lemma 17. *Under these identifications, the transpose of $\underline{A}$ is $2\pi i$ times the evaluation map ev.*

We assume this for the moment and proceed to establish Theorem 7. By definition, the kernel of ev is $H^0(K - D)$ and the exact sequence (8.1) implies that $h^0(D) = 1 + \dim \ker \underline{A}$. Thus the elementary formula (8.2) gives

$$h^0(D) - 1 = d - \dim H^{0,1} + h^0(K - D),$$

which is the result asserted in Theorem 7, since $\dim H^{0,1} = g$.

Now we prove Lemma 17. It is fairly clear that it suffices to consider the case of a single point p, around which we choose a local co-ordinate z. By our construction, the element $A = A(\partial/\partial z) \in H^{0,1}(X)$ is represented by the $(0,1)$-form $b = \bar{\partial}(\beta) \, 1/z$, where β is a cut-off function equal to 1 near the origin. We have to compute the pairing $\langle A, \theta \rangle$, where θ is a holomorphic 1-form on X, and by definition this is the integral

$$I = \int_X b \wedge \theta.$$

We write $\theta = g(z) \, dz$ in our local co-ordinate system. We can write I as an integral over $\mathbf{C}$,

$$\int_{\mathbf{C}} \bar{\partial}(\beta) \frac{1}{z} g(z) \, dz.$$

Choose a small circle γ around the origin, contained in the disc where $\beta = 1$. Then an application of Stokes' Theorem gives

$$I = \int_\gamma \beta \frac{1}{z} \theta = \int_\gamma \frac{1}{z} g(z)\, dz.$$

Now Cauchy's Residue Theorem gives $I = 2\pi i g(0)$, which is the desired statement.

There is a cleaner way of expressing the conclusion of this analysis. If f is a meromorphic function on X and θ is a holomorphic 1-form, the product $f\theta$ is a meromorphic 1-form and the residues of $f\theta$ at the poles are invariantly defined as complex numbers. Stokes' Theorem implies that the sum of the residues is 0, and this is precisely the assertion that $\sum \langle \mathrm{Res}_{p_i}\, f, \theta(p_i) \rangle = 0$, where $\mathrm{Res}_{p_i}\, f \in TX_{p_i}$ is the residue of f at p_i. This is a linear constraint on the collection of possible residues of f. Our result is the assertion that this necessary condition is also sufficient for the existence of a meromorphic function.

Exercises

1. Let $x_1, \ldots, x_n$ be distinct points in a compact Riemann surface Σ and let $w_1, \ldots w_n$ be distinct points in $\mathbf{C}$. Show that there is a meromorphic function on Σ which maps x_i to w_i for $i = 1, \ldots, n$.

2. Show that a compact Riemann surface admits a branched cover of the sphere with only simple branch points.

9 Proof of the Main Theorem

9.1 Discussion and motivation

We will now embark on the proof of our main analytical result, Theorem 5, for compact Riemann surfaces. Before getting to work, we give some preliminary discussion. The theorem consists of the three statements, on a compact connected Riemann surface X:

- If there is a solution to the equation $\Delta\phi = \rho$, then the integral of ρ is zero.
- If there is a solution, it is unique up to the addition of a constant.
- Conversely, if ρ is a form of integral 0, we can find a solution ϕ.

The first and second of these three statements are very easy to prove, so the real content is the third statement. The first statement follows immediately from Stokes' Theorem, since for any ϕ,

$$\int_X \Delta\phi = 2i \int_X \bar\partial\partial\phi = 2i \int_X d(\partial\phi) = 0.$$

The second statement is equivalent to the assertion that the only harmonic functions—solutions of the equation $\Delta f = 0$—are the constants. One can see this in two ways: either by the *maximum principle* for harmonic functions or by considering the *Dirichlet integral*. For the first, one considers a point in X where f is maximal, which exists by the compactness assumption, and one applies the maximum principle at that point. For the second, one writes

$$\int_X |df|^2 = \int_X f \Delta f = 0$$

when $\Delta f = 0$. Thus df vanishes everywhere on X, and f is a constant. These two proofs of the uniqueness are both very simple, but the two approaches are manifestations of two different approaches—via the maximum principle and the Dirichlet integral—which can be taken to the whole theory. To illustrate this, consider a problem which is closely related to that considered in our theorem, the solution of the Dirichlet boundary value problem. Here we consider a bounded domain $\Omega \subset \mathbf{C}$ with a smooth boundary $\partial\Omega$ and a given function g on the boundary. The problem is to solve the equation $\Delta\phi = 0$ in Ω with the boundary condition that ϕ has a continuous extension to $\overline{\Omega}$, equal to g on the boundary. Supposing that a solution ϕ exists, one can show that it is characterised by two different extremal properties:

1. For each x in Ω,

$$\phi(x) = \min\{\psi(x) : \psi|_{\partial\Omega} = g, \Delta\psi \geq 0\}.$$

2. The function ϕ minimises $\int_\Omega |\nabla\psi|^2$ over all functions ψ on $\overline{\Omega}$ with $\psi|_{\partial\Omega} = g$.

Conversely, one can prove the existence of a solution ϕ by showing that such extremal functions exist. Both of these approaches have their own merits, and they generalise in different ways. The line we will take in the proof of the Main Theorem, and the further results in the next chapter, will follow the Dirichlet integral approach, which is closest to the heuristic arguments originally employed by Riemann.

The equation $\Delta\phi = \rho$ is, in local co-ordinates, the *Poisson equation*, which may be familiar from potential theory in $\mathbf{R}^n$. In this vein, one can obtain some physical intuition into why the Main Theorem should be true, as follows. In this discussion we will anticipate a result proved in Chapter 13, that an oriented surface in $\mathbf{R}^3$ is naturally a Riemann surface. Suppose the Riemann surface X arises in this way. We get a standard area form on X, so we can identify 2-forms and functions. Think of this surface in $\mathbf{R}^3$ as being made of a thin metal sheet and the function ϕ as being the temperature distribution over the sheet. The function ρ represents some externally imposed source or removal of heat, varying over the surface. Then the Poisson equation $\Delta\phi = \rho$ is the equation for a steady-state temperature distribution, and the content of our theorem is that if the integral of ρ is zero—so there is no overall gain or loss of heat—then such a temperature distribution exists. One can also think about this in terms of the time-dependent heat equation

$$\frac{\partial\phi}{\partial t} = \rho - \Delta\phi.$$

Here ϕ is now a function on $X \times [0, \infty)$, while we still suppose that ρ is constant in time. One can prove analytically, in line with one's physical intuition, that if the integral of ρ is zero, then for any initial temperature distribution this heat equation has a solution which converges as $t \to \infty$ to a solution of the Poisson equation.

One case in which the theorem can be proved easily is that of a torus. Suppose, for example, that X is the quotient of $\mathbf{C}$ by the 'square' lattice $2\pi\mathbf{Z} \oplus 2\pi i\mathbf{Z}$. We take standard real angular co-ordinates θ_1, θ_2 on X and identify functions with 2-forms in the obvious way. Any smooth function on X can be written as a double Fourier series

$$f = \sum_{n,m} f_{nm} e^{in\theta_1 + im\theta_2}.$$

The Laplacian of such a function is

$$\Delta f = \sum (n^2 + m^2) f_{n,m} e^{in\theta_1 + im\theta_2},$$

and the integral of f with respect to the standard area form is $4\pi^2 f_{00}$. Thus, if ρ has integral zero, we can write down the solution to the Poisson equation in the form

$$\phi = \sum_{(m,n) \neq (0,0)} \frac{1}{m^2 + n^2} \rho_{mn} e^{im\theta_1 + n\theta_2},$$

where ρ_{mn} are the Fourier coefficients of ρ.

As a final remark, it is worth pointing out that our theorem fits into a wider setting of elliptic differential operators on compact manifolds. We shall not pause to explain what is meant by an elliptic operator: suffice it to say that this is a class of linear differential operators which includes the Laplace operator Δ as a particular case. If $\mathcal{L}$ is any linear differential operator over a compact manifold, and if we choose appropriate volume forms etc., there is a formal adjoint operator $\mathcal{L}^*$, characterised by the fact that for any f, g,

$$\langle \mathcal{L}f, g \rangle = \langle f, \mathcal{L}^* g \rangle,$$

where $\langle\, ,\, \rangle$ denotes the L^2 inner product. Then the main result is that if $\mathcal{L}$ is elliptic, one can solve the equation

$$\mathcal{L}\phi = \rho$$

if and only if ρ is orthogonal, in the L^2 sense, to the kernel of $\mathcal{L}^*$. In the Riemann surface situation of Theorem 5, if we choose an area form on the surface to identify functions and 2-forms, the formal adjoint of the Laplacian is the same operator, i.e. $\Delta^* = \Delta$, so the kernel of Δ^* consists of the constant functions, and the condition that ρ be orthogonal to this kernel is just that the integral of ρ is zero.

9.2   The Riesz Representation Theorem

We will now begin the proof. As we have said above, our approach will hinge on the Dirichlet integral, and an efficient way to build this into the argument uses the language of Hilbert spaces. Recall that we have defined the Dirichlet norm (or, more precisely, semi-norm) and inner product on functions on X. The norm and inner product are unchanged if we modify our functions by the addition of constants. We let $C^\infty(X)/\mathbf{R}$ be the vector space obtained by dividing out by the constant functions, so the norm and inner product descend to this quotient. Then we have the following proposition.

Proposition 27. *The Dirichlet norm and inner product make $C^\infty(X)/\mathbf{R}$ into a pre-Hilbert space.*

(Our notation will sometimes blur the distinction between a function on X and the equivalence class in $C^\infty(X)/\mathbf{R}$ which it represents. Note that if we fix a metric, i.e. an area form, on X, we can identify $C^\infty(X)/\mathbf{R}$ with the space of functions on X of integral zero.)

Now suppose that ρ is a 2-form on X. For any functions ϕ, ψ on X, we have

$$\int_X \psi(\rho - \Delta\phi) = \int_X \psi\rho - \int_X \psi\,\Delta\phi = \int_X \psi\rho - \int_X \nabla\phi.\nabla\psi = \int_X \psi\rho - \langle\phi,\psi\rangle_D.$$

By a simple, standard argument, the equation $\Delta\phi = \rho$ is equivalent to the condition that for all functions ψ we have

$$\int_X \psi(\rho - \Delta\phi) = 0.$$

Thus the content of Theorem 5 can be reformulated as the assertion that, when $\int_X \rho = 0$, there is a function ϕ such that

$$\int_X \psi\rho = \langle\psi,\phi\rangle_D$$

for all $\psi \in C^\infty(X)$. To write this more compactly, define

$$\hat\rho(\psi) = \int_X \rho\psi.$$

Then, if the integral of ρ is zero, this induces a linear map

$$\hat\rho : C^\infty(X)/\mathbf{R} \to \mathbf{R},$$

and our problem is to find a ϕ such that

$$\hat\rho(\psi) = \langle\psi,\phi\rangle_D \tag{9.1}$$

for all ψ.

In this formulation, our problem falls into the class covered by the well-known Riesz Representation Theorem from Hilbert space theory.

Theorem 8. *Let H be a real Hilbert space and let $\sigma : H \to \mathbf{R}$ be a bounded linear map (so there is a constant C such that $|\sigma(x)| \le C\|x\|$ for all $x \in H$). Then there is a $z \in H$ such that*

$$\sigma(x) = \langle z, x\rangle$$

for all x in H.

For a proof, see almost any elementary functional-analysis textbook.

With all this background in place, we can see that the proof of Theorem 5 divides into two parts. First, we want to put ourselves into the position where we can apply the Riesz Representation Theorem, and for this we need a Hilbert space. Thus we let H be the abstract *completion* of our pre-Hilbert space $C^\infty(X)/\mathbf{R}$, under the Dirichlet norm $\|\ \|_D$. A point of H is an equivalence class of Cauchy sequences (ψ_i) in $C^\infty(X)/\mathbf{R}$ under the equivalence relation $(\psi_i) \sim (\psi_i')$ if $\|\psi_i - \psi_i'\|_D \to 0$. The crucial thing we need now is the following theorem.

Theorem 9. *The functional $\hat\rho : C^\infty(X)/\mathbf{R} \to \mathbf{R}$ is bounded: there is a constant C such that $|\hat\rho(\psi)| \le C\|\psi\|_D$ for all ψ in $C^\infty(X)/\mathbf{R}$.*

Assuming this for the moment, it follows that $\hat\rho$ *extends* to a bounded linear map from H to $\mathbf{R}$, which we still denote by $\hat\rho$. (The proof is just to observe that for any Cauchy sequence (ψ_i) in $C^\infty(X)/\mathbf{R}$, the sequence $\hat\rho(\phi_i)$ is Cauchy in $\mathbf{R}$, so we can define the extension of $\hat\rho$ by taking the limit.) So we can apply the Riesz Representation Theorem, and we conclude that there is a ϕ in the completion H with $\hat\rho(\psi) = \langle\phi, \psi\rangle_D$ for all ψ. An object of this type is called a *weak solution* to our problem and the other part of the proof of Theorem 5 is to establish the following fact.

Theorem 10. *If ρ is a smooth 2-form on X of integral zero, then a weak solution ϕ in H of equation (9.1) is smooth, i.e. lies in the subset $C^\infty(X)/\mathbf{R}$ of H.*

9.3   The heart of the proof

The foundation of our proof of Theorem 9 will be a result from the calculus of two real variables. Suppose Ω is a bounded, convex, open set in $\mathbf{R}^2$. (For our applications, it suffices to consider the case of a circular disc.) Let A be the area of Ω and let d be its diameter.

Theorem 11. *Let ψ be a smooth function on an open set containing the closure $\overline{\Omega}$ and let $\overline{\psi}$ denote the average*

$$\overline{\psi} = \frac{1}{A}\int_\Omega \psi\, d\mu,$$

where $d\mu$ is the standard Lebesgue measure on $\mathbf{R}^2$. Then, for $x \in \Omega$, we have

$$|\psi(x) - \overline{\psi}| \le \frac{d^2}{2A}\int_\Omega \frac{1}{|x - y|}|\nabla\psi(y)|\, d\mu_y.$$

(Here the notation is supposed to indicate that the variable of integration on the right-hand side is $y \in \Omega$.)

To prove this, there is no loss in supposing that the point x is the origin in $\mathbf{R}^2$ (applying a translation in $\mathbf{R}^2$) and that $\psi(0)$ is zero (changing ψ by addition of a constant). We work in standard polar co-ordinates (r, θ) on the plane. Thus we can write

$$\overline{\psi} = \frac{1}{A} \int_0^{2\pi} \int_0^{R(\theta)} \psi(r, \theta) r \, dr \, d\theta,$$

where $R(\theta)$ is the length of the portion of the ray at angle θ lying in Ω. (Here we use the fact that Ω is convex.) Now, if we introduce another radial variable ρ, we can write, for each (r, θ),

$$\psi(r, \theta) = \int_0^r \frac{\partial \psi}{\partial \rho} \, d\rho,$$

using the fact that ψ vanishes at the origin. So now we have

$$\overline{\psi} = \frac{1}{A} \int_0^{2\pi} \int_0^{R(\theta)} \int_{\rho=0}^r r \frac{\partial \psi}{\partial \rho} \, d\rho \, dr \, d\theta.$$

We interchange the order of the r and ρ integrals, so

$$\overline{\psi} = \frac{1}{A} \int_0^{2\pi} \int_{\rho=0}^{R(\theta)} \left(\int_{r=\rho}^{R(\theta)} r \, dr \right) \frac{\partial \psi}{\partial \rho} \, d\rho \, d\theta.$$

The innermost integral is

$$\int_{r=\rho}^{R(\theta)} r \, dr = \frac{1}{2}(R(\theta)^2 - \rho^2),$$

which is positive and no larger than $R(\theta)^2/2$, while, by definition, $R(\theta) \leq d$. Thus

$$|\overline{\psi}| \leq \frac{d^2}{2A} \int_0^{2\pi} \int_0^{R(\theta)} \frac{1}{\rho} \left| \frac{\partial \psi}{\partial \rho} \right| \rho \, d\rho \, d\theta.$$

The modulus of the radial derivative $\partial \psi / \partial \rho$ is at most that of the full derivative $\nabla \psi$, so, switching back to a coordinate-free notation, we have

$$|\overline{\psi}| \leq \frac{d^2}{2A} \int_\Omega \frac{1}{|y|} |\nabla \psi_y| \, d\mu_y,$$

as required.

Corollary 6. *Under the hypotheses above,*

$$\int_\Omega |\psi(x) - \overline{\psi}|^2 \, d\mu_x \leq \left(\frac{d^3 \pi}{A} \right)^2 \int_\Omega |\nabla \psi|^2 \, d\mu.$$

To prove this, and for later use, we recall the notion of the *convolution* of functions on $\mathbf{R}^2$. The convolution of functions f, g is defined by

$$(f * g)(x) = \int_{\mathbf{R}^2} f(y)g(x - y) \, d\mu_y.$$

The operation $(f * g)$ is commutative and associative, and if $\| \ \|_T$ is any translation-invariant norm on functions on $\mathbf{R}^2$ we have

$$\|f * g\|_T \le \|f\|_{L^1} \|g\|_T,$$

where $\|f\|_{L^1}$ is the usual L^1 norm

$$\|f\|_{L^1} = \int_{\mathbf{R}^2} |f| \, d\mu.$$

In particular, this holds when $\| \cdot \|_T$ is the L^2 norm

$$\|g\|_{L^2}^2 = \int_{\mathbf{R}^2} |g|^2 \, d\mu.$$

(Strictly, we should specify what class of functions we are considering in the definition of the convolution, but this will be clear in the different contexts as they arise.)

To prove the corollary, we define

$$K(x) = \frac{d^2}{2A} \frac{1}{|x|} \quad \text{for } |x| < d,$$

and $K(x) = 0$ if $|x| \ge d$. This has a singularity at the origin but is nevertheless an integrable function, and

$$\|K\|_{L^1} = 2\pi \frac{d^2}{2A} \int_0^d dr = \frac{d^3 \pi}{A}.$$

We define a function g on $\mathbf{R}^2$ by

$$g(y) = |\nabla \psi(y)|$$

if $y \in \Omega$ and $g(y) = 0$ if $y \notin \Omega$. Then $K * g$ is a positive function on $\mathbf{R}^2$ and Theorem 11 asserts that for all $x \in \Omega$,

$$|\psi(x) - \overline{\psi}| \le |(K * g)(x)|.$$

It follows that

$$\int_\Omega |\psi(x) - \overline{\psi}|^2 \, d\mu_x \le \|K * g\|_{L^2}^2 \le \|K\|_{L^1}^2 \|g\|_{L^2}^2 \le \left(\frac{d^3 \pi}{A} \right)^2 \|\nabla \psi\|_{L^2}^2,$$

as asserted.

We can now prove Theorem 9. We begin with the case when ρ is supported in a single coordinate chart in our Riemann surface, which we identify with a bounded convex set Ω in $\mathbf{C} = \mathbf{R}^2$. Working in this local coordinate system, we use the Lebesgue area form to identify functions and 2-forms, so ρ can be regarded as a function of integral zero supported on Ω. Likewise, a function ψ on X can be regarded as defining a function, which we also call ψ, on a neighbourhood of Ω in $\mathbf{C}$, and we can write

$$\hat{\rho}(\psi) = \int_\Omega \rho\psi \, d\mu.$$

Now, since the integral of ρ is zero, we also have

$$\hat{\rho}(\psi) = \int_\Omega \rho(\psi - \overline{\psi}) \, d\mu,$$

and, by the Cauchy–Schwarz inequality,

$$\left| \int_\Omega \rho(\psi - \overline{\psi}) \, d\mu \right| \leq \|\rho\|_{L^2(\Omega)} \|\psi - \overline{\psi}\|_{L^2(\Omega)}.$$

Using Corollary 6, we then deduce that

$$|\hat{\rho}(\psi)| \leq C \|\nabla\psi\|_{L^2(\Omega)},$$

where $C = d^3 \pi A \|\rho\|_{L^2(\Omega)}$. Finally,

$$\|\nabla\psi\|_{L^2(\Omega)} \leq \|\nabla\psi\|_{L^2(X)} = \|\psi\|_D,$$

which completes the proof of Theorem 9 in this case.

To treat a general 2-form ρ on X, of integral zero, we recall from Chapter 5 that integration over X defines an isomorphism from $H^2(X)$ to $\mathbf{R}$, so we can write $\rho = d\theta$ for some 1-form θ on X. We fix a cover of X by a finite number of coordinate charts $U_\alpha \subset X$ of the kind considered above, and choose a partition of unity χ_α subordinate to this cover. Put $\rho_\alpha = d(\chi_\alpha\theta)$. Then each ρ_α is supported in the corresponding coordinate chart U_α, and

$$\int_X \rho_\alpha = \int_X d(\chi_\alpha\theta) = 0.$$

On the other hand,

$$\rho = d\theta = d\left(\left(\sum \chi_\alpha \right) \theta \right) = \sum \rho_\alpha.$$

Our previous argument shows that each of the linear maps $\hat{\rho}_\alpha$ is bounded, and so $\hat{\rho} = \sum \hat{\rho}_\alpha$ is also (as a finite sum of bounded linear maps).

9.4 Weyl's Lemma

Suppose that ϕ is an element of H which is a weak solution to our problem in the sense explained above. That is, we have a sequence of functions ϕ_i on X which is Cauchy with respect to the Dirichlet norm and, for any ψ,

$$\langle \phi_i, \psi \rangle \to \hat{\rho}(\psi)$$

as i tends to infinity. We want to see first that we can identify the abstract object ϕ with a function (up to a constant) on X, where initially this function will just be locally in L^2 (i.e. represented by an L^2 function, in the ordinary sense, in any local co-ordinate chart). To do this, we consider first any fixed co-ordinate chart, identified with $\Omega \subset \mathbf{C}$, as above. We can suppose, after changing the ϕ_i by the addition of suitable constants, that the integrals of the ϕ_i over Ω vanish and then, by Corollary 6, we have

$$\|\phi_i - \phi_j\|_{L^2(\Omega)} \leq C \|\phi_i - \phi_j\|_D.$$

Hence ϕ_i gives a Cauchy sequence in $L^2(\Omega)$ which converges to an L^2 limit by the completeness of L^2. We claim now that this same sequence ϕ_i converges locally in L^2 over *all* of X. We will give the argument in a form which will work equally well in the generalisation considered in the next chapter. Let A be the set of points x in X with the property that there is a co-ordinate chart around x in which ϕ_i converges to ϕ in L^2. Then A is non-empty by the preceding discussion, and A is open in X from the nature of its definition. Since X is connected, the complement of A is not open, unless it is the empty set, so either $A = X$ or there is a point y which is in the closure of A but not in A. But, in the latter case, we could find a coordinate neighbourhood Ω' about y and a sequence of real numbers c_i' such that $\phi_i - c_i'$ converges in L^2 over Ω'. But now there is a point x in $A \cap \Omega'$ and, on a small neighbourhood of x, both ϕ_i and $\phi_i - c_i'$ converge in L^2. This means that c_i' tends to 0 as $i \to \infty$, so in fact y is in A after all, a contradiction.

To sum up, we now have a function ϕ on X which is locally in L^2 and which is a *weak solution* to the equation $\Delta \phi = \rho$. We need to show that ϕ is smooth. Since smoothness is a local property, we can fix our attention on a single co-ordinate chart. What we need is a version of 'Weyl's Lemma', as stated in the following proposition.

Proposition 28. *Let Ω be a bounded open set in $\mathbf{C}$ and let ρ be a smooth 2-form on Ω. Suppose ϕ is an L^2 function on Ω with the property that, for any smooth function χ of compact support in Ω,*

$$\int_\Omega \Delta \chi \phi = \int_\Omega \chi \rho.$$

Then ϕ is smooth and satisfies the equation $\Delta \phi = \rho$.

The proof will involve a number of steps. The first step is to reduce the problem to the case when ρ is zero. Since smoothness is a local property, it suffices to prove that ϕ is smooth over any given interior set Ω', where we suppose that the ϵ-neighbourhood of Ω' is contained in Ω. Then we can choose a ρ' equal to ρ on a neighbourhood of the closure of Ω' and of compact support in Ω. Suppose we can find some smooth solution ϕ' of the equation $\Delta\phi' = \rho'$ over Ω. Then $\psi = \phi - \phi'$ will be a weak solution of the equation $\Delta\psi = 0$ on Ω'. If we can prove that ψ is smooth, then so will be ϕ.

To find the smooth solution ψ', we use the 'Newton potential' in two dimensions,

$$K(x) = \frac{1}{2\pi} \log |x|.$$

Of course, this is not defined at $x = 0$, but K is well defined as a locally integrable function on $\mathbf{C}$. For any smooth function f of compact support in $\mathbf{C}$, the convolution

$$K * f(x) = \int K(y) f(x - y) \, d\mu_y$$

is defined and $K * f$ is smooth.

Lemma 18.

- *If σ has compact support in $\mathbf{C}$ then $K * (\Delta\sigma) = \sigma$.*
- *If f has compact support, then $\Delta(K * f) = f$.*

This lemma essentially expresses the standard fact that convolution with K furnishes an inverse to the Laplace operator. To prove the first assertion we may, by translation invariance, calculate at the point $x = 0$. Then

$$(K * \Delta\sigma)(0) = \int \frac{1}{2\pi} \log(|y|)(\Delta\sigma)_y \, d\mu_y.$$

Now, $\Delta \log |y|$ vanishes on $\mathbf{C} \setminus \{0\}$. We write the integral as the limit as δ tends to zero of the integral over the set where $|y| \geq \delta$. We then use Green's identity to write this as a boundary integral and then take the limit as δ tends to zero. The argument is rather standard, so we do not give more details.

For the second part, we write

$$(K * f)(x) = \int K(y) f(x - y) \, d\mu_y.$$

When we take the Laplacian with respect to x, there is no problem in moving the differential operator inside the integral, since f is smooth and x does not appear inside the argument of K. Thus

$$\Delta(K * f) = \int K(y)\Delta_x f(x - y) \, d\mu_y,$$

where the notation means that we take the Laplcian with respect to x. But this is just the same as $K * \Delta f$, which is equal to f by the first part.

We have now reduced the problem to the case when $\rho = 0$ so, changing notation, let us suppose that ϕ is a weak solution of $\Delta \phi = 0$ on Ω and seek to prove that ϕ is smooth on the interior domain Ω', with the ϵ-neighbourhood of Ω' contained in Ω. The argument now exploits the mean-value property of smooth harmonic functions. This says that if ψ is a smooth harmonic function on a neighbourhood of a closed disc, then the value of ψ at the centre of the disc is equal to the average value on the circle boundary. Fix a smooth function β on $\mathbf{R}$ with $\beta(r)$ constant for small r and vanishing for $r \geq \epsilon$, and such that

$$2\pi \int_0^\infty r\beta(r) \; dr = 1.$$

Now let B be the function $B(z) = \beta(|z|)$ on $\mathbf{C}$. Then B is smooth and has integral 1 over $\mathbf{C}$ (with respect to the ordinary Lebesgue measure). Suppose first that ψ is a smooth harmonic function on a neighbourhood of the closed ϵ-disc centred at the origin. Then we have

$$\int_{\mathbf{C}} B(-z)\psi(z) \, d\mu_z = \int_0^\infty \int_0^{2\pi} r\beta(r)\psi(r,\theta) \, d\theta \, dr = \psi(0) \int_0^\infty r\beta(r) \, dr = \psi(0),$$

where we have switched to polar coordinates and used the mean-value property

$$\int_0^{2\pi} \psi(r,\theta) \, d\theta = 2\pi \psi(0).$$

Now the integral above is just that defining the convolution $B * \psi$ at 0. By translation invariance, we obtain the following result.

Proposition 29. *Let ψ be a smooth function on $\mathbf{C}$, and suppose that $\Delta \psi$ is supported in a compact set $J \subset \mathbf{C}$. Then $B * \psi - \psi$ vanishes outside the ϵ-neighbourhood of J.*

We see in particular from this that *if* our function ϕ on Ω is smooth we must have $B * \phi = \phi$ in Ω'. Conversely, for any L^2 function ϕ, the convolution $B * \phi$ is smooth. So, proving the smoothness of ϕ in Ω' is equivalent to establishing the identity $B * \phi = \phi$ in Ω'. To do this, we proceed as follows. It suffices to show that for any smooth test function χ of compact support in Ω' we have

$$\langle \chi, \phi - B * \phi \rangle = 0,$$

where we are writing $\langle \, , \, \rangle$ for the usual 'inner product'

$$\langle f, g \rangle = \int fg \; d\mu.$$

We use the fact that for any functions f, g, h in a suitable class,

$$\langle f, g * h \rangle = \langle g * f, h \rangle.$$

This follows by straightforward rearrangements of the integrals. We will not bother to spell out conditions on the functions involved, since the validity of the identity will be fairly obvious in our applications below.

Put $h = K * (\chi - B * \chi) = K * \chi - B * K * \chi$. Now $K * \chi$ is a smooth function on $\mathbf{C}$, and $\Delta K * \chi = \chi$ by the lemma above. Thus $\Delta K * \chi$ vanishes outside the support of χ, and hence, by the proposition above, $B * K * \chi$ equals $K * \chi$ outside the ϵ-neighbourhood of the support of χ. Thus h has compact support contained in Ω. So we can use h as a test function in the hypothesis that $\Delta\phi = 0$ weakly, i.e. we have

$$\langle \Delta h, \phi \rangle = 0.$$

But $\Delta h = \Delta K * (\chi - B * \chi) = \chi - B * \chi$ by the lemma above (since χ and $B * \chi$ have compact support). So we see that

$$\langle \chi - B * \chi, \phi \rangle = 0.$$

But, applying the identity above again, this gives

$$\langle \chi, \phi - B * \phi \rangle = 0,$$

as desired.

Exercises

1. For a compactly supported smooth function u on $\mathbf{C}$, we define

$$(Tu)(z) = \frac{1}{\pi} \int_{\mathbf{C}} \frac{u(w)}{w - z} \, d\mu_w,$$

where the notation $d\mu_w$ means the Lebesgue measure with respect to w. Show that $f = Tu$ solves the equation $\partial f / \partial \bar{z} = u$.

2. Show that there is a constant C such that for any smooth function f of compact support on the unit disc in $\mathbf{R}^2$ we have

$$\int |f|^2 \, d\mu \leq C \int |\nabla f|^2 \, d\mu.$$

(There are several ways of going about this. One is to use Fourier theory.)

3. Fix $p > 2$. Show that there is a constant C_p such that for f as in the previous question,

$$|f(0)| \leq C_p \int |\nabla f|^p \, d\mu.$$

4. We define a *harmonic* 1-form on a Riemann surface to be a real 1-form which is the real part of a holomorphic 1-form. Let X be a compact Riemann surface and let $\sigma : X \to X$ be an anti-holomorphic involution with no fixed points, so the quotient $Y = X/\sigma$ is a compact surface. Show how to define harmonic 1-forms on Y, and prove that each class in the de Rham cohomology $H^1(Y; \mathbf{R})$ has a unique harmonic representative (a case of the 'Hodge Theorem').

10 The Uniformisation Theorem

10.1 Statement

In this chapter, we prove the following theorem.

Theorem 12. *Let X be a connected, simply connected, non-compact Riemann surface. Then X is equivalent to either* **C** *or the upper half-plane H.*

Corollary 7. *Any connected Riemann surface is equivalent to one of the following:*

- *the Riemann sphere S^2;*
- **C** *or* $\mathbf{C}/\mathbf{Z} = \mathbf{C} \setminus \{0\}$ *or* $\mathbf{C}/\Lambda$ *for some lattice* Λ;
- *a quotient* H/Γ, *where* $\Gamma \subset PSL(2, \mathbf{R})$ *is a discrete subgroup acting freely on H.*

The corollary follows because any Riemann surface is a quotient of its universal cover by an action of its fundamental group, and we have seen that the only compact simply connected Riemann surface is the Riemann sphere.

Our proof of Theorem 12 will follow the same general pattern as the one we have already given to classify compact simply connected Riemann surfaces, but the non-compactness will require some extra steps. Most of our work goes into the proof of an analogue of the 'Main Theorem'. To state this, recall that if ϕ is real-valued function on a non-compact space X and c is a real number, we say that ϕ tends to c at infinity in X if for all $\epsilon > 0$ there is a compact subset K of X such that $|\phi(x) - c| < \epsilon$ if x is not in K. We say that ϕ tends to $+\infty$ at infinity in X if, for all $A \in \mathbf{R}$, there is a compact set $K \subset X$ such that $\phi(x) > A$ if x is not in K (and we say that ϕ tends to $-\infty$ at infinity in X if $-\phi$ tends to $+\infty$).

Theorem 13. *Let X be a connected, simply connected, non-compact Riemann surface. Then if ρ is a (real) 2-form of compact support on X with $\int_X \rho = 0$, there is a (real-valued) function ϕ on X with $\Delta\phi = \rho$ and such that ϕ tends to 0 at infinity in X.*

We now show that Theorem 13 implies Theorem 12. We choose a point $p \in X$ and a local complex co-ordinate z around p. Using the same notation as in Chapter 8, we put

$$A = \bar{\partial}\left(\frac{\beta}{z}\right),$$

where β is a cut-off function, so A is a $(0,1)$-form supported in an annulus around p. We put $\rho = \partial A$, so ρ is a complex-valued 2-form with integral zero, by Stokes' Theorem. By the result (applied to the real and imaginary parts of ρ), we can find a complex-valued function g with $\partial\bar{\partial}g = \rho$, and with the real and imaginary parts of g tending to 0 at infinity in X. Now let a be the real 1-form

$$a = (A - \bar{\partial}g) + (\overline{A} - \overline{\bar{\partial}g}).$$

By construction, $\partial(A - \bar{\partial}f) = 0$, and this means that $da = 0$. So, since $H^1(X) = 0$, there is a real-valued function ψ with $a = d\psi$. This means that $A = \bar{\partial}g + \bar{\partial}\psi$. Hence

$$\bar{\partial}\left(\frac{\beta}{z} - (g + \psi)\right) = 0$$

on $X \setminus \{p\}$. Hence $f = g + \psi$ is a meromorphic function on X with a simple pole at p, and the imaginary part of f tends to zero at infinity in X, since ψ is real. (Strictly, the imaginary part of f is not a function on X, since f has a pole, but the meaning should be clear—to be precise, we could say that the imaginary part of f tends to 0 at infinity on $X \setminus D$, where D is an open disc about p.)

The meromorphic function f is a holomorphic map from X to the Riemann sphere $S^2 = \mathbf{C} \cup \{\infty\}$. Let H_+ and H_- denote the (open) upper and lower half-planes in $\mathbf{C}$. Let $X_{\pm}$ be the pre-images $f^{-1}(H_{\pm})$ in X. So X_+ and X_- are open subsets of X, and $X_+ \cup X_-$ is dense in X, since f is an open map. The restrictions of f give holomorphic maps

$$f_{\pm} : X_{\pm} \to H_{\pm}.$$

We claim that f_+ and f_- are proper maps. For if B is a compact subset of H_+, say, there is an $\epsilon > 0$ such that $\mathrm{Im}(z) > \epsilon$ for all z in B. The fact that the imaginary part of f tends to zero at infinity implies that $f^{-1}(B)$ is a compact subset of X, but this is the same as $f_+^{-1}(B) \subset X_+$.

We know that f yields a local homeomorphism from a neighbourhood of p in X to a neighbourhood of ∞ in S^2. What we see first from this is that X_+, X_- are both non-empty. So we have degrees $d_+, d_- \geq 1$ of f_+, f_-. We claim that $d_+ = d_- = 1$. To see this, we apply the condition that $\mathrm{Im}(f)$ tends to zero at infinity to find a compact set K in X such that $\mathrm{Im}(f)(x) < 1$ if x is not in K.

Suppose the degree of f_+ is at least 2. Then, for each integer $n \geq 1$, we can find a pair of points $x_n, \tilde{x}_n \in X_+$ such that $f(x_n) = f(\tilde{x}_n) = in$ and either $x_n, \tilde{x}_n$ are distinct or $x_n = \tilde{x}_n$ and the derivative of f vanishes at x_n. The choice of K means that $x_n, \tilde{x}_n$ lie in this compact set, so we can find a subsequence $\{n'\}$ such that $x_{n'}$ and $\tilde{x}_{n'}$ converge to limits $x, \tilde{x}$. Since the points in, regarded as points of the Riemann sphere, converge to ∞, we must have $f(x) = f(\tilde{x}) = \infty$, and since f has just one pole, we must have $x = \tilde{x} = p$. But now we get a contradiction to the fact that f is a local homeomorphism, with non-vanishing derivative, on a neighbourhood of p.

So now we know that f maps $X_\pm$ bijectively to $H_\pm$ in $\mathbf{C}$. We claim next that f is an injection from X to S^2. For if x_1, x_2 are distinct points of X with $f(x_1) = f(x_2) = Z \in S^2$, we can find disjoint open discs D_1, D_2 about x_1, x_2 and a neighbourhood N of $Z \in S^2$ such that $f(D_1), f(D_2)$ each contain N. Now pick a point Z' in $N \cap H_+$. There are distinct points x_1', x_2' in D_1, D_2 which map to Z', and this contradicts the fact that f is injective on X_+.

Now we know that f maps X injectively to an open subset U of the Riemann sphere, containing $H_+ \cup H_- \cup \{\infty\}$. That is,

$$U = S^2 \setminus I$$

for some compact subset I of $\mathbf{R}$. Thus f yields an equivalence between X and this subset U. If I has more than one component, then $\pi_1(U)$ is non-trivial, which would contradict the fact that X is simply connected. So we conclude that either

- I is a proper closed interval $[a, b]$ for $a < b$ or
- I is a single point in $\mathbf{R}$.

This completes the proof, since in the first case it is easy to write down a holomorphic equivalence between $S^2 \setminus [a, b]$ and the upper half-plane, and in the second case it is obvious that the complement of any point in the Riemann sphere is equivalent to $\mathbf{C}$.

10.2   Proof of the analogue of the Main Theorem

10.2.1   Set-up

We now turn to the proof of Theorem 13. In the proof, we will make use of two facts which we state now.

Proposition 30. *Let X be a connected, simply connected, non-compact surface. Then, for any compact set $K \subset X$, the complement $X \setminus K$ has exactly one connected component whose closure is not compact.*

One says that a surface which satisfies the condition in the second sentence of this proposition has 'only one end'. Thus the statement is that a simply connected surface has only one end. We give a proof of this result in Section 10.2.4.

Our second fact involves calculus on surfaces. We recall from Chapter 5 the notion of the 'modulus' $|\rho|$ of a 2-form.

Proposition 31. *Let S be a smooth oriented surface and let $F : S \to \mathbf{R}^2$ be a smooth map. Then, for any compact set K in S,*

$$\mu(F(K)) \leq \int_S |F^*(dx_1 \, dx_2)|,$$

where x_1, x_2 are standard co-ordinates on $\mathbf{R}^2$ and μ denotes the Lebesgue measure on $\mathbf{R}^2$.

This is a variant of the usual change-of-variables formula. We will not give a proof here, although a proof is not hard given some general background in integration theory.

To prove Theorem 13, we adopt the same strategy as for the proof of the 'Main Theorem' in the compact case. There is one crucial additional step required (in Subsection 10.2.3 below), but to set the stage for this we will need to prove a number of elementary, but slightly delicate, preliminary results (in Subsection 10.2.2).

First, just as before, we can reduce the problem to the case when the 2-form ρ is supported in a co-ordinate disc D about p (this occurs anyway for our application). We consider the vector space Ω_c^0 of compactly supported real-valued functions on X with the Dirichlet norm

$$\| f \|_D^2 = i \int \overline{\partial} f \wedge \partial f.$$

Notice that this is now a genuine norm, since the constant functions do not have compact support. The proof that the functional $\hat{\rho}$ is bounded goes through just as before. We let H be the completion of Ω_c^0 under this norm, and the Riesz Representation Theorem gives an element, ϕ say, of H such that $\hat{\rho}(f) = \langle f, \phi \rangle_D$. Just as before, we can find a sequence f_i in Ω_c^0 converging to ϕ in H and a sequence of constants $c_i \in \mathbf{R}$ such that $\phi_i = f_i + c_i$ converges in L^2 over some co-ordinate disc. Again, the same argument as before shows that the sequence ϕ_i converges in L^2 over any co-ordinate disc. We should note, however, that ϕ_i need not have compact support. The same argument as before shows that ϕ is smooth and satisfies the desired equation $\Delta \phi = \rho$. What we achieve at this stage in the argument is summarised by the following proposition.

Proposition 32. *Let ρ be a 2-form of integral 0 supported in a co-ordinate disc D. Then there is a smooth function ϕ on X which satisfies the equation $\Delta\phi = \rho$, and a sequence ϕ_i of smooth functions on X which have the following properties:*

1. *There are real numbers c_i and compact sets $B_i \subset X$ such that $\phi_i = c_i$ outside B_i.*
2. *For any 1-form α of compact support on X, the norms $\|(\phi - \phi_i)\alpha\|$ tend to zero as i tends to infinity.*
3. *The norms $\|d\phi - d\phi_i\|$ tend to zero as i tends to infinity.*

What should be clear now is that the only essentially new thing required to prove Theorem 13 is to arrange that ϕ tends to zero at infinity in X. Of course, it is equally good to arrange that ϕ tends to a finite constant c at infinity, since we can replace ϕ by $\phi - c$. Notice here that there is an exceptional case when the derivative of ϕ vanishes outside a compact set. In that case, Proposition 30 implies immediately that ϕ tends to a constant at infinity and we are done. So we can suppose that the derivative of ϕ does not vanish on any open set in $X \setminus \text{supp}(\rho)$.

We also note the following lemma here.

Lemma 19. *Let J be any compact set in X. Then there is a sequence ϕ_i satisfying the conditions of Proposition 32 and with $\phi_i = \phi$ on J.*

To see this, let χ be a smooth function of compact support, equal to 1 on J. Given ϕ_i and ϕ as in Proposition 32, we define

$$\phi_i' = \chi\phi + (1 - \chi)\phi_i = \phi_i + \chi(\phi - \phi_i).$$

Thus $f_i' = \phi_i' - c_i$ have compact support, since $\phi_i = \phi_i'$ outside the fixed compact set $\text{supp}(\chi)$ and $\phi_i' = \phi$ on Γ. We have

$$d\phi_i' - d\phi_i = \chi(d\phi - d\phi_i) + (d\chi)(\phi - \phi_i),$$

and it follows that

$$\| f_i' - f_i \|_D \to 0$$

as $i \to \infty$. This means that the sequence ϕ_i' has the same properties (as required in Proposition 32) as ϕ_i.

10.2.2   Classification of behaviour at infinity

We begin with an elementary lemma, which applies to functions on any non-compact space.

Lemma 20. *Suppose ϕ is a continuous function on X; Then one of the following four statements holds:*

1. *There is a constant $c \in \mathbf{R}$ such that ϕ tends to c at infinity in X.*
2. *ϕ tends to $+\infty$ at infinity in X.*
3. *ϕ tends to $-\infty$ at infinity in X.*
4. *There are real numbers α, β with $\alpha < \beta$ such that $\phi^{-1}(-\infty, \alpha]$ and $\phi^{-1}[\beta, \infty)$ are both non-compact subsets of X.*

To prove this, let

$$A^- = \{t \in \mathbf{R} : \phi^{-1}(-\infty, t] \text{ is compact}\}.$$

Clearly, if t_- is in A^- then any $t < t_-$ is also in A^-. Symmetrically, put

$$A^+ = \{t \in \mathbf{R} : \phi^{-1}[t, \infty) \text{ is compact}\}.$$

Then, for any elements $t_\pm \in A_\pm$, we must have $t_- < t_+$, otherwise $\phi^{-1}(\mathbf{R}) = X$ would be compact, a contradiction.

Define $c_- = \sup A_-$ and $c_+ = \inf A_+$, with the convention that $c_- = -\infty$ if A_- is empty and $c_+ = +\infty$ if A_+ is empty. Thus, with obvious conventions involving $\pm\infty$, we have $c_- \le c_+$.

Suppose $c_- = c_+$ is a finite value $c \in \mathbf{R}$. Then, for any $\epsilon > 0$, we have $c \pm \epsilon \in A_\pm$, so $\phi^{-1}(-\infty, c - \epsilon]$ and $\phi^{-1}[c + \epsilon, \infty)$ are compact and $|\phi - c| \le \epsilon$ outside the union of these two compact sets, which is compact. Thus ϕ tends to c at infinity in X. Likewise, if $c_- = c_+ = \pm\infty$, we find that ϕ tends to $\pm\infty$ at infinity in X.

Suppose $c_- < c_+$. Then we can find real numbers α, β with $c_- < \alpha < \beta < c_+$ and, by definition, $\phi^{-1}(-\infty, \alpha]$ and $\phi^{-1}[\beta, \infty)$ are non-compact.

Now, in our situation, if ϕ tends to any real number c at infinity we can replace ϕ by $\phi - c$ to get a solution to our problem. Thus our task is to rule out the other three possibilities in Lemma 20.

Lemma 21. *The function ϕ does not tend to $+\infty$ at infinity in X.*

Suppose it did, and let C be the maximum value of ϕ on the compact set $\text{supp}(\rho)$. Then the set $K = \phi^{-1}(-\infty, C + 1]$ is compact. The points in $K \setminus \text{supp}(\rho)$ where the derivative of ϕ vanishes form a discrete set (since a harmonic function is locally the real part of a holomorphic function). It follows that we can find some $t_0 \in (1/2, 1)$ such that the derivative of ϕ does not vanish on $\phi^{-1}(t_0)$ (since we only need to avoid a finite number of points). Then $K_0 = \phi^{-1}(-\infty, C + t_0]$ is a compact surface-with-boundary in X, containing the support of ρ in its interior. Now, by Stokes' Theorem,

$$\int_{K_0} \Delta\phi = 2i \int_{\partial K_0} \partial\phi.$$

On the one hand, the integral on the left is zero, since it equals the integral of ρ. On the other hand, the boundary integral on the right is strictly *negative*. This is clear in local co-ordinates, or by thinking of the boundary integral in more classical notation as the 'flux' of the gradient of ϕ across the boundary.

Of course, the same argument shows that ϕ does not tend to $-\infty$, so our task is to rule out the fourth possibility in Lemma 20.

For $\alpha \in \mathbf{R}$, we write $Y_\alpha = \phi^{-1}(\infty, \alpha]$.

Lemma 22. *Any compact connected component of Y_α intersects* supp(ρ).

This is a simple consequence of the maximum principle. Suppose Z is a compact connected component which does not meet supp(ρ). There is a point x in Z which minimises ϕ over Z, and then the maximum principle implies that this can only happen if the derivative of ϕ vanishes near x, contrary to our assumption.

Remark One can also argue as follows. Suppose, for simplicity, that Z has a smooth boundary; then

$$\int_Z \overline{\partial}\phi \wedge \partial\phi = \int_{\partial Z} (\phi - \alpha)\, \partial\phi = 0.$$

It is not hard to handle the case when the boundary of Z is not smooth, but the maximum-principle argument is easier here.

Lemma 23. *For any point q in Y_α which is not in* supp(ρ) *and any neighbourhood N of q, there is an open disc D_q centred on q and contained in N such that $Y_\alpha \cap D_q$ is path-connected.*

This is rather obvious. We write the function $\phi - \alpha$ as the real part of a holomorphic function in a neighbourhood of q. Then this function is given in a suitable holomorphic co-ordinate as z^k, and the set

$$\{z \in \mathbf{C} : |z| < 1 : \mathrm{Re}(z^k) \leq 0\}$$

is path-connected.

Proposition 33. *Suppose Y_α is not compact. Then there is a non-compact path-connected subset of Y_α.*

In a nutshell, the argument is that any compact component must meet the support of ρ but these components have an obvious 'local finiteness' property. In more detail, let $D' \subset D$ be an open interior disc whose closure lies in D such that D' contains the support of ρ. In our local coordinate about p we can take D to correspond to $|z| < 1$ and D' to $|z| < r$, for some fixed $r < 1$. Let C be the circle corresponding to $|z| = (1 + r)/2$. It follows from Lemma 23 that we can write the intersection $Y_\alpha \cap C$ as the union of finitely many pieces $C_1, \ldots C_N$, say, such that any two points in the same C_j can be joined by a path in $Y_\alpha \setminus D'$. (We cover the compact set $Y_\alpha \cap C$ by finitely many small discs of the form D_q considered in the lemma.) Now consider the path-connected components of $Y_\alpha \setminus D'$. Notice that, by Lemma 23, these are the same as the connected components. If one of these components is non-compact, we are done. So, suppose all these components are compact. Since D' has compact closure and Y_α is not compact, there must be infinitely many different compact components of $Y_\alpha \setminus D'$. The circle C divides X into two connected components, one of which is a disc D'' containing p. Any compact component of $Y_\alpha \setminus D'$ must either intersect the circle C or lie entirely within the disc D'', for otherwise it lies in $X \setminus \overline{D''}$ and gives a compact component of Y_α which does not meet $\mathrm{supp}(\rho)$—contradicting Lemma 22. The union of the components lying in D'' is contained in a compact set—the closed disc $\overline{D''}$. So there must be infinitely many different components which intersect C. But two of these must meet the same subset C_j, and hence we get a contradiction since the points in C_j can be joined by paths in $Y_\alpha \setminus D'$.

Putting together our results from this subsection, we have the following proposition.

Proposition 34. *Either ϕ tends to a finite limit at infinity in X, or there are real numbers $\alpha < \beta$ and non-compact, closed, path-connected subsets $Z_\alpha, Z_\beta \subset X$ such that $\phi(x) \le \alpha$ for $x \in Z_\alpha$ and $\phi(x) \ge \beta$ for $x \in Z_\beta$.*

10.2.3   The main argument

Our task now is to show that the second alternative of Proposition 34 does not occur, so we suppose it does and argue for a contradiction. Fix points $x_\alpha \in Z_\alpha, x_\beta \in Z_\beta$ and a compact path-connected set Γ containing x_α and x_β (for example, the image of some path between the two points). By Lemma 19, there is no loss of generality in supposing that $\phi_i = \phi$ on Γ.

Our preparations are now complete. For each i, neither of the non-compact sets Z_α, Z_β can be contained in the compact set $\mathrm{supp}(f_i)$. Moreover, they must intersect a non-compact component of $X \setminus \mathrm{supp}(f_i)$ and, by Proposition 30, there is only one such component. Thus there is a path in $X \setminus \mathrm{supp}(f_i)$ joining the point y_α of Z_α to a point y_β of Z_β. Let $\gamma_i : [0, 1] \to X$ be a loop in X starting and ending at x_α of the following form (Figure 10.1):

- On the interval $0 \le t \le 1/4$, γ_i traces out a path from x_α to x_β in the set Γ.
- On the interval $1/4 \le t \le 1/2$, γ_i traces out a path in Z_β from x_β to y_β.
- On the interval $1/2 \le t \le 3/4$, γ_i traces out a path in $X \setminus \mathrm{supp}(\rho)$ from y_β to y_α.
- On the interval $3/4 \le t \le 1$, γ_i traces out a path in Z_α from y_α to x_α.

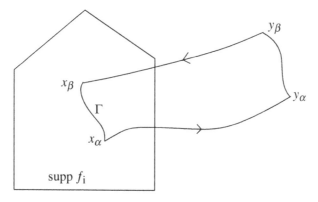

Figure 10.1 *The loop in X*

Since X is simply connected, this loop is contractible, so we can find a compact set $K_i \subset X$, containing the image of γ_i, such that γ_i is contractible in K_i.

Now consider the smooth map $F_i : X \backslash \mathbf{R}^2$ defined by

$$F_i(x) = (\phi(x), \phi_i(x)).$$

Then the composite $F_i \circ \gamma_i$ is a loop in $\mathbf{R}^2$ with the properties that:

- For $0 \le t \le 1/4$, $F_i \circ \gamma_i$ maps into the diagonal $\{(u, u)\}$ in $\mathbf{R}^2$.
- For $1/4 \le t \le 1/2$, $F_i \circ \gamma_i$ maps into the half-plane $\{(u_1, u_2) : u_1 \ge \beta\}$.
- For $1/2 \le t \le 3/4$, $F_i \circ \gamma_i$ maps into the horizontal line $\{(u_1, c_i)\}$.
- For $3/4 \le t \le 1$, $F_i \circ \gamma_i$ maps into the half-plane $\{(u_1, u_2) : u_1 \le \alpha\}$.

See Figure 10.2.

Let Z be the subset of $\mathbf{R}^2$ given by

$$Z = \{(u_1, u_2) : \alpha < u_1 < \beta, \min(u_1, c_i) < u_2 < \max(u_1, c_i)\}.$$

There are several possible pictures for Z, depending on whether c_i lies between α and β, above β or below α. However, it is elementary to see that in any case the area satisfies

$$\mu(Z) \ge \frac{1}{4}(\beta - \alpha)^2.$$

The conditions on our loop $F_i \circ \gamma_i$ imply that it does not meet Z. Now we have the following lemma.

Lemma 24. *If P is any point of Z, the winding number of $F_i \circ \gamma_i$ around P is equal to ± 1.*

This is a straightforward exercise in algebraic topology.

Corollary 8. *F_i maps the compact set $K_i \subset X$ onto Z.*

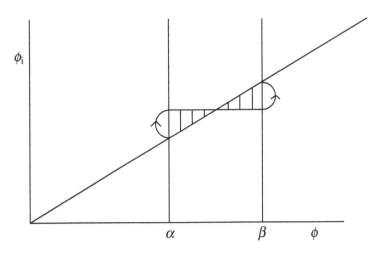

Figure 10.2   *The image of the loop, and the region Z*

This follows because if P is not in the image $F_i(K)$, the composite of F_i with a contraction of γ_i would give a contraction of $F_i \circ \gamma_i$ in $\mathbf{R}^2 \setminus \{P\}$, contradicting the homotopy invariance of the winding number.

Now we can apply Proposition 31 to deduce that the integral of the modulus of the form $F_i^*(du_1\, du_2)$ must be at least $\frac{1}{4}(\beta^2 - \alpha^2) = \delta$, say. Writing

$$F_i^*(du_1\, du_2) = d\phi \wedge d\phi_i,$$

we can state a conclusion to our preceding arguments as follows.

Corollary 9. *If ϕ does not tend to a finite limit at infinity in X, then there is some $\delta > 0$ such that for each i,*

$$\int_X |d\phi \wedge d\phi_i| \geq \delta.$$

Now we write

$$d\phi \wedge d\phi_i = (d\phi - d\phi_i) \wedge d\phi_i$$

and apply Lemma 13 from Chapter 5 to get

$$\int_X |d\phi \wedge d\phi_i| \leq \|d\phi - d\phi_i\|\,\|d\phi_i\|,$$

but this tends to zero by Proposition 32, and we have the desired contradiction. This completes the proof of the Uniformisation Theorem.

Remark In Chapter 3 we mentioned Rado's Theorem, which states that a (connected) Riemann surface always has a countable topology (it is paracompact). While we have not wanted to emphasise this somewhat specialised aspect of the theory, an analysis of our proof shows that the countable-topology condition is not needed, so we obtain a proof Rado's Theorem along the way. This is why we are careful only to integrate over compact sets. *A priori*, the Hilbert space we construct as an abstract completion might not be separable, but that does not matter.

10.2.4   Proof of Proposition 30

The statement is of course intuitively clear, and how one writes down a proof depends on the kind of background assumed. We will give a self-contained argument using the techniques developed in this book.

We suppose we have a connected oriented surface S and a compact subset $K \subset S$ such that $S \setminus K$ has two connected components U_0, U_1 whose closures are not compact. We want to show that S cannot be simply connected.

First, choose a finite number of open sets in S with compact closures which cover K. Let K^+ be the union of these sets, which is again compact. The complement of K^+ has two connected components whose closures are not compact and, further, these closures are disjoint. So, without loss of generality, to simplify the notation, suppose that K already has this property. The first step is to construct a smooth function f on S equal to 0 on U_0 and to 1 on U_1. We leave this as an exercise or refer readers to Spivak (1979). Now let $\theta = df$. This is a closed 1-form with compact support, but the hypotheses imply that its class in $H^1_c(S)$ is not zero, since there is no constant c such that $f + c$ has compact support. If we knew the general statement of Poincaré duality—that $H^1_c(S)$ is dual to $H^1(S)$ and hence isomorphic to the real cohomology of S—we would be finished immediately, so what we need to do is to establish a special case of this statement.

Pick a point $p_0 \in U_0$ and let σ_0 be a 2-form of compact support in U_0 and with integral 1, supported in a small disc about p_0. Choose σ_1 on U_1 in the same fashion. Then the σ_i can be regarded as forms on X in the obvious way and, by Proposition 12, there is a 1-form α on S, of compact support, such that $d\alpha = \sigma_1 - \sigma_0$. Now α is a closed 1-form on a neighbourhood N of the support of θ and we can choose N to be contained in K. We claim that we cannot have $\alpha = du$ on N. For

$$\int_S \alpha\theta = \int_S \alpha\, df = \int_S (\sigma_1 - \sigma_0)\, f = 1,$$

and if $\alpha = du$ on N then

$$\int_S \alpha\theta = \int_N \alpha\theta = \int_N du\,\theta = \int_N u\, d\theta = 0,$$

where we have used the fact that the support of θ lies in N. Now it follows that this neighbourhood N cannot be simply connected; we can choose a loop γ in N such that the integral of α around γ is 1. Suppose γ is contractible in the whole of S. The image of the contraction is a compact set $J \subset S$. By hypothesis, we can choose points p_0', p_1' in U_0, U_1, respectively, which are not contained in J. Let σ_i' be 2-forms supported in small neighbourhoods of these points, not meeting J, and of integral 1. Then $\sigma_0' - \sigma_0$ has compact support in U_0 and integral 0, so there is a compactly supported 1-form β_0 with $d\beta_0 = \sigma_0' - \sigma_0$. Likewise, we define β_1 on U_1. Now consider $\alpha' = \alpha - \beta_0 + \beta_1$. By construction, this has $d\alpha' = \sigma_1' - \sigma_0'$. Since the image of the contraction of γ does not meet the supports of σ_i' the integral of α' around γ is 0. On the other hand, the support of β_i is contained in U_i and so does not meet γ. Thus the integral of α' around γ is the same as that of α, which is 1. This gives the desired contradiction.

Exercises

1. Find a smooth function f on the punctured disc which is unbounded but such that

$$\int |\nabla f|^2 < \infty.$$

2. Let X be either $\mathbf{C}$ or the upper-half plane H. Given ρ of compact support and integral zero, find an explicit integral formula for a function ϕ with $\Delta\phi = \rho$, tending to zero at infinity. (*Hint:* In the case $X = \mathbf{C}$, we saw this in Chapter 9. In the case $X = H$, use the 'reflection principle'.) Show that the assumption that the integral of ρ is zero is not needed in the case when $X = H$ but is needed when $X = \mathbf{C}$.

3. Suppose $r(x)$ is a compactly supported function on $\mathbf{R}$ with integral zero. Show that there is a bounded solution to the equation $f''(x) = r(x)$ on $\mathbf{R}$ and that this tends to a limit at infinity (in the sense defined in this chapter) if and only if the integral of $xr(x)$ is zero.

4. Let $X = S^1 \times \mathbf{R}$. Show that it is not always possible to solve $\Delta\phi = \rho$ with ϕ tending to zero at infinity, even if the integral of ρ is zero.

5. Let $X = S^1 \times (0,1)$, viewed as a Riemann surface in the standard way. If ρ is a 2-form of compact support, find an integral formula for a function ϕ with $\Delta\phi = \rho$, tending to zero on each end of X. (*Hint:* Use separation of variables and Fourier series.)

6. Start with the strip $-1 \leq \operatorname{Im} z \leq 1$ in $\mathbf{C}$ and a sequence of numbers $r_n \in (0, 1/2)$, $n \geq 1$. Now remove closed discs of radius r_n centred on $\pm n$. Identify $n + \zeta$ with $-n + (r_n/2)\zeta^{-1}$, for integers $n \geq 1$ and $r_n < |\zeta| < 1/2$. Finally, identify the boundaries $\operatorname{Im} z = \pm 1$ in the obvious way. The result is a non-compact Riemann surface X.

(a) Convince yourself of this.

(b) Show that X has only one end.

(c) (Harder.) Suppose that $\sum_n \log r_n^{-1} < \infty$. Show that there is a 2-form ρ of compact support on X and with integral zero, and a ϕ with $\Delta\phi = \rho$ and finite Dirichlet norm but such that ϕ does not tend to a limit at infinity in X. Analyse what goes wrong in the proof of the main result in this chapter.

Part IV

Further developments

11 Contrasts in Riemann surface theory

11.1 Algebraic aspects

11.1.1 Fields of meromorphic functions

In Section 4.2.3, we saw how to associate a compact connected Riemann surface to any irreducible polynomial $P(z, w)$ in two variables. More precisely, we did this assuming two technical facts:

- There are only finitely points (z, w) where both P and $\partial P/\partial w$ vanish.
- The zero set $X \subset \mathbf{C}^2$ of P is connected.

(Recall that in the first item we exclude the rather trivial case $P = z - z_0$.)

In this subsection, we will establish the converse—any compact connected Riemann surface arises from a polynomial P. We will also prove the technical facts mentioned above. More generally, we explain how the whole theory can be cast in an algebraic form, involving fields of transcendence degree 1.

The fundamental idea is that we can associate to any compact connected Riemann surface Σ a field k_Σ of all meromorphic functions on Σ. It is clear that a non-constant holomorphic map $f : \Sigma_1 \to \Sigma_2$ induces an injection $k_{\Sigma_2} \to k_{\Sigma_1}$. We also know that the field $k_{\mathbf{CP}^1}$ associated to the Riemann sphere is the field $\mathbf{C}(z)$ of rational functions in one variable (Exercise 6 in Chapter 4). Recall that if $L \subset K$ are fields, the degree $[K, L]$ is defined to be the dimension (possibly infinite) of K as a vector space over L. By our fundamental results from Chapter 8, any compact Riemann surface Σ admits a non-trivial holomorphic map $f : \Sigma \to \mathbf{CP}^1$. Let d be the degree of f, as defined in Chapter 4. We regard $\mathbf{C}(z)$ as a subfield of k_Σ via the injection induced by f. (The notation becomes somewhat mixed here, since we could also write $\mathbf{C}(z)$ as $\mathbf{C}(f)$.) Thus we have an 'algebraic' degree $[k_\Sigma, \mathbf{C}(z)]$, and the first main result is the following theorem.

Theorem 14.

$$[k_\Sigma, \mathbf{C}(z)] = d.$$

The proof has two parts.

Part 1: $[k_\Sigma, \mathbf{C}(z)] \geq d$. Without loss of generality, we can suppose that $0 \in \mathbf{C} \subset \mathbf{CP}^1$ is a regular value of f so that $f^{-1}(0)$ is a set of d distinct points $p_1, \ldots, p_d$ in Σ. It follows easily from the existence theory of Chapter 8 that, for each i, we can find a meromorphic function g_i on Σ that has a simple pole at p_i but is holomorphic around p_j for each $j \neq i$. We claim that $g_1, \ldots, g_d$ are linearly independent over $\mathbf{C}(z)$. For, suppose there is a linear relation $\sum_j \lambda_j(z) g_j = 0$. Multiplying by a suitable power z^μ, we can assume that the λ_j are holomorphic around $z = 0$ and do not all vanish at $z = 0$. But now consider the identity $\sum \lambda_j g_j = 0$ around a point p_i. The only term which is not holomorphic around this point is $\lambda_i g_i$, and since g_i has a pole, we see that λ_i must vanish at 0, a contradiction.

Part 2: $[k_\Sigma, \mathbf{C}(z)] \leq d$. By the theorem of the 'primitive element', it suffices to show that any element g of k_Σ satisfies a polynomial identity $P(f, g) = 0$, where P has degree at most d in the second variable. Let z be a regular value of f, so that $f^{-1}(z) = \{p_1(z), \ldots, p_d(z)\} \subset \Sigma$. Then we have d points $w_i = g(p_i(z))$ in $\mathbf{CP}^1$. For all but finitely many z, the w_i all lie in $\mathbf{C} \subset \mathbf{CP}^1$, so for simplicity let us assume that z is such a point. Of course the *ordering* of the w_i is arbitrary, but we can form the elementary symmetric functions

$$a_1(z) = \sum w_i, \quad a_2(z) = \sum w_i w_j, \quad \ldots,$$

which do not depend on the ordering. So we have

$$\sum_{r=0}^{d} (-1)^r a_r(z) g^r = 0. \tag{11.1}$$

Clearly, the a_r are holomorphic functions on the complement of a finite subset Δ in $\mathbf{CP}^1$, and equation (11.1) is an identity among meromorphic functions on $f^{-1}(\mathbf{CP}^1 \setminus \Delta)$. It suffices then to prove that the a_r *extend* to meromorphic functions on $\mathbf{CP}^1$: for then, equation (11.1) becomes the desired identity $P(f, g) = 0$.

To see this extension property, we use a version of the Riemann Extension Theorem from elementary complex analysis. If a is a holomorphic function on a disc in $\mathbf{C}$ minus a single point z_0 and if $|a(z)| \leq C|z - z_0|^{-\nu}$ for some C, ν and all z in the punctured disc, then a extends as a meromorphic function over z_0. Now suppose that z_0 is a point in $\Delta \subset \mathbf{CP}^1$. There is clearly no loss in supposing that z_0 lies in $\mathbf{C}$. Let $f^{-1}(z_0) = \{q_1, \ldots q_n\} \subset \Sigma$, where f has ramification order d_i at q_i, so that $\sum d_i = d$. Suppose that $g(q_i) = \infty$ for $i \leq n_0$ and $g(q_i)$ is finite for $i > n_0$. Take local coordinates u_i on Σ centred at q_i. If z is sufficiently close to z_0, but not equal to z_0, then the set $f^{-1}(z)$ contains exactly d_i points close to q_i and, in fact, if p is one of these points its local co-ordinate u satisfies

$$C_0^{-1}|z - z_0|^{d_i} \le |u| \le C_0|z - z_0|^{d_i}.$$

Then $|g(p)| \le C_1|u|^{-\mu_i}$, where μ_i is the order of the pole of g at q_i (interpreted as zero if $i > n_0$). Thus

$$|g(p)| \le C_2|z - z_0|^{\mu_i d_i},$$

and it is clear that we get an estimate of the form $|a_r(z)| \le C|z - z_0|^{-\nu}$ for each symmetric function a_r.

Corollary 10. *If $f : \Sigma_1 \to \Sigma_2$ is a non-constant holomorphic map between compact connected Riemann surfaces, then the degree $[k_{\Sigma_1}, k_{\Sigma_2}]$ is equal to the degree of f.*

Here, of course, we are regarding k_{Σ_2} as a subfield of k_{Σ_1} via the inclusion induced by f. It would be more precise to write $[k_{\Sigma_1}, f^*k_{\Sigma_2}]$. There may be many different holomorphic maps from Σ_1 to Σ_2, and hence different inclusions giving different degrees.

To establish the above corollary, we consider a holomorphic map $f_0 : \Sigma_2 \to \mathbb{CP}^1$, so we have $\mathbb{C}(z) \subset k_{\Sigma_2} \subset k_{\Sigma_1}$. Now apply the previous result using the multiplicativity of the geometric and algebraic degrees (see Exercise 1 in Chapter 4), and

$$[k_{\Sigma_1}, \mathbb{C}(z)] = [k_{\Sigma_1}, k_{\Sigma_2}][k_{\Sigma_2}, \mathbb{C}(z)].$$

One can also argue more directly, without introducing f_0, by adapting the proof of Theorem 14.

Recall from the algebraic theory of fields that if K is a field containing $\mathbb{C}$, the *transcendence degree* (possibly infinite) of K is the supremum of the numbers n such that K contains a subfield isomorphic to $\mathbb{C}(z_1, \ldots, z_n)$ (the field of rational functions in n variables).

Corollary 11. *If Σ is a compact connected Riemann surface, the transcendence degree of k_Σ is 1.*

In one direction, we know we have a $\mathbb{C}(z) \subset k_\Sigma$, so that the transcendence degree is at least 1. In the other direction, suppose we had $\mathbb{C}(z_1, z_2) \subset k_\Sigma$. Then z_1 corresponds to a non-constant holomorphic map $f : \Sigma \to \mathbb{CP}^1$. By Theorem 14, we know that $[k_\Sigma, \mathbb{C}(z_1)]$ is finite and hence $[\mathbb{C}(z_1, z_2), \mathbb{C}(z_1)]$ is finite, which is a contradiction.

Next, we have a converse.

Theorem 15. *If K is any field containing $\mathbb{C}$ and of transcendence degree 1, then there is a compact connected Riemann surface Σ with $k_\Sigma = K$.*

More precisely, what we shall show is that any subfield $\mathbb{C}(z) \subset K$ corresponds to a pair consisting of a Riemann surface Σ and a holomorphic map

$f : \Sigma \rightarrow \mathbf{CP}^1$. At the same time, we shall establish the technical facts assumed in Section 4.2.3 and recalled at the beginning of this chapter.

In the course of the proof we make use of 'Gauss's Lemma'. This states that if R is a unique factorisation domain with field of fractions F and if p is a polynomial in $R[z]$ which factors as $p = q_1 q_2$ for $q_1, q_2 \in F[z]$, then there is a non-zero $\lambda \in F$ such that $\lambda q_1, \lambda^{-1} q_2$ are in $R[z]$. Thus we have a factorisation $p = (\lambda q_1)(\lambda^{-1} q_2)$ in $R[z]$.

If K has transcendence degree 1, then by the theorem of the primitive element it can be written as

$$K = \mathbf{C}(z)[w]/P,$$

where $\mathbf{C}(z)[w]$ consists of polynomials in w with coefficients which are rational functions of z, and $P \in \mathbf{C}(z)[w]$ is irreducible in $\mathbf{C}(z)[w]$ (i.e. cannot be written as a non-trivial product $P_1 P_2$ for $P_i \in \mathbf{C}(z)[w]$). Multiplying by a suitable polynomial in z, there is no loss in supposing that P is in $\mathbf{C}[z, w]$, i.e. is a polynomial in both variables. Gauss's Lemma shows that P is also irreducible regarded as an element of $\mathbf{C}[z, w]$, and it is easy to see that K is the field of fractions of the integral domain $\mathbf{C}[z, w]/ < P >$, where $< P >$ is the ideal in $\mathbf{C}[z, w]$ generated by P.

At this point, we want to recall the theory of the 'resultant'. Suppose Π_1, Π_2 are elements of $\mathbf{C}[z, w]$ with no common factor. By Gauss's Lemma, they are coprime, regarded as elements of $\mathbf{C}(z)[w]$. By the Euclidean algorithm in $\mathbf{C}(z)[w]$, we can find $\lambda, \mu \in \mathbf{C}(z)[w]$ such that $\lambda \Pi_1 + \mu \Pi_2 = 1$. Clearing denominators, we get $\tilde{\lambda}, \tilde{\mu} \in \mathbf{C}[z, w]$ such that

$$\tilde{\lambda} \Pi_1 + \tilde{\mu} \Pi_2 = \rho(z),$$

for a non-zero polynomial $\rho \in \mathbf{C}[z]$. The set of polynomials ρ which can be realised in this way forms an ideal in $\mathbf{C}[z]$ which is generated by some $\rho_0(z)$. This is the *resultant* of Π_1, Π_2. (Which we can normalise, for example by specifying that it is a monic polynomial.) Clearly, if Π_i both vanish at some (z_0, w_0), then $\rho_0(z_0) = 0$. In fact, the converse is true: if $\rho_0(z_0) = 0$, then there is a w_0 such that $\Pi_i(z_0, w_0) = 0$. But we do not need this converse.

Now, in particular, we see that if P is an irreducible polynomial in $\mathbf{C}[z, w]$, then for any other polynomial Q not identically zero, there are only finitely many common roots of P, Q in $\mathbf{C}^2$. Taking $Q = \partial P/(\partial w)$, we immediately get the first of the statements left over from Section 4.2.3. So now, starting with an irreducible polynomial P, we can run the construction described in Section 4.2.3 to produce a compact Riemann surface Σ. The only small issue is that we do not at this stage know that Σ is connected: in fact, we shall show that the zero set of P in $\mathbf{C}^2$ is connected. To see this, suppose that the zero set had two components. Then Σ is a disjoint union $\Sigma_1 \cup \Sigma_2$ of compact Riemann surfaces. By construction, we have holomorphic maps f, g from Σ_i to $\mathbf{CP}^1$, and Theorem 14 implies that there are polynomials P_1, P_2 such that $P_1(f, g) = 0$ on Σ_1 and $P_2(f, g)$ on Σ_2. Suppose the polynomial P had degree d in w. Then we know that, for z_0 outside a finite set in $\mathbf{C}$, there are exactly d solutions of the equation $P(z_0, w) = 0$ for w. If d_i of these solutions correspond to Σ_i, then $d_1 + d_2 = d$ and $P_i(z_0, w)$ is a polynomial with these roots. So we see that, outside a finite set, $P(z, w) = a(z) P_1(z, w) P_2(z, w)$. By considering the leading term in w we see that a is meromorphic, so we have a non-trivial factorisation of P in $\mathbf{C}(z)[w]$, and by Gauss's Lemma this contradicts the irreducibility of P.

Finally, we claim that k_Σ is isomorphic to the field K we started with. Clearly, there is a natural inclusion $K \subset k_\Sigma$. Let P have degree d in the variable w. Then $f : \Sigma \to \mathbf{CP}^1$ has geometric degree d, and so $[k_\Sigma, \mathbf{C}(z)] = d$ by Theorem 14. On the other hand, algebraically, we know that $[K, \mathbf{C}(z)] = d$, so $[k_\Sigma, K] = 1$ and $k_\Sigma = K$.

11.1.2 Valuations

A more precise and satisfactory statement of this relation between compact Riemann surfaces and fields is to say that there is an *equivalence of categories* between

- the category of compact connected Riemann surfaces and non-constant holomorphic maps;
- the category of fields of transcendence degree 1 and field inclusions.

In plain language, what this means is that we should have a definite procedure for associating a compact Riemann surface Σ_K to a field K which is inverse to our association of k_Σ to Σ. We also want a field inclusion $K_2 \subset K_1$ to induce a holomorphic map from Σ_{K_1} to Σ_{K_2} with the obvious compatibility with the association we have of a field inclusion to a holomorphic map. A neat way of achieving this is through the notion of a 'valuation' on a field.

Definition 9. *Let K be a field. A valuation on K is a surjective map $v : K \to \mathbf{R} \cup \{\infty\}$ with the following properties:*

1. $v^{-1}(\infty) = \{0\}$.
2. $v(f + g) \geq \min(v(f), v(g))$ for all $f, g \in K$.
3. $v(fg) = v(f) + v(g)$ for all $f, g \in K$.

If v is a valuation, then so is cv for any real number $c > 0$. We say that v and cv are *equivalent*. The *trivial* valuation is defined by mapping all non-zero elements in K to 0. It is clear that if $L \subset K$ is a subfield, then a valuation on K restricts to a valuation on L. We say that v is a valuation on K over L if its restriction to L is trivial.

If v is a valuation, then $R_v = \{f \in K : v(f) \geq 0\}$ is a subring of K, the *valuation ring*. It is easy to check from the axioms that the subset $I_v = \{f \in K : v(f) > 0\}$ is a maximal ideal in R_v, so R_v/I_v is a field, called the *residue class field* of the valuation.

Given a compact connected Riemann surface Σ, we let $\mathrm{Val}(\Sigma)$ be the set of equivalence classes of non-trivial valuations on the field k_Σ over the subfield $\mathbf{C}$.

Suppose x is a point in Σ. We define a valuation v_x on k_Σ by setting $v_x(f)$ to be the order of vanishing of f at x. In other words, $v_x(f) = v$ if, in a local co-ordinate ζ centred at x, the Laurent series of f begins with a non-zero multiple of ζ^v. The valuation ring associated to v_x consists of meromorphic functions which are holomorphic around x; the ideal I consists of functions vanishing at x and $R/I = \mathbf{C}$.

We now have a map of sets from Σ to $\mathrm{Val}(\Sigma)$ taking x to the equivalence class of v_x.

Theorem 16. *This map is a bijection between Σ and* $\mathrm{Val}(\Sigma)$.

We begin by establishing the theorem in the case of the Riemann sphere. So we have to consider a non-trivial valuation v on the field of rational functions $\mathbf{C}(z)$, over the subfield $\mathbf{C}$. Suppose first that $v(z) \geq 0$. Then it follows from the axioms that $v(p) \geq 0$ for any polynomial p, that is, $\mathbf{C}[z] \subset R$. If $v(p) = 0$ for all p, then $v(p/q) = v(p) - v(q) = 0$, which contradicts the assumption that v is non-trivial. So $\mathbf{C}[z] \cap I$ is a non-trivial prime ideal in $\mathbf{C}[z]$. This means that $\mathbf{C}[z] \cap I = \mathbf{C}[z](z - z_0)$ for some $z_0 \in \mathbf{C}$. Write $v(z - z_0) = \gamma$. For any polynomial π coprime to $z - z_0$, we have $\pi \notin I$, so $v(q) = 0$. Writing a general element of $\mathbf{C}(z)$ as $(z - z_0)^n (\pi_1/\pi_2)$ and applying axiom 3 in Definition 9, we see that v is γ times the valuation v_{z_0}.

Now suppose that $v(z) < 0$. Then $v(z^{-1}) > 0$. There is an automorphism of $\mathbf{C}(z)$ taking z^{-1} to z, so we can reduce this case to the previous case, and in fact we see that v is equivalent to the valuation determined by the point at infinity in $\mathbf{CP}^1$.

Next, consider a general compact connected Riemann surface Σ and choose a holomorphic map $f : \Sigma \to \mathbf{CP}^1$. Let v be a non-trivial valuation on k_Σ over $\mathbf{C}$. This induces a valuation on $\mathbf{C}(z)$ regarded as a subfield of k_Σ. We want to see first that this valuation on $\mathbf{C}(z)$ is non-trivial. Choose $g \in k_\Sigma$ such that $v(g) = \gamma > 0$. We know that g satisfies an equation $\sum a_r g^r = 0$ for $a_r \in \mathbf{C}(z)$, and we can obviously suppose that $a_0 \neq 0$. Thus

$$-a_0 = \sum_{r \geq 1} a_r g^r.$$

Then, if v were trivial on $\mathbf{C}(z)$, we would have $v(a_r g^r) = r\gamma$ and

$$v(-a_0) \geq \min(\gamma, 2\gamma, \dots) = \gamma > 0,$$

which is a contradiction. So, by the previous case, we know that the restriction of v to $\mathbf{C}(z)$ is equivalent to the valuation defined by some point z_0 in $\mathbf{CP}^1$, which we can obviously suppose is the point $0 \in \mathbf{C} \subset \mathbf{CP}^1$.

Now, the pre-image $f^{-1}(0)$ is a finite set $p_1, \ldots, p_n$ in Σ. We choose another meromorphic function g on Σ which maps these points to distinct points $w_1, \ldots, w_n$ in $\mathbf{C}$ and which generates k_Σ over $\mathbf{C}(z)$ (see Exercise 1 in Chapter 8). Now we can apply the same argument as before to g and the embedding of the field of rational functions in k_Σ induced by g. Let us call this $\mathbf{C}(w) \subset k_\Sigma$. We know that the valuation defines a particular $w_0 \in \mathbf{CP}^1$, which we may assume is 0. That is to say, we can suppose $v(g) > 0$. Now we claim that one of the w_i must be zero. For if not, we have a polynomial $P(z, w)$ in two variables such that $P(f, g) = 0$ and $P(0, 0) = a \neq 0$. Then we have an equation in k_Σ

$$-a = \sum_{i+j>0} P_{ij} f^i g^j,$$

which implies that $v(a) \geq \min(v(f^i), v(g^j)) > 0$, contradicting the fact that v is trivial on the constants $\mathbf{C} \subset k_\Sigma$.

Now we can suppose $w_1 = 0$ and write $p = p_1$. Clearly, we should try to show that v is equivalent to the valuation v_p. We do this in steps:

1. Suppose $h \in k_\Sigma$ and $h(p) = 0$. We claim that $v(h) > 0$. For we can write

$$h = \sum_{r \geq 0} b_r(f) g^r$$

   for polynomials b_r, and the condition $h(p) = 0$ means that b_0 vanishes at 0. Then, since $v(f), v(g) > 0$, we get $v(h) > 0$ as before.
2. If $h \in k_\Sigma$ and $h(p) = \infty$, then $v(h) < 0$. This follows by replacing h by h^{-1} in the statement above.
3. If $v(h) \geq 0$, then $h(p)$ is finite. This is logically equivalent to the statement in step 2.
4. Now we show the converse to the statement in step 1: if $v(h) > 0$, then $h(p) = 0$. For the statement in step 3 shows that evaluation at p defines a homomorphism from the ring R_v to $\mathbf{C}$. This is non-trivial, since it is the identity on the constants. So the kernel is a maximal ideal in R. But the statement in step 1 asserts that kernel of the evaluation map is contained in the proper ideal I_v, so the two must be equal, by maximality. This is the desired statement.
5. Now, replacing h by h^{-1}, we see that $v(h) < 0$ if and only if $h(p) = \infty$. So we know that $v(h) = 0$ if and only if $h(p)$ is finite and non-zero. So we have established that the signs (i.e. $>, <, = 0$)) of $v(h), v_p(h)$ are the same for all h.

6. Using the multiplicative property, we see that $v(h_1) > v(h_2)$ if and only if $v_p(h_1) > v_p(h_2)$ and that $v(h_1) = v(h_2)$ if and only if $v_p(h_1) = v_p(h_2)$.

7. From this, it is straightforward to show that v maps $k_\Sigma \setminus \{0\}$ onto a subgroup $c\mathbf{Z}$ (for some $c > 0$) of $\mathbf{R}$ and is equal to cv_p.

To give a fully satisfactory development of the theory, we should do the following. Start from any field K of transcendence degree 1. Then we have a set $\mathrm{Val}(K)$ of equivalence classes of non-trivial valuations on K over $\mathbf{C}$. We should define a Riemann surface structure on this set in such a way that if $L \subset K$, the natural induced map of sets, which we already have, is holomorphic. Of course, if we start with a Riemann surface Σ, we also want the bijection we have defined above to be a holomorphic isomorphism. But we will not carry this programme through in detail, since it would largely be to repeat the work we have already done in different language. Suffice it to say that one can show, either from first principles or by going through the existence result that we have, that the residue class fields defined by valuations on K are all canonically isomorphic to $\mathbf{C}$. Now, for any non-constant $f \in K$ and valuation v, we define $e_v(f) \in \mathbf{C} \cup \{\infty\}$ by the following:

- If $v(f) < 0$, then $e_v(f) = \infty$.
- If $v(f) \geq 0$, so f is in R_v, we set $e_v(f)$ to be the image of f in R_v/I_v under the identification of this field with $\mathbf{C}$.

This means that an element $f \in K \setminus \mathbf{C}$ defines a map of sets

$$\tilde{f} : \mathrm{Val}(K) \to \mathbf{CP}^1,$$

with $\tilde{f}(v) = e_v(f)$. The Riemann surface structure on $\mathrm{Val}(K)$ is characterised by the property that all these maps are holomorphic.

The upshot of all this is that the study of compact Riemann surfaces is formally equivalent to an algebraic problem: the study of fields of transcendence degree 1 over $\mathbf{C}$. The same questions can be studied either geometrically or algebraically. To illustrate this, recall two concepts.

- Given fields $L \subset K$, say of characteristic zero, we have a *Galois group* $\mathrm{Gal}(K : L)$.
- Given a non-constant holomorphic map $f : \Sigma_1 \to \Sigma_2$, we have a monodromy $\pi_1(\Sigma_2 \setminus \Delta) \to S_d$. We call the image of this the monodromy group of the map. (It is well defined up to conjugacy in S_d.)

Now $f : \Sigma_1 \to \Sigma_2$ corresponds to a field extension $k_{\Sigma_2} \subset k_{\Sigma_1}$, and we have the following theorem.

Theorem 17. *The monodromy group of f is isomorphic to the Galois group* $\mathrm{Gal}(k_{\Sigma_2} : k_{\Sigma_1})$.

We now outline the proof, leaving the interested reader to fill in some details. Suppose that a finite group Γ acts effectively on a compact Riemann surface Σ. The quotient Σ/Γ is naturally a Riemann surface Σ', and meromorphic functions on Σ' can be identified with Γ-invariant meromorphic functions on Σ. The degree d of the natural map $p : \Sigma \to \Sigma'$ is the order of Γ. Let $f \in k_\Sigma$ be a generator of k_Σ over $k_{\Sigma'}$, the zero of an irreducible polynomial P with coefficients in $k_{\Sigma'}$. Then, for each $\gamma \in \Gamma$, the element $f \circ \gamma \in k_\Sigma$ is also a root of P, and since the degree of P is d, the polynomial P splits in k_Σ. It follows that the extension $k_{\Sigma'} \subset k_\Sigma$ is normal.

Now we go back to the general situation of $f : \Sigma_1 \to \Sigma_2$, as above, and let $\Gamma \subset S_d$ be the image of the monodromy $\rho : \pi_1(\Sigma_2 \setminus \Delta) \to S_d$. Write $\underline{d}$ for the standard set $\{1, \dots, d\}$ of d elements. Then Γ acts transitively on $\underline{d}$ (since Σ_1 is connected). Write $\underline{\Gamma}$ for the set of elements of Γ. Then Γ acts transitively on $\underline{\Gamma}$ on the left and the right, and the left and right actions commute. Let $e : \underline{\Gamma} \to \underline{d}$ be the map $e(\gamma) = \gamma(1)$. This is an equivarient map for the left action of Γ on $\underline{\Gamma}$. Now ρ induces a homomorphism from $\pi_1(\Sigma_2 \setminus \Delta)$ to the permutations of the set $\underline{\Gamma}$, so the general existence theorem gives a compact Riemann surface Σ with a holomorphic map $\hat{f} : \Sigma \to \Sigma_2$ of degree equal to $|\Gamma|$. The map of sets e induces a factorisation of $\hat{f}$ as

$$\Sigma \to \Sigma_1 \to \Sigma_2,$$

and the right action of Γ on $\underline{\Gamma}$ induces an action of Γ on Σ such that $\Sigma_2 = \Sigma/\Gamma$. Now one sees that k_Σ is the normal closure of the extension $k_{\Sigma_2} \subset k_{\Sigma_1}$ and that the group of automorphisms of k_Σ fixing k_{Σ_2} is precisely Γ. By definition, this is the Galois group of $k_{\Sigma_2} \subset k_{\Sigma_1}$.

Remark

1. This theory—connecting algebra and geometry—is in some ways a proto-type for the theory of 'schemes'. However, the (complex) one-dimensional case has special features which mean that one can work with fields and valuations as opposed, in the more general situation, to sheaves, rings and the 'spectra' of rings.

2. We have explained in Chapter 4 that an irreducible polynomial P in two variables defines an algebraic curve in the projective plane $\mathbf{CP}^2$. The Riemann surface of P maps holomorphically onto this curve but, typically, the curve has singularities and the map is not one-to-one. It is not hard to show that any compact Riemann suface can be *embedded* as a complex submanifold in $\mathbf{CP}^N$ for some N, cut out by the zeros of a finite set of homogeneous polynomial equations. That is, a compact Riemann surface can be realised as a smooth algebraic curve. In fact, one can take $N = 3$. One can also show that a compact Riemann surface can be mapped into the projective plane in such a way that the image has only 'ordinary double

point' singularities. These are modelled locally on the solutions of the equation $zw = 0$. Two branches of the Riemann surface cross in $\mathbf{CP}^2$.

11.1.3   Connections with algebraic number theory

There are well-known analogies between the ring $\mathbf{C}[z]$ of polynomials in one variable and the ring of integers $\mathbf{Z}$. Both are Euclidean domains. Here we replace the field $\mathbf{C}(z)$ (the field of fractions of $\mathbf{C}[z]$) by the rational numbers $\mathbf{Q}$ (the field of fractions of $\mathbf{Z}$) and consider a finite extension k of $\mathbf{Q}$, that is, an 'algebraic number field'. What are the valuations on k? First consider the case $k = \mathbf{Q}$. Given a prime $p > 0$, we write any rational number x in the form $x = p^v q/r$, where q and r are coprime to p, and set $v_p(x) = v$. This is just the same as the construction of the valuation on $\mathbf{C}(z)$ associated to a point $z_0 \in \mathbf{C}$, with the prime p in place of $(z - z_0)$ (which is a prime in $\mathbf{C}[z]$). The basic fact is that any non-trivial valuation on $\mathbf{Q}$ is equivalent to one of the v_p. One difference between this case and the previous one is that the residue class fields vary. The valuation ring associated to a prime p consists of fractions $p^v q/r$ with $v \geq 0$; the valuation ideal consists of such fractions with $v > 0$ and the quotient is the finite field $\mathbf{Z}/p$. Notice that a rational number x is an integer if and only if $v(x) \geq 0$ for all valuations v.

Now consider an extension k of $\mathbf{Q}$ and a non-trivial valuation v on k. The restriction of v to $\mathbf{Q}$ must be some v_p for a prime p (after normalising by a constant multiple). So we have to study the possible extensions of v_p on $\mathbf{Q}$ to k, just as we studied the extensions of a valuation determined by a point z_0 in $\mathbf{C}$ to a finite extension K of $\mathbf{C}(z)$. Let us think through the procedure in the latter case.

1. Suppose K is generated by w, where $P(z, w) = 0$ for an irreducible polynomial P of degree d in w. For simplicity, suppose we are in a case when the coefficient of w^d is 1. We solve the equation $P(z_0, w) = 0$ for w in $\mathbf{C}$. Suppose first that we are in the case when there are d distinct roots $w_1, \ldots, w_d$.
2. Fix a root w_i. We find a holomorphic function $\phi(z)$ defined in a neighbourhood of z_0 such that $P(z, \phi(z)) = 0$ and $\phi(z_0) = w_i$. This is the basic fact we established in Chapter 1.
3. Now we define a valuation on K as follows. Write $g \in K$ as $Q(z, w)$ where Q is a polynomial in w with co-efficients in $\mathbf{C}(z)$. Substitute $w = \phi(z)$ to get a meromorphic function $\gamma(z) = Q(z, \phi(z))$ of z. Now define $v(g)$ to be the order of vanishing of γ at z_0.
4. If the equation $P(z_0, w) = 0$ has multiple roots, we generalise the construction using Puiseaux series, solving the equation $P(z_0, \phi((z - z_0)^{1/r})) = 0$ (see equation (11.2)).

Guided by this, we proceed as follows.

1. Suppose our algebraic number field k is $\mathbf{Q}(\theta)$, where $f(\theta) = 0$ for an irreducible polynomial f of degree d with integer coefficients. For simplicity, suppose we are in a case when f is monic (i.e. the coefficient of θ^d is 1). Now, given a fixed prime number p, we factorise the reduction of the polynomial f to a polynomial with co-efficients in the associated *residue class field* $\mathbf{Z}/p$. Since this field is not algebraically closed, there is a wider variety of possibilities. For simplicity, the factors are all linear and distinct, that is, the equation $f(x) = 0$ has d distinct roots modulo p. We choose integers $\theta_1, \ldots, \theta_d$ representing these roots modulo p.

2. We fix a root $\theta_i \in \mathbf{Z}$. This is a solution of the equation $f(\theta_i) = 0 \bmod p$. It is a fact (Hensel's Lemma) that we can 'promote' this to a solution modulo any power of p. Thus we can find an $a_1 \in \mathbf{Z}$ such that

$$f(\theta_i + a_1 p) = 0 \bmod p^2,$$

and then an a_2 such that

$$f(\theta_i + a_1 p + a_2 p^2) = 0 \bmod p^3,$$

and so on. We encode all these solutions in a formal series

$$\phi = \theta_i + a_1 p + a_2 p^2 + a_3 p^3 + \ldots \tag{11.2}$$

(If we choose the representatives $\theta_i, a_1, a_2, a_3, \ldots$ to lie between 0 and $p - 1$, then they are unique.) Of course, this is the analogue of the holomorphic function ϕ in the previous case, viewed as a power series in $(z - z_0)$.

3. Now we define a valuation v on k as follows. We write an element of k as $g(\theta)$ for a polynomial g with rational coefficients and substitute $\theta = \phi$. Since we have not yet attached a precise meaning to ϕ, we can interpret this as saying that we take some sufficiently large integer N and substitute a finite sum $\phi_N = \theta_i + a_1 p + \cdots + a_N p^N$. Then $g(\phi_N)$ is a rational number. We define $v(g(\theta))$ to be the 'number of times p divides' this rational number or, more precisely, $v_p(g(\phi_N))$. (Of course, we need to check that this does not vary with N, once N is sufficiently large.)

4. To complete the story we would need to extend this discussion to the case when the polynomial f does not factorise into distinct linear factors modulo p, but for this we refer to texts on algebraic number theory.

This sketch indicates how one can describe the valuations on a number field 'lying over' the primes in $\mathbf{Z}$ in analogy to the points of the Riemann surface of an algebraic function. The ring of *algebraic integers* in k is the set of elements g such that $v(g) \geq 0$ for all valuations v, and valuations are another language for taking about ideals in this ring, but we will not go further into this.

One of the important differences between the field $\mathbf{Q}$ and $\mathbf{C}(z)$ is that in the latter case one has a valuation at the point at infinity in the Riemann sphere $\mathbf{CP}^1$. The analogy is made more precise by introducing the notion of an *absolute value* on a field.

Let F be any field. An absolute value on F is a map $x \mapsto |x|$ from F to $\mathbf{R}$ such that for all x, y in F,

- $|xy| = |x|\,|y|$,
- $|x + y| \leq |x| + |y|$.

Of course, the usual absolute value, which we write as $|\cdot|_\infty$, is an absolute value on $\mathbf{C}$ and hence on $\mathbf{Q}$. A valuation v defines an absolute value by

$$|x| = e^{-v(x)}$$

(where, of course, we set $e^{-\infty} = 0$). In this case we have the stronger inequality

$$|x + y| \leq \max(|x|, |y|).$$

Absolute values of this kind are called *non-Archimedean*. A theorem of Ostrowski asserts that the only absolute values on $\mathbf{Q}$ are the standard one $|\cdot|_\infty$ and those obtained from valuations.

Remark

1. Suppose one is presented with a calculation in elementary arithmetic such as

$$845,213 \times 597,161 = 504,728,240,293,$$

and one wants to check if it is probably correct. Then there are two standard kinds of test. One is by 'order of magnitude':

$$845,213 \sim 8 \times 10^5; \quad 597,161 \sim 6 \times 10^5 \Rightarrow \text{product} \sim 5 \times 10^{11}.$$

The other is by divisibility, reducing modulo an integer, which (by the Chinese Remainder Theorem) we can take to be a prime power. For example, using the 'rule of three', we have

$$845,213 = 8 + 4 + 5 + 2 + 1 + 3 = 2 \bmod 3; \quad 597,161 = 5 + 9 + 7 + 1 + 6 + 1$$

$$= 2 \bmod 3 \Rightarrow \text{product} = 1 \bmod 3.$$

This agrees with

$$5 + 0 + 4 + 7 + 2 + 8 + 2 + 4 + 0 + 2 + 9 + 3 = 1 \bmod 3.$$

The two kinds of valuations on $\mathbf{Q}$ can be seen as developing these two familiar ideas.

2. Terminology differs in the literature. Some authors (Cassels 1986) define a valuation to be what we have called an absolute value. Our terminology follows Cohn (1991).

If we work with absolute values, we get an analogue of the 'point at infinity' in the Riemann sphere, furnished by $|\cdot|_\infty$. It is convenient to normalise the multiplicative constants by defining

$$|x|_p = p^{-v},$$

if $x = p^v q/r$ for q, r coprime to p. This gives us a *product formula*

$$\prod_p |x|_p = 1, \tag{11.3}$$

where the product runs over the $|\cdot|_p$ and $|\cdot|_\infty$. The analogue holds in $\mathbf{C}(z)$, where it is the assertion that the number of zeros of a rational function is equal to the number of poles. The absolute values on a number field k comprise the non-Archimedean ones, corresponding to the valuations discussed above, and Archimedean ones extending $|\cdot|_\infty$ on $\mathbf{Q}$. The latter are analogous to the points of a Riemann surface lying over the point at infinity in the Riemann sphere. One also has a product formula like equation (11.3).

An absolute value on a field F defines a metric, and we can form the *completion* in the usual sense of metric space theory. This is again a field. The completion of $\mathbf{Q}$ in $|\cdot|_\infty$ is $\mathbf{R}$. The completion of $\mathbf{Q}$ with respect to $|\cdot|_p$ is the field $\mathbf{Q}_p$ of p-adic numbers. The closure of $\mathbf{Z}$ in $\mathbf{Q}_p$ is a subring $\mathbf{Z}_p$. Elements of $\mathbf{Z}_p$ can be written as formal power series

$$a_0 + a_1 p + a_2 p^2 + \ldots,$$

where the a_i are integers between 0 and $p - 1$. Thus we can attach a precise meaning to equation (11.2) in $\mathbf{Z}_p$. This ring corresponds to the ring of formal power series $\sum a_r (z - z_0)^r$ about z_0.

11.2 Hyperbolic surfaces

11.2.1 Definitions

In this section, we need some ideas from *Riemannian geometry*. We start with some very basic definitions, taking an informal point of view. A Riemannian metric on a smooth surface is given by specifying a positive definite quadratic form on each tangent space, varying 'smoothly' as we move around the surface. We can make this precise in local coordinates x_1, x_2. At each point (in the domain of this coordinate chart), we have a basis for the tangent space

$\partial_i = \partial/\partial x_i$, and a quadratic form is specified by the inner products $g_{ij} = \langle \partial_i, \partial_j \rangle$. We need the symmetric matrix

$$\begin{pmatrix} g_{11} & g_{12} \\ g_{12} & g_{22} \end{pmatrix} \tag{11.4}$$

to be positive definite in the usual sense and we need the g_{ij} to be smooth functions of the co-ordinates x_1, x_2. The traditional notation is

$$ds^2 = g_{11}\,dx_1^2 + 2g_{12}\,dx_1\,dx_2 + g_{22}\,dx_2^2.$$

A Riemannian metric on a surface S defines the *lengths* of smooth paths in S and the *area* of open sets in S. In local co-ordinates, a path is given by functions $x_1(t), x_2(t)$ of a real variable $t \in [a,b]$, say, and the length is defined to be

$$\int_a^b \left(\sum_{ij} g_{ij}\dot{x}_i\dot{x}_j \right)^{1/2} dt.$$

The *distance* between two points of a connected Riemannian surface is defined to be the infimum of the lengths of the paths joining the points. This distance function makes the surface into a metric space.

The *area* of a region U contained in the co-ordinate chart is

$$\int_U \det(g_{ij})\,dx_1\,dx_2.$$

One needs to check that these definitions are independent of the choice of local co-ordinates and, in the case of path length, of the choice of parametrisation of the path.

Let $S, \tilde{S}$ be smooth surfaces with Riemannian metrics. A diffeomorphism $f : S \to \tilde{S}$ is called an *isometry* if at each point $p \in S$ the derivative $df_p : TS_p \to T\tilde{S}_{f(p)}$ is an isometry of Euclidean vector spaces.

Now suppose that Σ is a Riemann surface. Each tangent space is a one-dimensional complex vector space, and a Riemannian metric on Σ is *compatible* with the Riemann surface structure if multiplication by i preserves lengths of tangent vectors. Another way of expressing this is that at a point of a Riemann surface the *angle* between tangent vectors is intrinsically defined, and the compatibility condition is that this is the same as the angle furnished by the Euclidean geometry of the tangent space. In terms of a local holomorphic co-ordinate $z = x_1 + ix_2$, the compatibility condition is that the metric takes the form

$$ds^2 = V(dx_1^2 + dx_2^2),$$

for a positive function V, i.e. the matrix (11.4) is V times the identity matrix. We will sometimes use the shorthand $V|dz|^2$ to denote this metric. Let $\Sigma, \tilde{\Sigma}$ be two Riemann surfaces with compatible metrics and let $f : \Sigma \to \tilde{\Sigma}$ be a holomorphic diffeomorphism. Choose a holomorphic co-ordinate z around a point of Σ and w around the corresponding point of $\tilde{\Sigma}$. Then, in terms of these co-ordinates, we can write $w = f(z)$. The metrics are given by $V(z)|dz|^2$ and $\tilde{V}(w)|dw|^2$, say. The condition that f be an isometry is

$$V(z) = |f'(z)|^2 \, \tilde{V}(f(z)). \qquad (11.5)$$

11.2.2   Models of the hyperbolic plane

Now we introduce three 'models' of the hyperbolic plane: that is, three partic-ular surfaces with Riemannian metrics which will turn out to be isometric.

The Poincaré (disc) model. The surface is simply the unit disc $D = \{w \in \mathbf{C} : |w| < 1\}$ in $\mathbf{C}$. The metric, in the notation introduced above, is

$$\frac{4}{(1 - |w|^2)^2}|dw|^2.$$

The Beltrami (half-plane) model. The surface is the upper half-plane $H = \{z \in \mathbf{C} : \mathrm{Im}(z) > 0\}$, and the metric is

$$\left(\frac{1}{\mathrm{Im}(z)}\right)^2 |dz|^2.$$

These metrics are clearly compatible with the obvious Riemann surface structures. By contrast, the third model is not obviously a Riemann surface.

The Klein model. We write $\mathbf{R}^{2,1}$ for $\mathbf{R}^3$ equipped with a standard symmetric bilinear form B of signature $- + +$. So B has the matrix

$$\begin{pmatrix} -1 & 0 & 0 \\ 0 & 1 & 0 \\ 0 & 0 & 1 \end{pmatrix}.$$

The set $\{u \in \mathbf{R}^{3,1} : B(u,u) = -1\}$ is a hyperboloid of two sheets, and we let Q be the sheet where the first co-ordinate is positive. This is obviously a smooth surface, diffeomorphic to $\mathbf{R}^2$. The tangent space TQ_p of Q at a point $p \in Q$ is a two-dimensional subspace of $\mathbf{R}^{2,1}$. So we get a quadratic form on TQ_p by restricting B. As the reader can check, and we will see presently, this restriction is always positive definite, so it defines a Riemannian metric on Q.

Now we have the following theorem.

Theorem 18.   *The three Riemannian surfaces described above are isometric.*

Let f be the Möbius map $f(z) = (z - i)/(z + i)$. It is clear that z lies in H if and only if $|z - i| < |z + i|$, and so if and only if $|f(z)| < 1$. Thus f is a holomorphic diffeomorphism from H to D. We have $f'(z) = 2i/(z + i)^2$, so

$$|f'(z)|^2 = \frac{4}{|z + i|^2}.$$

On the other hand, if $w = f(z)$,

$$1 - |w|^2 = 1 - \frac{z\bar{z} + i(z - \bar{z}) + 1}{z\bar{z} - i(z - \bar{z}) + 1} = \frac{4 \operatorname{Im}(z)}{|z + i|^2}.$$

So

$$\frac{4}{(1 - |w|^2)^2} = \frac{1}{|f'(z)|^2} \frac{1}{\operatorname{Im}(z))^2},$$

which is the isometry condition (11.5).

Now we construct an isometry from the Klein model Q to the disc D. We can clearly parametrise Q by 'polar' co-ordinates

$$(\cosh \rho, \ \sinh \rho \cos \theta, \ \sinh \rho \sin \theta).$$

Then (away from the co-ordinate singularity where $\rho = 0$) we have a basis for the tangent spaces of Q given by $\partial_\rho, \partial_\theta$. These correspond to vectors in $\mathbf{R}^{2,1}$;

$$\partial_\rho = (\sinh \rho, \ \cosh \rho \cos \theta, \cosh \rho \sin \theta),$$

$$\partial_\theta = (0, \ -\sinh \rho \sin \theta, \sinh \rho \cos \theta).$$

Computing inner products using the indefinite form B, we obtain

$$\langle \partial_\rho, \partial_\rho \rangle = 1, \quad \langle \partial_\rho, \partial_\theta \rangle = 0, \quad \langle \partial_\theta, \partial_\theta \rangle = \sinh^2 \rho.$$

So the metric on Q expressed in the ρ, θ co-ordinates is

$$ds^2 = d\rho^2 + \sinh^2 \rho \ d\theta^2.$$

Now we define a map g from Q to D which takes the point with polar co-ordinates ρ, θ in Q to $w = \tanh(\rho/2)e^{i\theta}$ in D. In other words, if r, θ are the standard polar co-ordinates in D, our map is defined by $r = \tanh(\rho/2)$. Then

$$\frac{dr}{d\rho} = \frac{1}{2} \frac{1}{\cosh^2(\rho/2)} = \frac{1 - r^2}{2}.$$

This means that

$$d\rho^2 + \sinh^2 \rho \ d\theta^2 = \frac{4}{(1 - r^2)^2} dr^2 + \sinh^2 \rho \ d\theta^2.$$

Now, $\sinh\rho = 2\sinh(\rho/2)\cosh(\rho/2) = 2r\cosh^2(\rho/2) = 2r/(1-r^2)$,   so   the metric is

$$\frac{4}{(1-r^2)^2}\left(dr^2 + r^2\,d\theta^2\right),$$

which is the same as the metric on D, written in polar coordinates. This calculation shows that g is an isometry.

We can call any one of these three manifolds 'the hyperbolic plane'.

11.2.3   Self-isometries

The definition of an isometry $f : S \to \tilde{S}$ allows, in particular, the case when $S = \tilde{S}$ (as Riemannian surfaces), and clearly these self-isometries form a group. For our purposes, we restrict attention to oriented surfaces and orientation-preserving isometries, so we speak of the oriented isometry group. A 'typical' Riemannian metric has no non-trivial isometries: the distinctive feature of the hyperbolic plane is that this isometry group is 'large'.

Recall that the group of *holomorphic automorphisms* of the upper half-plane H is $PSL(2,\mathbf{R})$, acting as Möbius maps.

Proposition 35. *The oriented isometry group of H, with the metric given above, is $PSL(2,\mathbf{R})$.*

In one direction, consider a Möbius map $f(z) = (az+b)/(cz+d)$ with a,b,c,d real and $ad - bc = 1$. Then

$$f'(z) = \frac{ad - bc}{(cz+d)^2} = \frac{1}{(cz+d)^2}.$$

If $w = f(z)$, then

$$2i\,\mathrm{Im}(w) = \frac{az+b}{cz+d} - \frac{a\bar{z}+b}{c\bar{z}+d},$$

which simplifies to

$$2i\,\mathrm{Im}(w) = \frac{(ad-bc)(z-\bar{z})}{|cz+d|^2} = 2i\,\mathrm{Im}(z)\frac{1}{|cz+d|^2} = 2i\,\mathrm{Im}(z)\,|f'(z)|^2.$$

This is the desired identity (11.5), which shows that f is an isometry.

In the other direction, it is in general the case that an oriented isometry of a Riemannian metric which is compatible with a Riemann surface structure must be holomorphic. For the isometry condition implies that the map preserves angles between tangent vectors and so is a *conformal map*, and an orientation-preserving conformal map is holomorphic. Since we know the holomorphic automorphism group of H, there can be no other oriented isometries.

In fact, a stronger statement is true. Let $f : N \to \tilde{N}$ be a diffeomorphism from an open neighbourhood of a point $p \in H$ to a neighbourhood of another point $\tilde{p}$. Then, if f is an oriented isometry of the hyperbolic metric, it is the restriction of a map defined by an element of $PSL(2,\mathbf{R})$.

Of course, it follows from Theorem 18 that the oriented isometry groups of our three models of the hyperbolic plane are isomorphic. For the disc model, the oriented isometries are the Möbius maps

$$f(w) = \frac{aw + b}{\bar{b}w + \bar{a}},$$

which are the holomorphic automorphisms of the disc. For the Klein model, we start with the group $O(2,1)$ of linear automorphisms of $\mathbf{R}^{2,1}$ which preserve the quadratic form B. This has an index-2 subgroup $O^+(2,1)$ which maps S to S. The oriented isometry group is the index-4 subgroup $SO^+(2,1) \subset O^+(2,1) \subset O(2,1)$ of elements of $O^+(2,1)$ with determinant 1. This is the identity component of $O(2,1)$.

11.2.4   Hyperbolic surfaces

Let S be a smooth surface. A *hyperbolic structure* on S is a Riemannian metric on S which is locally isometric to the hyperbolic plane. That is, each point in S has a neighbourhood which is isometric to an open set in H. An equivalent definition is that there is an atlas of charts for S with overlap functions given by the restrictions of elements of $PSL(2,\mathbf{R})$. Let $\Gamma \subset PSL(2,\mathbf{R})$ be a discrete group which acts freely on H. We have seen that the quotient $\Sigma = H/\Gamma$ is a Riemann surface. Since Γ preserves the metric on H, there is an obvious induced metric on Σ. This is a hyperbolic metric on Σ, compatible with the Riemann surface structure. It is a fact that these hyperbolic surfaces are precisely those which are oriented and *complete*, in the sense of metric space theory. (Note that a metric on a compact surface is always complete.)

We know that, with a very few exceptions (the complex plane, the Riemann sphere, tori and the once-punctured plane), all Riemann surfaces arise as such quotients H/Γ. Thus, with these few exceptions, all Riemann surfaces have compatible complete hyperbolic metrics. In particular, all compact Riemann surfaces of genus 2 or more have compatible hyperbolic metrics. Moreover, it follows from the discussion above that the metric is unique. So the study of compact Riemann surfaces is essentially equivalent to the study of compact oriented hyperbolic surfaces.

11.2.5   Geodesics

Let S be a connected smooth surface with a Riemannian metric. A *geodesic* in S can be defined as a path $\gamma : I \to S$ (for some interval $I \subset \mathbf{R}$), parametrised by arc length, such that for all t_0 in the interior of I, there is a $\delta > 0$ with the property that if $t_0 - \delta < t_1 < t_2 < t_0 + \delta$ then the length of the path given by restricting γ to the interval $[t_1, t_2]$ is equal to the distance from $\gamma(t_1)$ to $\gamma(t_2)$. Of course, when S is $\mathbf{R}^2$ with the Euclidean metric, the geodesics are just line segments. In general, geodesics are solutions of a second-order ordinary differential equation—the Euler–Lagrange equation for the length functional. However, we do not need this theory if we confine our attention to hyperbolic surfaces.

Strictly, as we have defined it, a geodesic is a parametrised curve: a map from an interval rather than the image of the map. We will sometimes use language imprecisely, calling this image a geodesic where no confusion will be caused.

Consider the half-plane model H and a pair of points on the imaginary axis iy_1, iy_2, with $y_1 < y_2$. Write $z = x + iy$ and consider any path $(x(t), y(t))$ between these points. The path $(0, y(t))$ has the same end points and, clearly, its length is no greater. It follows that the infimum is realised by the path which runs along the imaginary axis from iy_1 to iy_2. So the imaginary axis is a geodesic. Recall that Möbius maps preserve the set of 'circlines'—circles and straight lines in $\mathbf{C}$. Thus an element of $PSL(2, \mathbf{R})$ maps the imaginary axis to a circline C. But since Möbius maps preserve angles, C must meet the real axis orthogonally. That is, C is either a circle with its centre on the real axis or a line parallel to the imaginary axis. It is easy to verify, using the action of the isometries, that the intersections of these circlines with H are precisely the geodesics in H. By the same reasoning, the geodesics in D are arcs of circles which meet the unit circle orthogonally or line segments through the origin.

Geodesics are also easy to describe in the Klein model. Let $Q^* \subset \mathbf{R}^{2,1}$ be the set of vectors v with $B(v, v) = 1$. For $v \in Q^*$, let

$$v^\perp = \{u \in Q : B(u, v) = 0\}.$$

Suppose u, u' are distinct points of Q. They span a two-dimensional subspace Π of $\mathbf{R}^{2,1}$. The orthogonal complement $\Pi^\perp$ of Π with respect to the indefinite form B is a one-dimensional subspace; let v be a basis vector. We cannot have $B(v, v) < 0$, for that would mean that v, u span a two-dimensional subspace on which B is negative definite, and we cannot have $B(v, v) = 0$, for then the form would be degenerate. So $B(v, v) > 0$, and we can suppose that $B(v, v) = 1$. Then v is in Q^* and $u, u' \in v^\perp$. The sets $v^\perp$ are exactly the geodesics in Q. (See Exercise 5 at the end of this chapter). Clearly, if $v \in Q^*$ then so is $-v$ and the sets $v^\perp, (-v)^\perp$ are the same. So the set of geodesics in Q can be identified with the quotient $Q^*/\pm 1$. Points of Q^* can be viewed as oriented geodesics: a geodesic with a choice of direction.

Let Δ be the unit disc in $\mathbf{R}^2$ and let $\pi : Q \to \Delta$ be the linear projection map

$$\pi(\cosh \rho, \sinh \rho \cos \theta, \sinh \rho \sin \theta) = (\tanh \rho \cos \theta, \tanh \rho \sin \theta).$$

Then π takes geodesics in Q to ordinary straight line segments in Δ.

The metric geometry of the hyperbolic plane makes clear some of the constructions and assertions that we have performed and made above. In the Klein model, the hyperbolic distance from the point $(1,0,0)$ to $(\cosh\rho, \sinh\rho\cos\theta, \sinh\rho\sin\theta)$ is ρ. In the disc model, the distance from the origin to the point $re^{i\theta}$ is

$$\int_0^r \frac{2}{1-t^2}\,dt = 2\tanh^{-1}(r).$$

This makes evident the formula, which we produced before out of a hat, for the isometry between the Klein and disc models. Given a point p in the hyperbolic plane and a radius s, we can consider the *hyperbolic disc* centred at p, of radius s. This is the set of points whose hyperbolic distance to p is less than s. Clearly, in the disc model when p is the origin, the hyperbolic disc is the ordinary Euclidean disc with radius $\tanh(s/2)$. Using the fact that Möbius maps preserve the class of circlines, we see that, in either the disc or the half-space model, hyperbolic discs are Euclidean discs, but in general the hyperbolic centre is not the same as the Euclidean centre. The fact that the local isometries of the hyperbolic plane are the same as the global ones becomes clear when one considers polar co-ordinates on a hyperbolic disc.

Cutting and pasting, collar neighbourhoods

We now turn to geodesics in hyperbolic surfaces. Since the geodesic condition is local, these are locally modelled on the geodesics in the hyperbolic plane discussed above. The crucial feature of geodesics in hyperbolic surfaces is that they can be used in cutting-and-gluing constructions. We will present this informally. The starting point is the following lemma.

Lemma 25. *The oriented isometry group of the hyperbolic plane acts simply transitively on pairs consisting of a point p and an oriented geodesic containing p.*

That is, if γ is an oriented geodesic in the hyperbolic plane through p and γ' is an oriented geodesic through another point p', there is a unique oriented isometry taking p to p' and γ to γ'. Now suppose that Σ, Σ' are two hyperbolic surfaces, that γ, γ' are geodesics in Σ, Σ', respectively, that p is a point of γ and that p' is a point of γ'. Then there are neighbourhoods N, N' of p, p' and a unique isometry from N to N' taking p to p' and $\gamma \cap N$ to $\gamma' \cap N'$. In short, geodesics in hyperbolic surfaces are locally identical.

Definition 10. *A simple closed geodesic of length l in a hyperbolic surface S is a geodesic (parametrised by arc length) $\gamma : \mathbf{R} \to S$ such that $\gamma(t_1) = \gamma(t_2)$ if and only if $t_1 - t_2$ is an integer multiple of l.*

Thus the image of a simple closed geodesic is an embedded circle in the surface (and we will often refer to this image as a simple closed geodesic).

Given $l > 0$, set $a = e^l$ and let G_l be the subgroup of $PSL(2, \mathbf{R})$ generated by the map $z \mapsto az$. The imaginary axis is preserved by G_l, and its image yields a simple closed geodesic γ_0 of length l in the quotient $\Sigma_l = H/G_l$. If γ is a simple closed geodesic of length l in any hyperbolic surface Σ, a neighbourhood of γ in Σ is isometric to a neighbourhood of γ_0 in Σ_l. As a consequence, *any two simple closed geodesics in hyperbolic surfaces, of the same length, have isometric neighbourhoods*. Now suppose, for example, that γ divides Σ into two components Σ^+, Σ^-. Suppose we have another such situation, with a simple closed geodesic of the same length l in another hyperbolic surface Ξ which divides Ξ into components Ξ^+, Ξ^-. Then we can cut Σ, Ξ along the geodesics and glue Σ^+ to Ξ^- (say); then the resulting space has a natural hyperbolic structure. In fact, we have a choice in doing this. If we fix a point p on γ, we can make the gluing in such a way as to identify p with any given point on the given geodesic in Ξ.

As a variant of the same idea, consider the 'ideal hyperbolic triangle' P in H defined by the inequalities $\operatorname{Im} z > 0, -0 \leq \operatorname{Re} z \leq 1, |z - 1/2| \geq 1/2$. This is a 'hyperbolic surface with a geodesic boundary'. If we take two copies of P, we can glue them along the boundary geodesics to get a hyperbolic surface, homeomorphic to the sphere punctured at three points. Again we have a choice of ways of doing this, but there is one natural choice, in which the boundaries of the two components are identified by the identity map. This is called the *double* of P. When we make this choice, the surface we construct is complete and can be described slightly differently, as follows (Figure 11.1). We take the subgroup Γ of $PSL(2, \mathbf{R})$ generated by

$$z \mapsto z + 2, \quad z \mapsto \frac{z}{2z + 1}.$$

A fundamental domain for the action of Γ on H is given by the union of P and its translate by -1. The action of $z \mapsto z + 2$ glues the edge $\operatorname{Re}(z) = -1$ to the edge $\operatorname{Re}(z) = 1$, and the action of $z \mapsto z/2z + 1$ glues the point $-1/2 + e^{i\theta}/2$ on one semicircular boundary component to $1/2 - e^{-i\theta}/2$ on the other. So the double of P is the quotient H/Γ. In fact, Γ is the group Γ_2 discussed in Section 3.2.3.

We can use the same approach to give a different picture of the model hyperbolic surface $\Sigma_l = H/G_l$, containing a simple closed geodesic of length l. We take the region in H between the two geodesics $|z| = 1$ and $|z| = e^l$ and glue these two boundaries by $z \mapsto e^l z$.

11.2.6   Discussion

The topic of this section fits into two different more general theories. On the one hand, we can consider general Riemannian metrics, or just metrics on surfaces. Suppose such a metric is compatible with a Riemann surface structure

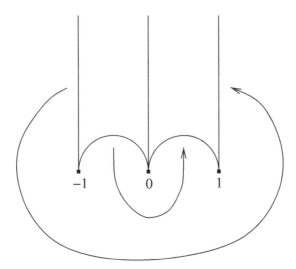

Figure 11.1   *Gluing two ideal triangles*

and that it is written in a local co-ordinate $z = x + iy$ as $e^{\phi}|dz|^2$. The *Gauss curvature* of the metric can be defined to be

$$K = \frac{1}{2}e^{-\phi}(\phi_{xx} + \phi_{yy}),$$

where subscripts denote partial derivatives with respect to x, y. Of course, we have to check that this formula is independent of the local co-ordinate used. Hyperbolic metrics are precisely those with $K = -1$.

On the other hand, we can fit our discussion into the general theory of Lie groups and homogeneous spaces. The three models of the hyperbolic plane yield the descriptions $SL(2, \mathbf{R})/SO(2)$, $SU(1, 1)/U(1)$, $SO(2, 1)^+/SO(2)$. Here $SU(1, 1)$ is the group of complex matrices of determinant 1 which preserve an indefinite Hermitian form. In fact, the three groups $SL(2, \mathbf{R})$, $SU(1, 1)$, $SO(2, 1)^+$ are isomorphic. But these isomorphisms are special to this low dimension. In addition, we can view $SL(2, \mathbf{R})$ either as the matrices of determinant 1, generalising to $SL(n, \mathbf{R})$, or as the matrices which preserve the standard *skew-symmetric* form on $\mathbf{R}^2$, generalising to $Sp(2n, \mathbf{R})$ on $\mathbf{R}^{2n}$. We can think of the different models of the hyperbolic plane as leading on to four different families of homogeneous spaces:

- the *n*-dimensional real hyperbolic spaces $SO(n, 1)^+/SO(n)$;
- the Siegel upper half-spaces $Sp(2n, \mathbf{R})/U(n)$;
- the complex hyperbolic spaces $SU(n, 1)/U(n)$;
- the spaces $\mathcal{H}_n = SL(n, \mathbf{R})/SO(n)$.

The last of these, $\mathcal{H}_n = SL(n, \mathbf{R})/SO(n)$, can be identified with the set of positive definite symmetric $n \times n$ matrices of determinant 1. When $n = 2$,

$$\det \begin{pmatrix} X_0 + X_1 & X_2 \\ X_2 & X_0 - X_1 \end{pmatrix} = X_0^2 - X_1^2 - X_2^2$$

and we see $\mathcal{H}_2$ as a sheet of a hyperboloid.

11.2.7   The Gauss–Bonnet Theorem

Consider a geodesic in the upper half-plane—a semicircle centred on the real axis parametrised by the hyperbolic arc length s. Let $\theta(s)$ be the angle that the tangent vector makes with the real axis. Then elementary geometry gives the identity

$$\frac{d\theta}{ds} - \frac{1}{y}\frac{dx}{ds} = 0. \tag{11.6}$$

Now suppose that $\Gamma(s) = x(s) + iy(s)$ is any curve in H parametrised by hyperbolic arc length, and define $\Theta(s)$ in the same way (Of course, there the usual ambiguity of factors of 2π in the definition of this angle.) We define the *geodesic curvature* of the curve to be

$$\kappa(s) = \frac{d\Theta}{ds} - \frac{1}{y}\frac{dx}{ds}.$$

Although it is not obvious from this definition, this quantity is a hyperbolic invariant. To see this, we give an alternative, intrinsic definition. Without loss of generality, consider the parameter value $s = 0$. Let γ be the oriented geodesic through $\Gamma(0)$ with the same tangent vector as Γ and let $d(s)$ be the 'signed distance' from $\Gamma(s)$ to γ. That is, for small s, there is a unique geodesic through $\Gamma(s)$ which cuts γ orthogonally at a point $p(s)$, and we set $d(s)$ to be $\pm$ the distance from $\gamma(s)$ to $p(s)$, with the sign determined by whether $p(s)$ is on the left- or right-hand side of γ, viewed with the given orientation.

Lemma 26. *The geodesic curvature $\kappa(0)$ is the second derivative $d''(0)$.*

To see this, consider for simplicity a case where $|\theta(0)| < \pi/2$ so $x'(0)$ is positive. For small s, we can parametrise γ as $x(s) + i(y(s) + h(s))$, where $h(0) = h'(0) = 0$. This parametrisation of γ is not by arc length on γ, but clearly the identity (11.6) holds for any parametrisation of a geodesic. So if $\Theta(s)$ is the angle made by the tangent vector of γ, we have

$$\frac{d\Theta}{ds} - \frac{1}{y+h}\frac{dx}{ds} = 0.$$

On the other hand, since $\tan\Theta = (y' + h')/x'$ and $\tan\theta = y'/x'$, we have, at $s = 0$,

$$\Theta' = \theta' + \frac{h''}{x'}.$$

Also, it is easy to see that

$$d(s) = \cos\theta \frac{h(s)}{y} + o(s^2),$$

so we get

$$d''(0) = \frac{\cos\Theta(0)}{y} h''(0).$$

Now the condition that Γ is parametrised by arc length means that

$$x'(0) = \cos\Theta(0)\, y,$$

which gives

$$\Theta' = \theta' - d''$$

and the result follows, using the identity (11.6) for the geodesic γ.

There are many other ways of giving an intrinsic definition of the geodesic curvature. Another approach is just to verify by calculation that the formula is invariant under the action of $PSL(2,\mathbf{R})$.

Now suppose that we have a simple *closed* curve Γ in H, parametrised anticlockwise by arc length. Then

$$\int_\Gamma \kappa\, ds = \int_\Gamma \frac{d\Theta}{ds} ds - \int_\Gamma \frac{1}{y}\frac{dx}{ds}\, ds.$$

The first integral is equal to 2π, since in traversing Γ the tangent vector rotates once in a positive direction (with our sign conventions). The second integral can be written in terms of a 1-form

$$\int_\Gamma \frac{dx}{y}.$$

By Stokes' Theorem, if Γ is the boundary of a compact region Δ, we have

$$\int_\Gamma \frac{dx}{y} = \int_\Delta \frac{dx\, dy}{y^2},$$

which is the hyperbolic area of Δ. So we have our first form of the *Gauss–Bonnet Theorem*,

$$\int_{\partial\Delta} \kappa\, ds = 2\pi + \text{Hyperbolic Area}(\Delta).$$

Of course, this should be compared with the formula in plane Euclidean geometry that the integral of the (Euclidean) geodesic curvature of a simple closed curve with respect to the Euclidean arc length is equal to 2π. (*Remark:* In fact, it is not much harder to extend the discussion to general Riemannian surfaces, where one obtains

$$\int_{\partial \Delta} \kappa \, ds = 2\pi - \int_{\Delta} K \, dA$$

when Δ is a topological disc.)

Now consider a hyperbolic polygon P whose edges are geodesic segments with internal angles α_i at the vertices. We choose a family of approximations Γ_ϵ by smooth curves, rounding off the vertices in a small ϵ-neighbourhood. The contribution to the integral of the geodesic curvature from such a neighbourhood is approximately $\pi - \alpha_i$ and, taking a limit, we obtain the second form of the Gauss–Bonnet Theorem,

$$\sum (\pi - \alpha_i) = 2\pi - \text{Hyperbolic Area}(P).$$

So, if there are N vertices, we have

$$\sum \alpha_i = (N - 2)\pi - \text{Hyperbolic Area}(P).$$

More generally, for a polygon Q whose edges E_j are general smooth curves and with internal angles α_i at the vertices, we have

$$\sum_i (\pi - \alpha_i) + \sum_j \int_{E_j} \kappa \, ds = 2\pi - \text{Hyperbolic Area}(Q). \tag{11.7}$$

Now consider a general compact hyperbolic surface Σ and assume that we can choose a triangulation, decomposing Σ as a union of polygons Q_k isometric to polygons in H. Each edge is in the boundary of two of these polygons, and the signs work out in such a way that the geodesic curvatures viewed from different sides are opposite. We sum over k the identity (11.7) for the Q_k, viewed as polygons in H. The integrals over the edges cancel and the sum of the internal angles at each vertex is 2π. Rearranging, we get

$$-2\pi (n_0 - n_1 + n_2) = \text{Hyperbolic Area}(\Sigma),$$

where n_0, n_1 and n_2 are, respectively, the numbers of vertices, edges and triangles. In other words, the hyperbolic area of a compact hyperbolic surface is -2π times the Euler characteristic.

11.2.8   Right-angled hexagons

In the remainder of this section, we establish some more technical results which will be important in Chapter 14. We begin here with the 'hexagon lemma'.

Lemma 27. *Given $l_1, l_2, l_3 > 0$, there is a hexagon $ABCDEF$ in the hyperbolic plane with all internal angles right angles and with prescribed side lengths $AB = l_1$, $CD = l_2$, $EF = l_3$ (see Figure 11.2). Moreover, this hexagon is unique up to isometries of the hyperbolic plane.*

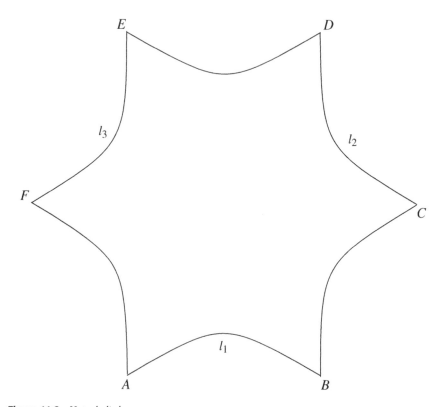

Figure 11.2 *Hyperbolic hexagon*

This is easiest to see using the Klein model of the hyperbolic plane. Recall that a vector $v \in Q^*$ defines a geodesic with a choice of orientation. It is convenient to regard this orientation as a choice of one of the two components of the complement of the geodesic, the component Ω_v given by the inequality $B(u, v) > 0$. Now we have the following lemma.

Lemma 28.

1. *If $v_1, v_2 \in S^*$, the corresponding geodesics intersect in Q if and only if $|B(v_1, v_2)| \geq 1$.*
2. *If $B(v_1, v_2) < -1$, the sets $\Omega_{v_1}, \Omega_{v_2}$ are disjoint. There is a unique geodesic which cuts $v_1^\perp, v_2^\perp$ orthogonally, and the distance between the two intersection points is $\cosh^{-1}(-B(v_1, v_2))$.*

We leave the proof as an exercise.

Now write $\alpha_i = \cosh l_i > 1$ for $i = 1, 2, 3$. It is clear that our problem comes down to finding three vectors $v_1, v_2, v_3 \in S^*$ such that

$$B(v_1, v_2) = -\alpha_3, \quad B(v_2, v_3) = -\alpha_1, \quad B(v_1, v_3) = -\alpha_2.$$

To see that this can be done, we use the classification of symmetric forms. Consider a three-dimensional vector space V with basis e_1, e_2, e_3, and define a bilinear form b on V by $b(e_i, e_i) = 1$, $b(e_2, e_3) = -\alpha_1$ etc. In other words, the form is given by the symmetric matrix

$$\begin{pmatrix} 1 & -\alpha_3 & -\alpha_2 \\ -\alpha_3 & 1 & -\alpha_1 \\ -\alpha_2 & -\alpha_1 & 1 \end{pmatrix}.$$

The determinant of this matrix is

$$1 - (\alpha_1^2 + \alpha_2^2 + \alpha_3^2) - 2\alpha_1\alpha_2\alpha_3,$$

which is less than $-(\alpha_1^2 + \alpha_2^2 + 2\alpha_1\alpha_2\alpha_3)$, since $\alpha_3 > 1$. Clearly, then, the determinant is strictly negative, which implies that the quadratic form has signature $(-++)$ (since the diagonal entries are positive). By the classification of quadratic forms, there is a linear isomorphism $A : V \to \mathbf{R}^{2,1}$ taking this form on V to the standard one B on $\mathbf{R}^{2,1}$. Then we can solve our problem by taking $v_i = A(e_i)$.

11.2.9   Closed geodesics

Throughout this subsection, we let Σ be a compact hyperbolic surface (although some of the definitions and statements apply more generally).

Recall that a pair of continuous maps ρ_0, ρ_1 from the circle to a space X (or 'loops in X') are said to be *freely homotopic* if there is a continuous map $R : S^1 \times [0, 1] \to X$ which restricts to ρ_0 on $S^1 \times \{0\}$ and to ρ_1 on $S^1 \times \{1\}$. This notion should be distinguished from that of *based homotopy*, where we fix base points and require that all our maps preserve them. It is the based homotopy classes which define the fundamental group $\pi_1(X, x_0)$. If X is connected, then the free homotopy classes of loops in X can be identified with the *conjugacy classes* in $\pi_1(X, x_0)$.

Proposition 36. *Any non-trivial free homotopy class of loops in Σ which contains an embedded representative contains a unique simple closed geodesic. If two such classes contain disjoint representatives, then the simple closed geodesics are disjoint.*

(Here the condition 'non-trivial' is needed only because the statement is false for the free homotopy class of a constant map.)

To reduce the possibility of confusion, let us emphasise that there are three slightly different notions:

- a *geodesic loop* based at a point x_0 in Σ, of length L, is a geodesic $\gamma : \mathbf{R} \to \Sigma$ with $\gamma(0) = \gamma(L) = x_0$.

- a *closed geodesic* of length L is a geodesic $\gamma : \mathbf{R} \to \Sigma$ which is periodic with period L.
- a *simple closed geodesic* of length L is a closed geodesic such that $\gamma(t_1) \neq \gamma(t_2)$ if $t_1 - t_2$ is not a multiple of L.

Of course, we can view any of these as a map from the circle to Σ, restricting to $[0, L]$ and identifying the end points.

Another notion which we will use, here and in Chapter 14, is that of the *injectivity radius*. We can identify the universal cover of Σ with the unit disc in such a way that 0 in the disc maps to p. This identification is uniquely determined by p, up to the action of rotations of the disc. Thus we have a map $e_p : D \to \Sigma$ which is a local hyperbolic isometry mapping 0 to p, and e_p is uniquely determined up composition with rotations of the disc. (This map e_p is essentially what is called the 'exponential map' in general Riemannian geometry.) For $s > 0$, let $D_s \subset D$ be the open disc centred at 0 with hyperbolic radius s. For small s, the map e_p gives an embedding of D_s in Σ, and this condition is obviously unaffected by rotations of the disc. The injectivity radius of Σ at p, namely $i(\Sigma, p)$, is defined to be the supremum of such radii s. This is clearly finite, since the area of Σ is. The *injectivity radius* of Σ is defined to be the infimum over $p \in \Sigma$ of the $i(\Sigma, p)$. It is clear that since Σ is compact, the injectivity radius is strictly positive.

Lemma 29. *If $i(\Sigma, p) = r$, then there is a geodesic loop of length $2r$ based at p and there is no non-trivial geodesic loop based at p of length strictly less than $2r$.*

Suppose that $s < r$, so the hyperbolic disc D_s of radius s centred at p is embedded by e_p. Then any loop in Σ starting and ending at p and of length strictly less than $2s$ cannot reach the boundary of $e_p(D_s)$, and so can be regarded as a loop in this disc. Clearly, there are no non-trivial geodesic loops of this kind. Conversely, it follows from the definition of $i(\Sigma, p)$ that e_p must be an embedding of the *open* disc D_r but must map two points w_1, w_2 on the boundary of D_r to the same point q of Σ. We then have two one-dimensional subspaces of the tangent space $T\Sigma_q$ given by the images under the derivative of e_p of the tangent spaces to the boundary of D_r at these two points. It is also clear that from the definition that these one-dimensional subspaces must in fact be equal, i.e. that the image under e_p of the boundary circle is a loop in Σ which passes twice through q with the same tangent space. This implies that the images under e_p of the geodesics in D from 0 to w_1 and from 0 to w_2 match up at q and give a geodesic loop of length $2r$.

We have a very similar result for the injectivity radius.

Lemma 30. *If $i(\Sigma) = r$, then there is a simple closed geodesic of length $2r$.*

We leave the proof as an exercise. The point is that minimising $i(\Sigma, p)$ over the points p in Σ forces the tangents of the geodesic loop constructed above to match up at p. Notice that it follows immediately from the above lemma that no simple closed geodesic can have length strictly less than $2r$.

Now we begin the proof of Proposition 36. First, we consider based homotopy classes with a fixed base point x_0, that is, elements of $\pi_1(\Sigma, x_0)$. We can suppose that $0 \in D$ maps down to x_0. Elements of $\pi_1(X)$ are identified with the pre-image of x_0 in H. Each such point can be joined to 0 by a geodesic in H, and the image of this in Σ yields a geodesic loop in the corresponding based homotopy class.

Now fix a non-trivial free homotopy class of loops in Σ, which is the same as a conjugacy class in $\pi_1(\Sigma)$. For each base point and each representative of the conjugacy class, we have a non-trivial geodesic loop. We know that the lengths of these loops must exceed twice the injectivity radius, so the infimum L_0 of the lengths is strictly positive. We claim that this infimum is attained. For we can choose a *minimising sequence* of points in Σ and unit tangent vectors at these points which generate geodesic loops in the given free homotopy class with lengths tending to L_0. By the compactness of Σ, we can suppose that these sequences converge. Taking the limit, we get a geodesic loop of length L_0 in some free homotopy class. But then it is clear that the only geodesic loops close to this limit are in the same homotopy class, so we have achieved a minimiser in the original class. The argument that this minimiser is actually a closed geodesic is the same as that for Lemma 30 above. (If the tangent vectors at the end points do not match up, we can move the base point a little to get a slightly shorter loop.)

Now we have found a closed geodesic γ in our free homotopy class of length L. The next task is to show that if this class contains any embedded loop, the closed geodesic is simple. Equivalently, we shall show that if γ has a self-intersection point, then any other representative must do too. Consider the pre-image of Γ in the universal covering D. This is a union of geodesics in D. It follows from the definitions that γ is simple if and only if these geodesics are disjoint. So, suppose that two components $\tilde{\gamma}_1, \tilde{\gamma}_2$ in D intersect. Pick a base point p on γ. The pre-image of p in $\tilde{\gamma}_1$ is a set of points $\tilde{p}_n$ ($n \in \mathbf{Z}$) on $\tilde{\gamma}_1$, with consecutive points separated by distance L. Let δ be any loop in Σ free-homotopic to γ. Pick a path from p to a point q of δ. Let λ be the length of this path. The lift of this path determines a component $\tilde{\delta}_1$ of the pre-image of δ in H, and the pre-image of q gives a set of points $\tilde{q}_n$ ($n \in \mathbf{Z}$) in $\tilde{\delta}_1$. The hyperbolic distance from $\tilde{q}_n$ to $\tilde{p}_n$ is at most λ. Moreover, the portion of $\tilde{\delta}_1$ between $\tilde{q}_n$ and $\tilde{q}_{n+1}$ has a fixed hyperbolic diameter M, say. Of course, we also obtain another component $\tilde{\delta}_2$, which has the same relation to $\tilde{\gamma}_2$ as $\tilde{\delta}_1$ has to $\tilde{\gamma}_1$.

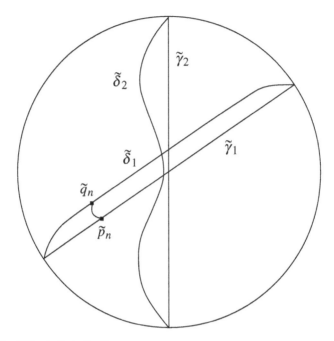

Figure 11.3 *Lifting paths to the disc*

Now extend the curves $\tilde{\gamma}_i$ to the closed disc $\overline{D}$ in the obvious way, includ-ing end points on the boundary. The discussion above, and the comparison between hyperbolic distances, shows that $\tilde{\delta}_i$ can also be extended to the images of continuous maps from a closed interval to $\overline{D}$ with the same end points as the $\tilde{\gamma}_i$. Now the hypothesis that the $\tilde{\gamma}_i$ intersect in D implies that the four end points are 'linked'. It follows that the $\tilde{\delta}_i$ must also intersect in D and, in turn, this shows that δ is not embedded. See Figure 11.3.

Now we have achieved a simple closed geodesic γ, and we want to prove uniqueness in the free homotopy class. We switch to the upper-half-plane model and suppose, without loss of generality, that the imaginary axis maps to γ. The covering transformation corresponding to the period L is $z \mapsto e^L z$, where $a = e^L$. So we have set of points $\tilde{p}_n = a^n i$ $(n \in \mathbf{Z})$ which map to the same point p in γ. Suppose now that δ is another simple closed geodesic in the same homotopy class as γ. The homotopy yields a path in Σ from p to δ. As before, we consider the lifts of this path to H starting from $\tilde{p}_n$. The end points of these lifts give a set of points $\tilde{q}_n$, which are just $\tilde{q}_n = a^n \tilde{q}_0$. But these points must all lie on a geodesic $\tilde{\delta}$ in H. This is only possible if $\tilde{q}_0$ lies on the imaginary axis and $\tilde{\delta} = \tilde{\gamma}$, which implies that $\delta = \gamma$.

The proof of the remaining statement—that the simple closed geodesics do not intersect if there are any disjoint representatives—is a similar argument, which we leave as an exercise for the reader.

Exercises

1. Let K be a field with $\mathbf{C} \subset K \subset \mathbf{C}(z)$. Prove that K is isomorphic (as a field) to $\mathbf{C}(z)$ (Lüroth's Theorem).

2. Consider the field $\mathbf{Q}(\theta)$, where $\theta^2 + 1 = 0$. Fix the prime $p = 5$, so that there are two solutions ± 2 of the equation $\theta^2 + 1 = 0$ modulo 5. Verify that there are solutions in $\mathbf{Z}_5$

$$\pm(2 + (1.5) + (2.5^2) + \ldots),$$

   and hence find the valuations in $\mathbf{Q}(\theta)$ lying over the prime 5.

3. Consider the Riemann surface defined by the polynomial

$$P(z, w) = w^5 + 2w^2 z + 3wz^2 + z^5.$$

   Show that there are solutions $w = \phi_1(z)$, $w = \phi_2(z)$, $w = \phi_3(z^{1/3})$ for power series ϕ_1, ϕ_2, ϕ_3 convergent around the origin.

4. (For those with some background in differential geometry.) Find a surface of revolution in $\mathbf{R}^3$ isometric to the quotient of the region $\operatorname{Im} z > a$ (for suitable $a > 0$) in the upper half-plane by the action of translations $z \mapsto z + n$ for integers n.

5. Verify that the geodesics in Q have the form stated in Section 11.2.5.

6. Show that the Q^* has a natural *Lorentzian metric*, i.e. quadratic forms on the tangent spaces with signature $+-$, as considered in general relativity.

7. Prove Lemma 28.

8. Show that the distance from a point $z = x + iy$ to the imaginary axis in H is $\sinh^{-1}(|x|/y)$.

9. Show that when we construct a hyperbolic surface by gluing two copies of the ideal triangle P, the surface will not in general be complete, if we do not use the canonical identifications of the boundaries.

10. Prove Lemma 30.

11. Find a formula for the lengths of the other three sides of the right-angled hexagon with lengths l_1, l_2, l_2.

12 Divisors, line bundles and Jacobians

12.1 Cohomology and line bundles

There is a choice, in developing Riemann surface theory, in the extent to which one uses sophisticated modern language and the techniques of sheaves, cohomology and so forth. On one hand, these are certainly not necessary: all the main results were developed long before the creation of these techniques, and from that point of view they are an inessential abstraction. On the other hand, the modern language makes many of the results much clearer. Also, when one goes on to study higher-dimensional complex manifolds, for example, these techniques become essential, and it is very useful to see first how they work in the simpler setting of Riemann surfaces. Thus, in this book, we take a middle course. We have developed the core of the theory without these abstract ideas, but now we give a very condensed outline of how they fit in.

12.1.1 Sheaves and cohomology

We motivate the notion of *sheaf cohomology* in two ways.

Čech cohomology

Consider a topological space X and an open cover $X = \bigcup_\alpha U_\alpha$. For simplicity, suppose X is compact and the cover is finite. The *nerve* of the cover is an abstract simplicial complex with one vertex (a 0-simplex) σ_α for every open set U_α and a p-simplex $\sigma_{\alpha_0,\dots,\alpha_p}$ for each non-empty intersection $U_{\alpha_0} \cap \cdots \cap U_{\alpha_p}$. Now, given an abelian group A, we can form the simplicial cochain complex of this nerve. For example, a 0-cochain is the assignment of an element $g_\alpha \in A$ to each U_α. A 1-cochain assigns an element $g_{\alpha\beta} \in A$ to each non-empty intersection $U_\alpha \cap U_\beta$, and a 2-cochain assigns an element $g_{\alpha\beta\gamma}$ to each non-empty triple intersection $U_\alpha \cap U_\beta \cap U_\gamma$. We have cochain groups C^p for $p \geq 0$. There is a standard coboundary map $\delta : C^p \to C^{p+1}$. For example, a 1-cochain $\underline{g} = (a_\alpha)$ has $\delta \underline{g} = \underline{h}$, where

$$\underline{h}_{\alpha\beta} = g_\alpha - g_\beta,$$

and a 2-cochain $\underline{h}$ has $\delta\underline{h} = \underline{k}$, where

$$\underline{k}_{\alpha\beta\gamma} = h_{\alpha\beta} - h_{\alpha\gamma} + h_{\beta\gamma}.$$

(*Note:* There is an issue about how one handles orientations here; see any textbook on algebraic topology. One way of dealing with this is to say that the cochains depend on a choice of ordering of the vertices of a simplex and change sign according to the sign of permutations.)

Now we can form the cohomology of this complex $H^*(X;\mathcal{U}, A)$, say, where $\mathcal{U}$ denotes the open cover. Suppose $X = \bigcup V_i$ is another cover $\mathcal{V}$. We say that $\mathcal{V}$ is a *refinement* of $\mathcal{U}$ if we can choose a map of index sets $\alpha(i)$ such that V_i is contained in $U_{\alpha(i)}$. In this case, we get a natural induced map on cochain complexes and hence a map $H^*(X;\mathcal{U}, A) \to H^*(X;\mathcal{V}, A)$. The Čech cohomology $H^*(X, A)$ of X with co-efficients in A is defined by taking the 'inverse limit'. That is, an element of $H^*(X, A)$ is represented by an element of $H^*(X;\mathcal{U}, A)$ for some cover $\mathcal{U}$, but two elements represented by $\omega_1 \in H^*(X;\mathcal{U}_1, A), \omega_2 \in H^*(X;\mathcal{U}_2, A)$ are equal in $H^*(X; A)$ if there is some common refinement $\mathcal{V}$ of $\mathcal{U}_1, \mathcal{U}_2$ such that ω_1, ω_2 have the same image in $H^*(X;\mathcal{V}, A)$. Thus $H^*(X, A)$ is a topological invariant of X and, at least for reasonable spaces such as manifolds, this is in fact isomorphic to the singular cohomology with co-efficients in A. In practice, one rarely needs to perform the limiting process explicitly if one works with a sensible cover.

Now, from this standpoint, the generalisation to 'sheaf cohomology' is very easy and natural. To keep things simple, for the moment suppose one has a class S of functions on the space X. For example, these could be continuous functions or, if X is a manifold, smooth functions or, if X is a Riemann surface, holomorphic functions. The only property we require is that for each non-empty open set $U \subset X$ there is a corresponding set $S(U)$ of functions on U of the given class, and if $V \subset U$ we have a restriction map from $S(U)$ to $S(V)$. Then we generalise the Čech construction by allowing $g_{\alpha_0,...,\alpha_p}$ to be an element of $S(U_0 \cap \cdots \cap U_p)$. The differential δ is defined by the same formula, but restricting functions on $p+1$-fold intersections to p-fold intersections. Then we get cohomology groups $H^*(X;\mathcal{U}, S)$ and, taking a limit over coverings, we arrive at an invariant $H^*(X, S)$. For example, a 0-cochain $\underline{g}$ is given by a function g_α on U_α for each α, and the cocycle condition $\delta\underline{g} = \overline{0}$ asserts that $g_\alpha = g_\beta$ on any non-empty intersection $U_\alpha \cap U_\beta$. This just means that the g_α are restrictions of a single function g on X. Thus $H^0(X;\mathcal{U}, S) = S(X)$.

Meromorphic functions on Riemann surfaces

(This discussion follows Griffiths and Harris (1994), Section 0.2)

We go back to the problem considered in Chapter 8 of the existence of a meromorphic function on a Riemann surface Σ with prescribed poles. For simplicity, consider a single pole at $p \in \Sigma$ and choose a local co-ordinate z centred at p. We seek a holomorphic function f on $\Sigma \setminus \{p\}$ which, near p, is equal to $z^{-1} + g(z)$ in this co-ordinate system, where g is holomorphic across 0. Let $D \subset \Sigma$ be a small disc centred at p, and let h be the function on $D \setminus \{p\}$ given by z^{-1} in our co-ordinate system. Now we have an open cover of Σ by two open sets $U_1 = \Sigma \setminus \{p\}$ and $U_2 = D$ and we can regard h as a Čech cocycle $\underline{h}$, defined on the intersection $U_1 \cap U_2$. Here we are taking the class of functions to be the holomorphic functions $\mathcal{O}(U)$. A moment's thought shows that we can find the desired meromorphic function if and only $h = f_1 - f_2$, where f_i is holomorphic on U_i, that is, if and only if the cohomology class of $\underline{h}$ in $H^1(\Sigma; \mathcal{U}, \mathcal{O})$ vanishes.

In the approach we adopted before, we defined a class in $H^{0,1}(\Sigma)$ which vanished if and only if there is a meromorphic function of the required kind. This suggests that the spaces $H^{0,1}(\Sigma)$ and $H^1(\Sigma; \mathcal{U}, \mathcal{O})$ should be related and, in fact, we will see in a moment that they are isomorphic. The theory of *sheaf cohomology* gives a universal language for discussing this kind of existence problem. One should think of the partial-differential-equation approach leading to $H^{0,1}(\Sigma)$ and the Čech approach as two different representations of the same more fundamental object, the cohomology group. There could be other ways of expressing our model problem, but these would lead to other representations of the same underlying cohomology group.

With this motivation, we will now proceed to a more formal treatment. A *sheaf* $\mathcal{S}$ on a space X is an assignment of a set $\mathcal{S}(U)$ to each open set $U \subset X$ with 'restriction maps' $\rho_{UV} : \mathcal{S}(U) \to \mathcal{S}(V)$ when $V \subset U$. These restriction maps should satisfy the obvious property when we have $W \subset V \subset U$. Also, if U is covered by open sets V_i and if we have $f_i \in \mathcal{S}(V_i)$ such that whenever $V_i \cap V_j$ is non-empty f_i and f_j have the same image under restriction to $V_i \cap V_j$, then there is a unique element f of $\mathcal{S}(U)$ such that $\rho_{UV_i} f = f_i$ for every i. The 'classes of functions' we considered above obviously define sheaves, but the notion of a sheaf is more general.

We will always consider sheaves with extra algebraic structure, at least that of abelian groups. This means that each $\mathcal{S}(U)$ is an abelian group and the restriction maps are group homomorphisms.

A homomorphism of sheaves $a : \mathcal{S} \to \mathcal{S}'$, over a fixed space X, is given by a homomorphism $a_U : \mathcal{S}(U) \to \mathcal{S}'(U)$ for each open $U \subset X$, compatible with restriction maps. We say a is injective if all the a_U are injective. We say a is surjective if, for each point, each open $U \subset X$, each point $p \in U$ and each h in $\mathcal{S}'(U)$, there is an open set $V \subset U$ containing p and $g \in \mathcal{S}(V)$ such that h restricts to $a_V(g)$ on V. It is important to realise that this is not the same as saying that all the maps a_U are surjective. Any sheaf homomorphism a can be factored as $a = i \circ \pi$, where i is injective and π is surjective. Regarding i as an inclusion, this defines the image of a, a subsheaf of $\mathcal{S}'$. Thus we have a notion of an exact sequence of sheaf homomorphisms.

Definition 11. *A sheaf $\mathcal{S}$ over M is called* fine *if, for each locally finite cover $\{U_i\}$ of M by open sets, there exists for each i an endomorphism e_i of $\mathcal{S}$ such that*

- $\mathrm{supp}(e_i) \subset U_i$,
- $\sum_i e_i = \mathrm{id}$.

By supp *(the* support *of e_i), we denote the closure of the set of points $m \in M$ for which $e_i \mid \mathcal{S}_m$ is not zero. We call $\{e_i\}$ a* partition of unity *for $\mathcal{S}$ subordinate to the cover $\{U_i\}$ of M.*

Now the basic fact is that there is a way to define cohomology groups $H^p(X, \mathcal{S})$ for any sheaf over a space X. Since X is fixed, we will often drop it from the notation and just write $H^p(\mathcal{S})$. These can be characterised by the following axioms.

1. $H^0(X, \mathcal{S}) = \mathcal{S}(X)$, the space of 'global sections' of $\mathcal{S}$.
2. A sheaf homomorphism $a : \mathcal{S} \to \mathcal{S}'$ induces homomorphisms from $H^p(X, \mathcal{S})$ to $H^p(X, \mathcal{S}')$.
3. A short exact sequence of sheaves $0 \to \mathcal{S} \to \mathcal{S}' \to \mathcal{S}'' \to 0$ induces a long exact sequence in cohomology

$$\cdots \to H^p(\mathcal{S}') \to H^p(\mathcal{S}'') \overset{\delta}{\to} H^{p+1}(\mathcal{S}) \to H^{p+1}(\mathcal{S}') \to \ldots,$$

where the boundary maps δ have an obvious naturality property with respect to homomorphisms of exact sequences.
4. If $\mathcal{S}$ is fine, then $H^p(\mathcal{S}) = 0$ for $p > 0$.

One approach is to use the Čech construction outlined above to define the groups $H^p(\mathcal{S})$ and then show uniqueness from the axioms. But there are several other approaches, and we refer to more specialised texts for a proper treatment. The point is that however we go about things, we arrive at the same cohomology groups.

To see how all this works, we return to our model problem of finding a meromorphic function with a single pole at a point p of a Riemann surface Σ. Let $\mathcal{O}$ be the sheaf of holomorphic functions on Σ and let $\mathcal{O}[p]$ be the meromorphic functions with at worst a pole at p. Let $\mathbf{C}_p$ be the 'skyscraper sheaf' defined by assigning $\mathbf{C}$ to an open set which contains p, and 0 otherwise. Then, using a local co-ordinate around p as before, we have a short exact sequence

$$0 \to \mathcal{O} \to \mathcal{O}[p] \to \mathbf{C}_p \to 0,$$

where the second map takes a function (in local co-ordinates) $a_{-1}z^{-1} + a_0 + a_1z = \dots$ to the 'residue' a_{-1}. The long exact sequence goes

$$0 \to H^0(\mathcal{O}) \to H^0(\mathcal{O}[p]) \to \mathbf{C} \overset{\delta}{\to} H^1(\mathcal{O}) \to \dots, \qquad (12.1)$$

and our problem is to find an element of $H^0(\mathcal{O}[p])$ which maps to 1 in $\mathbf{C}$. We can do this if and only if the image $\delta(1) \in H^1(\mathcal{O})$ vanishes.

To match this up with the partial-differential-equation point of view, consider the sheaf homomorphisms

$$0 \to \mathcal{O} \to \Omega^0 \to \Omega^{0,1} \to 0.$$

We claim that this is an exact sequence. This comes down to the fact that holomorphic functions are the solutions of the Cauchy–Riemann equation $\bar{\partial}f = 0$ and the fact that we can solve locally the inhomogeneous equation $\bar{\partial}f = \alpha$. (See Exercise 1 in Chapter 9.) The sheaves $\Omega^0, \Omega^{0,1}$ are fine, so the long exact sequence goes

$$\Omega^0 \to \Omega^{0,1} \to H^1(\mathcal{O}) \to 0,$$

which asserts that $H^1(\mathcal{O})$ is isomorphic to the cokernel $H^{0,1}$ of the $\bar{\partial}$-operator.

We have now understood from several points of view the fact that the obstruction to constructing meromorphic functions with prescribed poles lies in the sheaf cohomology group $H^1(\mathcal{O})$. Once one becomes accustomed to the language, it is usually easier to work directly with the cohomology, exact sequences etc. rather than detailed constructions and specific representatives.

12.1.2   Line bundles

Definition 12. *A rank-r vector bundle over a Riemann surface Σ is given by:*

1. *An $r+1$-dimensional complex manifold E and a holomorphic map $p : E \to \Sigma$.*
2. *The structure of an r-dimensional complex vector space on each 'fibre' $p^{-1}(x)$ for $x \in \Sigma$.*
3. *This data is required to be locally trivial in that any point $x \in \Sigma$ is contained in an open set $U \subset \Sigma$ such that we have a holomorphic isomorphism from $U \times \mathbf{C}^r$ to $p^{-1}(U)$ compatible with the projection maps to U and the vector space structures on the fibres.*

We can regard E as a family of vector spaces E_x parametrised by $x \in \Sigma$ and 'varying holomorphically' with x. Given a vector bundle E, we obtain another one E^*, the dual bundle, by taking the vector spaces dual to the fibres of E.

There are at least three ways of thinking of a holomorphic vector bundle. One is through the total space, as above. A second is through the sheaf of holomorphic sections. Given an open set $U \subset \Sigma$, a holomorphic section of E

over U is a holomorphic map $s : U \to E$ such that $p \circ s$ is the identity map on U. Thus s picks out an element $s(x)$ of each fibre $p^{-1}(x)$, and this 'varies holomorphically' with $x \in U$. We get a sheaf over Σ of holomorphic sections of E. This is a sheaf of modules over the sheaf of rings $\mathcal{O}$ and is locally free of rank r; that is, locally isomorphic to $\mathcal{O} \oplus \cdots \oplus \mathcal{O}$. Conversely, any such sheaf arises from a vector bundle.

A third approach is through 'transition functions'. Given a holomorphic vector bundle E over Σ, we can cover Σ by open sets U_α over which we have local trivialisations, as in item 3 of the definition. Over a non-empty intersection $U_\alpha \cap U_\beta$, we have two trivialisations which differ by a holomorphic map from $(U_\alpha \cap U_\beta) \times \mathbf{C}^r$ to itself, covering the identity on $U_\alpha \cap U_\beta$ and linear on the fibres. Such a map is given by a holomorphic function $g_{\alpha\beta}$ on $U_\alpha \cap U_\beta$ with values in the general linear group $GL(r, \mathbf{C})$. By definition, we have $g_{\beta\alpha} = g_{\alpha\beta}^{-1}$ and, on a triple intersection $U_\alpha \cap U_\beta \cap U_\gamma$, we have $g_{\alpha\gamma} = g_{\alpha\beta}g_{\beta\gamma}$. Conversely, if we have an open cover $\{U_\alpha\}$ of Σ and $GL(r, \mathbf{C})$-valued functions $g_{\alpha\beta}$ defined on intersections, satisfying these conditions, then we can construct a holomorphic vector bundle by gluing together products $U_\alpha \times \mathbf{C}^r$.

In this subsection, we will just consider *line bundles*, so $r = 1$. The set of isomorphism classes of holomorphic line bundles forms a group with multiplication operation the fibrewise tensor product and inverses given by duals. This is the *Picard group* $\mathrm{Pic}(\Sigma)$. Some examples are given below:

1. The tangent bundle of Σ is a line bundle; sections are holomorphic vector fields.
2. The cotangent bundle of Σ is a line bundle; sections are holomorphic 1-forms. The cotangent bundle is the dual of the tangent bundle.
3. Let V be a two-dimensional complex vector space. Recall that the complex projective line $\mathbf{P}(V)$ is defined to be the set of one-dimensional linear subspaces of V. Let E be the set of pairs (v, ζ) in $V \times \mathbf{P}(V)$ such that v lies in the one-dimensional subspace ζ and let $p(v, \zeta) = \zeta$. This makes E a line bundle over $\mathbf{P}(V)$; it is called the tautological bundle because the fibre over $\zeta \in \mathbf{P}(V)$ is ζ. Let H be the dual bundle. An element a of V^* defines a holomorphic section of H over $\mathbf{CP}^1$.

Let p be a point in the Riemann surface Σ. There is a line bundle $L_{[p]}$ over Σ, unique up to isomorphism, with the property that the sheaf of sections of $L_{[p]}$ is isomorphic to the sheaf $\mathcal{O}[p]$ of functions with at worst a simple pole at p. One way of seeing this is to observe that $\mathcal{O}[p]$ is a locally free sheaf of rank 1 over $\mathcal{O}$. Another way is via transition functions. We take a cover of Σ by a $U_1 = \Sigma \setminus \{p\}$ and a disc $U_2 = D$, as above. Then a line bundle is specified by a transition function g mapping $U_1 \cap U_2$, which is the punctured disc, to $GL(1, \mathbf{C}) = \mathbf{C}^*$. That is, our bundle will be covered by two open sets

$$E_1 = (\Sigma \setminus \{p\}) \times \mathbf{C}, \quad E_2 = D \times \mathbf{C},$$

and for $x \in D \setminus \{p\}$ and $\lambda \in C$, the point (x, λ) in E_1 is equal to the point $(x, h(x)\lambda)$ in E_2. To define $L_{[p]}$, we choose a local co-ordinate z centred on p and simply set $h = z$. There is a section of $L_{[p]}$ over $\Sigma \setminus \{p\}$ defined by

$$x \mapsto (x, 1) \in E_1 \subset L_{[p]},$$

and a section of $L_{[p]}$ over D defined by

$$x \mapsto (x, z(x)) \in E_2 \subset L_{[p]}.$$

These agree on the intersection, and so define a section over the whole of Σ which obviously vanishes at p. One can check that up to isomorphism, the bundle is independent of the choices made.

Definition 13. *A* divisor *on a compact Riemann surface Σ is a map from Σ to Z which is zero for all but finitely many points.*

There is a partial ordering on divisors defined by saying that $D \geq E$ if $D(p) \geq E(p)$ for all p in Σ. We can also regard a divisor as a formal sum $D = \sum n_i p_i$, where p_i are points of Σ. The *degree* of the divisor is defined to be $\sum n_i$. The set of divisors forms an abelian group in an obvious way, and the degree yields a homomorphism to Z. We associate a divisor (f) to a meromorphic function f by

$$(f) = \sum n_i p_i - \sum m_j q_j,$$

where the p_i are the poles of f, with multiplicities n_i, and the q_j are the zeros, with multiplicities m_j. Divisors of this kind are called principal divisors. Obviously, $(fg) = (f) + (g)$, so the principal divisors are a subgroup of the group of all divisors. The quotient is called the *divisor class group* $\mathrm{Cl}(\Sigma)$. Since the degree of a principal divisor vanishes, we get an induced homomorphism from $\mathrm{Cl}(\Sigma)$ to Z.

Now we relate divisors to line bundles. To any divisor $D = \sum_{i=1}^{k} n_i p_i$ we associate a line bundle L_D by

$$L_D = L_{p_1}^{n_1} \otimes \cdots \otimes L_{p_k}^{n_k},$$

where we use the duals to define negative powers as above. The properties of this construction are:

- The holomorphic sections of L_D correspond to meromorphic functions f on Σ with $(f) \leq D$.
- Two divisors D_1, D_2 define isomorphic line bundles L_{D_1}, L_{D_2} if and only if $D_1 - D_2 = (f)$ for some meromorphic function f.

We leave the proofs as an exercise.

The second property above implies that the assignment $D \mapsto L_D$ induces a homomorphism from the divisor class group $\mathrm{Cl}(\Sigma)$ to the group $\mathrm{Pic}(\Sigma)$ of

isomorphism classes of holomorphic line bundles. The first property above
implies that this is injective. A much more substantial fact is that the map is
also surjective: any line bundle arises from a divisor. We postpone the proof of
this until Section 12.2.3.

Line bundles give an efficient, alternative language for talking about meromorphic functions on a Riemann surface. In particular, they give an efficient
way of formulating a general version of the Riemann–Roch formula, a simple
version of which we stated in Chapter 8. To begin this, let us look once again
at our model problem, the existence of a meromorphic function with a single
pole at a point $p \in \Sigma$. Now we consider the line bundle $L_{[p]}$ which has a
section s vanishing at p. We identify the fibre of $L_{[p]}$ at p with $\mathbf{C}$, so we have
an evaluation map from the sheaf $\mathcal{O}(L_{[p]})$ of holomorphic sections of $L_{[p]}$ to
the skyscraper sheaf $\mathbf{C}_{[p]}$. Multiplication by s defines a map from the sheaf of
functions $\mathcal{O}$ to $\mathcal{O}(L_{[p]})$, and we have an exact sequence

$$0 \to \mathcal{O} \to \mathcal{O}(L_{[p]}) \to \mathbf{C}_p \to 0.$$

Our problem is equivalent to finding another holomorphic section s' of $L_{[p]}$
which maps to 1 in $\mathbf{C}_p$. Of course, we consider the long exact sequence in
cohomology

$$H^0(\mathcal{O}(L_{[p]})) \to \mathbf{C} \to H^1(\mathcal{O}) \to H^1(\mathcal{O}(L_{[p]})) \cdots ,$$

and the familiar obstruction in $H^1(\mathcal{O})$. Moreover, what we are doing here is
exactly the same as in equation (12.1), but in a different notation.

We will now formulate the Riemann–Roch Theorem for a general holomorphic line bundle L over our compact Riemann surface Σ. This requires the
notion of the degree of a line bundle, which we postpone until Section 12.2.2.
Suffice it to say here that the degree of a line bundle L_D is equal to the degree
of D. To simplify the notation, we write $H^i(L)$ for the cohomology groups of
the sheaf of holomorphic sections of L over Σ.

Theorem 19. *For $p > 1$, the cohomology groups $H^p(L)$ are trivial. The cohomology groups $H^0(L), H^1(L)$ are finite-dimensional complex vector spaces,
and*

$$\dim H^0(L) - \dim H^1(L) = d - g + 1,$$

where g is the genus of Σ.

We shall prove this for line bundles arising from divisors. We know the result
holds for the trivial bundle. It suffices then to show that if it holds for a line
bundle L, it also holds for $L \otimes L_{[p]}$ for any $p \in \Sigma$. Now we have an exact
sequence of sheaves

$$0 \to L \to L \otimes L_{[p]} \to \mathbf{C}_p \to 0,$$

where we have identified the fibre of $L \otimes L_{[p]}$ at p with **C**. The statements all follow from the long exact sequence in cohomology.

This formulation looks different from that (in the case of a positive divisor) of Chapter 8. The two are related by the *Serre Duality Theorem*. To state this, we need some more background.

We have explained that the sheaf cohomology group $H^1(\mathcal{O})$ can be identified with $H^{0,1}$, the cokernel of $\bar{\partial} : \Omega^0 \to \Omega^{0,1}$. We also have a $\bar{\partial}$-operator

$$\bar{\partial} : \Omega^{1,0} \to \Omega^2,$$

which we will write here as $\bar{\partial}_K$. We can apply the same argument to this. We have an exact sequence of sheaves

$$0 \to \mathcal{O}(K) \to \Omega^{1,0} \to \Omega^2 \to 0,$$

from which it follows that the cokernel of $\bar{\partial}_K$ is $H^1(K)$. But we have seen in Chapter 8 that this cokernel is identified with **C** by the map induced by integration of forms over Σ. (Of course, we already knew that $H^0(K)$—the space of holomorphic 1-forms—is the kernel of $\bar{\partial}_K$.)

More generally, consider any holomorphic line bundle $L \to \Sigma$.

Lemma 31. *There is a unique way to define $\bar{\partial}_L : \Omega^0(L) \to \Omega^{0,1}(L)$ such that*

$$\bar{\partial}_L(fs) = f\bar{\partial}_L s + (\bar{\partial} f)s,$$

and the sheaf of holomorphic sections is the kernel of $\bar{\partial}_L$.

In a local trivialisation, $\bar{\partial}_L$ is given by the usual $\bar{\partial}$-operator, so it follows as before that we have a short exact sequence of sheaves

$$0 \to \mathcal{O}(L) \to \Omega^0(L) \to \Omega^{0,1}(L) \to 0,$$

and the cokernel of $\bar{\partial}_L$ can be identified with $H^1(L)$. One approach to the theory would be to use this as the definition of $H^1(L)$.

What is clear from this description of cohomology is that if L_1, L_2 are holomorphic line bundles over Σ, we have a product map

$$H^0(L_1) \otimes H^1(L_2) \to H^1(L_1 \otimes L_2).$$

This is induced by the product map from sections and forms with values in the bundles. In particular, we have a product map

$$H^0(L^* \otimes K) \otimes H^1(L) \to H^1(K) = \mathbf{C}.$$

Theorem 20. (Serre duality). *The above map defines a dual pairing between $H^1(L)$ and $H^0(L^* \otimes K)$. In particular, these spaces have the same dimension.*

Notice that when L is the trivial bundle, this is just what we proved in Chapter 8.

This Serre duality is the basic reason why the theory of Riemann surfaces could be developed without explicit mention of cohomology, as it did historically. Rather than using the H^1's, one formulates things dually using H^0's. In particular, we get another version of the Riemann–Roch formula,

$$\dim H^0(L) - \dim H^0(K \otimes L^*) = d - g + 1.$$

If $L = L_D$ for a divisor $D = p_1 + \cdots + p_d$, with p_i distinct, then $H^0(L)$ is the space of meromorphic functions with at worst simple poles at the p_i, and $H^0(K \otimes L^*)$ is the space of holomorphic 1-forms vanishing at the p_i, so we recover the earlier version.

Now we move on to discuss the proof of the Serre Duality Theorem. The approach that would fit best into the general scheme of the heart of this book (Chapters 8 and 9) would be to extend the analysis of the $\bar{\partial}$-operator to the more general class $\bar{\partial}_L$. To do this, one introduces a Hermitian metric on L and thus an inner product on sections of $\Omega^{0,1}(L)$. Then it is easy to show that the orthogonal complement of the image of $\bar{\partial}_L$, with respect to this inner product, can be identified with the dual of $H^0(K \otimes L^*)$. In a finite-dimensional situation, the orthogonal complement of the image can be identified with the cokernel. The essence of the analytical problem is to show that the same conclusion holds in this infinite-dimensional situation. The arguments are variants of those which we have developed in Chapter 9, but we will not develop them in more detail here. The advantage of this approach is that it also leads to a proof of the fundamental fact that any line bundle arises from a divisor: see Exercise 3 at the end of this chapter.

What we will do here is to give a proof of Serre duality for line bundles which arise from divisors. Let us say that a line bundle L 'satisfies Serre duality' if the natural pairings we have defined yield an isomorphism

$$H^1(L) = H^0(L^* \otimes K)^* \quad and \quad H^1(L^* \otimes K) = H^0(L)^*.$$

So, by definition, L satisfies Serre duality if and only if $L^* \otimes K$ does. As in the proof of the Riemann–Roch formula, it suffices to show that if L satisfies Serre duality then so does $L \otimes L_{[p]}$ for any $p \in \Sigma$.

We have a long exact sequence

$$0 \to H^0(L) \to H^0(L \otimes L_{[p]}) \to L_p \otimes K_p^* \to H^1(L) \to H^1(L \otimes L_p) \to 0.$$

Here, the middle term is a one-dimensional complex vector space, but it is important to identify it invariantly. This identification comes about because a section of $L \otimes L_{[p]}$ can be viewed as a meromorphic section of L and its residue at p lies in $L_p \otimes K_p^*$. Replacing L by $L^* \otimes L_{[p]}^*$, we have another exact sequence,

$$0 \to H^0(L^* \otimes L^*_{[p]} \otimes K) \to H^0(L^* \otimes K) \to L^*_p \otimes K_p \to H^1(L^* \otimes L^*_{[p]} \otimes K)$$
$$\to H^1(L^* \otimes K) \to 0.$$

We take the dual of the second sequence. Then we have defined 'vertical' maps between all the terms in the two sequences, giving a big diagram of maps, consisting of the two exact sequences and vertical maps between corresponding terms.

Now we have the following lemma.

Lemma 32. *The above diagram commutes.*

The proof of this, unwinding the definitions, amounts to the 'residue formula' identification we established in Section 8.2. We leave it as an exercise for the reader. By the 'Five Lemma', it follows that if L satisfies Serre duality then so does $L \otimes L_{[p]}$.

12.1.3   Line bundles and projective embeddings

Line bundles give an efficient language for discussing maps from a Riemann surface into projective spaces. In a sense, we already know how to construct these. Let $f_1, \ldots, f_N$ be non-constant meromorphic functions on Σ and let Δ be the union of their poles, a finite subset of Σ. Then we have

$$\underline{f} : \Sigma \setminus \Delta \to \mathbf{C}^N,$$

and it is not hard to see that this extends to a holomorphic map from Σ to $\mathbf{CP}^N$, regarded as a compactification of $\mathbf{C}^N$. The image of this map is a 'complex curve' in the projective space. But it is a little complicated to analyse the behaviour of the image of this map 'at infinity' in $\mathbf{CP}^N$ from this point of view, and the language of line bundles makes things much clearer.

Suppose $L \to \Sigma$ is a holomorphic line bundle and $s_0, \ldots, s_N$ are linearly independent holomorphic sections of L. If there is a point $q \in \Sigma$ where all the sections s_i vanish, then these sections can be regarded as sections of $L \otimes L^*_q$. Repeating this process if necessary, we can reduce the situation to the case when there is no such common zero. Now write V for the linear subspace of $H^0(L)$ spanned by the s_i. For each point p of Σ, we have an evaluation map

$$e_p : V \to L_q,$$

where L_q denotes the fibre of L at q. By hypothesis, this is surjective. Dually, the transposed map

$$e_p^T : L_q^* \to V^*$$

is injective, with image a one-dimensional subspace of V^*. By definition, this is a point of the projective space $\mathbf{P}(V^*)$. Now we define a map from Σ to $\mathbf{P}(V^*)$ which takes p to this one-dimensional subspace.

To make this more concrete, let us return to our basis s_i of V. If we identify a fibre L_p with $\mathbf{C}$, then we have a non-zero vector

$$(s_0(p), \ldots, s_N(p))$$

in $\mathbf{C}^{N+1}$. Changing the identification with $\mathbf{C}$ changes each co-ordinate by multiplying by the same factor, so there is a well-defined element $[s_0, \ldots, s_N]$ in $\mathbf{CP}^N$. Said in another way, consider the ratios s_i/s_0 for $i = 1, \ldots, N$. These are meromorphic functions f_i on Σ, and our map can be obtained from these meromorphic functions as above.

There is a converse of this construction. Any holomorphic map from Σ to a projective space arises from a line bundle L and a subspace $V \subset H^0(L)$ with no common zero. But, although nothing difficult is involved, we will not take up space to develop this here.

Riemann surfaces of genus 2 and 3

Let Σ be a Riemann surface of genus $g \geq 2$. Then the space of holomorphic 1-forms has dimension g. If there are no common zeros, we get a map from Σ to $\mathbf{CP}^{g-1}$ by the above procedure (taking $V = H^0(K)$). We will analyse these maps when $g = 2, 3$. This provides a good illustration of the techniques developed in this chapter.

First we show that, indeed, there are no common zeros.

Lemma 33. *Let Σ be a Riemann surface of genus $g \geq 1$. Then there is no point of Σ where all holomorphic 1-forms vanish.*

For if p were such a point, then the Riemann–Roch theorem shows that there is a meromorphic function with a simple pole at p and no other poles, and we have seen that this implies that Σ is isomorphic to the Riemann sphere.

Now we have a well-defined map $f_\Sigma : \Sigma \to \mathbf{CP}^{g-1}$.

Lemma 34. *If Σ has genus 2, then f_Σ has degree 2 and is a double branched cover with six branch points.*

Corollary 12. *Any Riemann surface of genus 2 is isomorphic to the Riemann surface of a function $P(z, w) = w^2 - f(z)$, where f is a polynomial of degree 6 with distinct roots in $\mathbf{C}$. Moreover, two such Riemann surfaces, defined by polynomials f_1 and f_2, are isomorphic if and only there is a Möbius mapping taking the roots of f_1 to the roots of f_2.*

Proposition 37. *If Σ has genus 3, then either f_Σ embeds Σ as a smooth quartic curve in $\mathbf{CP}^2$ or f_Σ is a degree-2 map onto a smooth conic C in $\mathbf{CP}^2$ and exhibits Σ as a double branched cover of $C \equiv \mathbf{CP}^1$ with eight branch points.*

First, suppose Σ is a Riemann surface of genus 3 which can be represented as a double branched cover of the Riemann sphere. Thus Σ is the Riemann surface of the function $\sqrt{p}$, where p is a polynomial of degree 8. A basis for the holomorphic forms on Σ is given by $\theta_0, \theta_1 = z\theta_0, \theta_2 = z^2\theta_0$, where $\theta_0 = dz/\sqrt{p}$. Taking this basis, the map $f_\Sigma : \Sigma \to \mathbf{CP}^2$ is seen to be the composite of the double cover $\Sigma \to \mathbf{CP}^1$ and the map $z \mapsto [1, z, z^2]$ from $\mathbf{CP}^1$ to $\mathbf{CP}^2$. The image in $\mathbf{CP}^2$ is the conic with equation $Z_1^2 = Z_0 Z_2$. Now, suppose that Σ is any Riemann surface of genus 3 and that f_Σ is not an embedding. Thus there are two different points $p, q \in \Sigma$ which have the same image in $\mathbf{CP}^2$. This is the same as saying that any section of K_Σ which vanishes at p also vanishes at q. The space of sections of K_Σ has dimension 2, so we see that $H^0(K_\Sigma(-p-q))$ has dimension 2. Set $L = K_\Sigma \otimes \mathcal{L}_{-p-q}$. Then L has degree 2, and the two-dimensional space of sections defines a double cover of $\mathbf{CP}^1$. So now we know that if Σ is not a double cover, the map $f_\Sigma : \Sigma \to \mathbf{CP}^2$ is injective. Since K_Σ has degree 4, the image is a quartic curve. The fact that this curve is smooth can be seen in various ways, although they all go a little beyond the theory we have actually developed. One way is to repeat the argument above for a point q which is 'infinitely near' to p. That is, we show that for each point $p \in \Sigma$ there is a section of K_Σ which vanishes at p to order exactly one. This implies that the image is a smooth curve.

We see then that there is a dichotomy for Riemann surfaces of genus 3: either they are double covers (known as *hyperelliptic curves*) or smooth quartics. But (to anticipate slightly the ideas we will develop in the next two chapters) one can deform from one class to the other. Let Q be a non-degenerate quadratic form in three variables, so the equation $Q = 0$ defines a conic C in $\mathbf{CP}^2$. Let F be a typical polynomial of degree 4 and let Σ_ϵ be the curve defined by the equation $Q^2 + \epsilon F = 0$. Thus, when $\epsilon = 0$, we get Σ_0, which is, as a set, the same as C. Let Z be the zero set of F. Thus, for typical F, the intersection $Z \cap Q$ consists of $2 \cdot 4 = 8$ points $c_1, \ldots, c_8 \in C$. One can show that for small but non-zero ϵ the curve Σ_ϵ is smooth. So we can regard f_{Σ_ϵ} as the inclusion map of Σ_ϵ in $\mathbf{CP}^2$. If c is a typical point of C, then, for small but non-zero ϵ, the part of Σ_ϵ near to c consists of two 'parallel copies' of C, but when c is close to one of the c_i, these two 'sheets' come together. So, informally, Σ_ϵ is approximately a double cover of C branched over c_i.

To make this more precise, consider the projection from a typical point of $\mathbf{CP}^2$ to a line L. So we have $\pi_C : C \to L$ and degree-4 maps $\pi_\epsilon : \Sigma_\epsilon \to L$ for $\epsilon \neq 0$. We could suppose that, in affine co-ordinates z, w, the conic C is defined by $z^2 - w^2 = 1$ and the projection is $(z, w) \mapsto z$. There are two critical points $(1, 0), (0, 1)$ of π_C. There are 12 critical points of π_ϵ (for typical ϵ), and these fall into two sets. One set, with eight elements, consists of points $c_i(\epsilon) \in \Sigma_\epsilon$ close to c_i. The other set, with four elements, is made up of two pairs, one pair close to $(1, 0)$ and one pair close to $(0, 1)$. These 12 points map, for

typical parameters ϵ, to 12 distinct points in C, say $z_i(\epsilon)$ for $i = 1, \ldots, 8$ and
$1 + \delta_1, 1 + \delta_2, -1 + \delta_3, -1 + \delta_4$, where $\delta_i = \delta_i(\epsilon)$. As $\epsilon \to 0$, the critical values
$z_i(\epsilon)$ converge to distinct limits z_i, while $\delta_i(\epsilon) \to 0$. On the other hand, if we
start with the Riemann surface Σ_0 defined as a double cover of C branched
over c_i and then compose with π_C we get a 4–1 map from Σ_0 to L which has
12 critical points, eight lying over the z_i, one pair lying over +1 and one pair
lying over -1. So, using this description as four-sheeted branch covers, we can
see Σ_0 as the limit of the Σ_ϵ in a natural way. The only change is that when
$\epsilon = 0$, two pairs of critical points map to common critical values.

Examples with large symmetry

In Chapter 11, we saw that compact Riemann surfaces can be described in two
ways: as algebraic curves defined by polynomial equations or as hyperbolic
surfaces obtained by pasting edges of hyperbolic polygons or, which is much
the same, as quotients of the upper half-plane by discrete groups. Each of these
descriptions is in a sense explicit, but only rarely can one simultaneously find
an explicit description of both kinds for the same Riemann surface. The cases
when one can do this usually have large symmetry groups. Here we give two
examples.

A Riemann surface of genus 2

Let Σ be the Riemann surface associated to the equation $w^2 = f(z)$, where
$f(z) = z^6 - 1$. Then, if $\lambda = e^{\pi i/3}$, we have $f(\lambda z) = f(z)$, so there is an automorphism $\alpha : \Sigma \to \Sigma$ defined (on the dense open set of Σ which is contained in
$\mathbf{C}^2$) by $\alpha(z, w) = (\lambda z, w)$. Let D be the open unit disc in C, so there are two
lifts D_+, D_-, say, of D in Σ, and these are interchanged by the hyperelliptic
involution. Clearly, α preserves D_+ and D_-. The map α also preserves the
hyperbolic metric on Σ. Restrict the metric to the closure $\overline{D_+}$ of D_+. It is clear
that the metric has corners at the six branch points on the boundary of $\overline{D_+}$.
It follows then from symmetry that the hyperbolic metric on $\overline{D_+}$ is that of a
regular hexagon Δ with all internal angles equal to $\pi/2$—this is unique up to
isometries of hyperbolic space.

Of course, the same discussion holds for D_-. Further, the same holds if we
replace D by the interior of its complement in the Riemann sphere. So we see
that Σ can be constructed from four copies of Δ by gluing edges appropriately.
In fact, the automorphism group of Σ has order 24. It acts transitively on the
four discs, and the stabiliser of a disc is the cyclic group of order 6.

The 'Klein curve'

Now we consider a very famous example, of genus 3. In Chapter 3 and in
Section 7.2.4, we discussed modular curves. Now take the particular case

associated to the subgroup $\tilde{\Gamma} \subset SL(2, \mathbf{R})$ of matrices with integer entries and equal to 1 modulo the prime 7. This has image $\Gamma = \tilde{\Gamma} / \pm 1$ in $PSL(2, \mathbf{R})$, and Γ acts freely on the upper half-plane. The quotient H/Γ is non-compact, but we saw in Chapter 7 that it can be compactified by adding points corresponding to 'cusps'. So we get a compact Riemann surface Σ. Let $\tilde{G}$ be the finite group $SL(2, \mathbf{Z}/7)$. The order of $\tilde{G}$ is $(7^2 - 1)(7^2 - 7)/6 = 336$. The element -1 is in the centre of $\tilde{G}$, and the quotient $G = \tilde{G}/\pm 1 = PSL(2, \mathbf{Z}/7)$ has order 168. As we saw in Chapter 7, the group G acts on Σ, and the quotient Σ/G can be identified with the compactification of $H/PSL(2, \mathbf{Z})$ which is the Riemann sphere.

In Section 7.2.4, we saw that the genus of Σ is

$$1 + \frac{(7 - 6)(7^2 - 1)}{24} = 3.$$

So G acts on the three-dimensional complex vector space $H^0(\Sigma, K)$ of holomorphic 1-forms on Σ. We shall see that this is enough to determine Σ explicitly, as an algebraic curve. In fact, we do not need to use the full symmetry. Let $\tilde{P}$ be the group of matrices

$$\begin{pmatrix} a & b \\ 0 & a^{-1} \end{pmatrix}$$

with entries in $\mathbf{Z}/7$. So a runs over the six invertible non-zero elements of $\mathbf{Z}/7$, and b is arbitrary in $\mathbf{Z}/7$. Thus $\tilde{P}$ has order 42 and is a subgroup of $\tilde{G}$. Let $P \subset G$ be the quotient $\tilde{P}/\pm 1$. So P has order 21 and fits into an exact sequence of group homomorphisms

$$1 \to \mathbf{Z}/7 \to P \to \mathbf{Z}/3 \to 1.$$

Let $\tilde{\alpha}, \tilde{\beta}$ be the two elements of $\tilde{P}$

$$\tilde{\alpha} = \begin{pmatrix} 3 & 0 \\ 0 & 5 \end{pmatrix}, \quad \tilde{\beta} = \begin{pmatrix} 1 & 1 \\ 0 & 1 \end{pmatrix}.$$

Then $\tilde{\alpha}$ has order 6, $\tilde{\beta}$ has order 7 and one finds that $\tilde{\alpha}\tilde{\beta}\tilde{\alpha}^{-1} = \tilde{\beta}^2$. Let α, β be the images of $\tilde{\alpha}, \tilde{\beta}$ in G, so α has order 3, β has order 7 and $\alpha\beta\alpha^{-1} = \beta^2$. A few moments, thought shows that α, β generate G and that the group has a presentation

$$G = \langle \alpha, \beta : \alpha^3 = 1, \beta^7 = 1, \alpha\beta\alpha^{-1} = \beta^2 \rangle.$$

Now consider a representation of G on a complex vector space V. Let A, $B \in GL(V)$ correspond to α, β. Suppose that B *is not the identity*. This means that there is an eigenvector $e \in V$ with $Be = \lambda e$, where λ is a seventh. root of unity. Then the relation $ABA^{-1} = B^2$ implies that

$$B^2(Ae) = ABA^{-1}(Ae) = ABe = \lambda\, Ae,$$

and then, since $B^7 = 1$,

$$B(Ae) = (B^2)^4 Ae = \lambda^4\, Ae.$$

In the same way, one sees that

$$B(A^2 e) = \lambda^2\, Ae.$$

Since $A^3 = 1$, the three vectors e, Ae, $A^2 e$ form a basis for an invariant subspace of V. Thus we see that there are exactly two irreducible representations of P on which β acts non-trivially. They are both three-dimensional and complex conjugate, corresponding to different choices of the seventh. root of unity (i.e. the two decompositions of the roots of unity into subsets $\lambda, \lambda^4, \lambda^2$ and $\lambda^{-1}, \lambda^{-4}, \lambda^{-2}$).

Now suppose that V is this one of these irreducible representations. We ask what quartic polynomials on V are invariant, up to a scalar, under G. If x, y, z are the linear functions corresponding to the basis above, we have a basis for the quartic polynomials

$$\begin{array}{ccccc} & & y^4 & & \\ & zy^3 & xy^3 & & \\ & y^2z^2 & xy^2z & x^2y^2 & \\ yz^3 & xyz^2 & x^2yz & x^3y & \\ z^4 & xz^3 & x^2z^2 & x^3z & x^4 \end{array}$$

The action of B on a polynomial $x^p y^q z^r$ is multiplication by $\lambda^{p+4q+2r}$, so we get the following weights for the basis above:

$$\begin{array}{ccccc} & & 2 & & \\ & 0 & 6 & & \\ & 5 & 4 & 3 & \\ 3 & 2 & 1 & 0 & \\ 1 & 0 & 6 & 5 & 4 \end{array}$$

If P is our invariant polynomial, it must be built of monomials in the same weight space. On the other hand, it must be invariant under the cyclic permutation of x, y, z. One sees from the above table that the only possibility is with weight 0 and, up to a scalar,

$$P(x, y, z) = x^3 y + y^3 z + z^3 x.$$

(Notice that it does not matter whether we work with V or its conjugate.)

To sum up, we have the following proposition.

Proposition 38. *If Σ' is a Riemann surface of genus 3 with an action of P and if β does not act trivially on $H^0(\Sigma', K)$, then Σ' is the plane quartic curve defined by the equation $x^3 y + y^3 z + z^3 x = 0$.*

It remains to check that indeed β does not act trivially on $H^0(\Sigma, K)$ for the modular curve Σ above. In fact, we have the following result.

Proposition 39. *Let Σ' be any Riemann surface of genus 2 or more. If $a : \Sigma' \to \Sigma'$ is a holomorphic automophism of finite order which acts trivially on $H^0(\Sigma', K)$, then a is the identity.*

In fact, the hypothesis that a has finite order is unnecessary, since this is always the case, but we shall not take the time to prove it.

To prove the proposition, consider the quotient $\Sigma'' = \Sigma'/A$, where A is the cyclic group generated by a. Then there is a holomorphic map π from Σ' to Σ''. If a is not the identity, the Riemann–Hurwitz formula shows that the genus of Σ'' is strictly less than the genus of Σ'. But if a acts trivially on a holomorphic 1-form θ' on Σ', there is a form θ'' on Σ'' such that $\theta' = \pi^*(\theta'')$. It follows that if a acts trivially on $H^0(\Sigma', K)$, then the spaces $H^0(\Sigma', K)$, $H^0(\Sigma'', K)$ are isomorphic, which is a contradiction to the strict genus inequality.

(There are many other ways to prove this proposition: for example, one can use the Lefschetz fixed-point formula and the identification of the holomorphic forms with the real cohomology.)

A great deal of information about this 'Klein curve' can be found in the book edited by Levy (1999).

12.1.4   Divisors and unique factorisation

We want now to explain how divisors are related to questions of unique factorisation. Let $z_1, \ldots, z_k$ be distinct points in $\mathbf{C}$, and consider the affine curve $\Sigma_0 \subset \mathbf{C}^2$ defined by the equation $P(z, w) = 0$, where

$$P(z, w) = w^2 - (z - z_1) \ldots (z - z_k).$$

Let R be the quotient of $\mathbf{C}[z, w]$ by the ideal generated by P, so elements of R can be viewed as holomorphic functions on Σ_0. We consider the question of whether R is a unique factorisation domain (UFD). More specifically, we ask whether the equation $w^2 = \prod(z - z_i)$, which holds in R by construction, contradicts unique factorisation into irreducibles, up to multiplication by units.

When $k = 2$, the equation does not contradict unique factorisation and, in fact, R is a UFD. Without loss of generality, take $z_1 = 1, z_2 = -1$, so the equation is $w^2 - z^2 = -1$. We parametrise the curve by a variable $t \in \mathbf{C} \setminus \{0\}$ with

$$z = \frac{t + t^{-1}}{2}, \quad w = \frac{t - t^{-1}}{2},$$

so $t = w + z, t^{-1} = w - z$. Our identity is

$$\frac{((t + 1)(t - 1))^2}{t^2} = \frac{(t + 1)^2}{t} \frac{(t - 1)^2}{t},$$

and this does not contradict unique factorisation, *because t is a unit in R.*
The complete factorisation into irreducibles involves products of the elements
$g_+ = (t + 1) = z + w + 1, g_- = t - 1 = z + w - 1$ in R, so

$$w^2 = \frac{1}{4t^2} g_+^2 g_-^2,$$

while

$$z + 1 = \frac{1}{2t} g_+^2, \quad z - 1 = \frac{1}{2t} g_-^2.$$

For $k > 2$, the situation is different and R is not a UFD. For simplicity,
suppose that k is odd. Then the compact Riemann surface Σ associated to
P is obtained by adding a single point q at infinity to Σ_0. Let p_i be the points
in Σ corresponding to $z_i \in \mathbf{C}$. Then the divisor of the meromorphic function w
on Σ is

$$(w) = (p_1 + \cdots + p_k) - kq,$$

while

$$(z - z_i) = 2 p_i - 2q.$$

The identity $w^2 = \prod(z - z_i)$ corresponds to the additive identity

$$2(w) = \sum(2 p_i - 2q).$$

The source of the difficulty is now clear. At the level of divisors, we have
a 'unique factorisation', writing each side as sums of the divisors, $(p_i - q)$.
But we can only have unique factorisation in R if these divisors correspond
to elements of R and hence, in particular, to meromorphic functions on Σ.
But if we have a meromorphic function on Σ with divisor $p_i - q$, for any i,
this would imply that Σ has genus 0, since the function has a single pole.
More generally, we can think of the quotient Cl_0 of the divisors of degree 0
by those arising from meromorphic functions as a measure of the failure of
unique factorisation in a suitable ring.

We can formulate the definition of divisors in a more algebraic fashion. Let
F be a field, and suppose that

• all non-trivial valuations on F map to a discrete subgroup of $\mathbf{R}$;
• given $x \in F^*$, there are only many inequivalent valuations v such that
$v(x) \neq 0$.

The first property means that we can normalise each (non-trivial) valuation, within its equivalence class, by requiring that the image be $\mathbf{Z} \subset \mathbf{R}$. We define the group of divisors to be finite formal sums $\sum n_i v_i$ of such normalised valuations. By the second property, a non-zero element x in F defines a 'principal' divisor

$$(x) = \sum_v v(x)v.$$

Then we can take the quotient group of divisors modulo principal divisors. Clearly, when $F = k_\Sigma$ for a compact Riemann surface Σ, this agrees with the divisor class group $\mathrm{Cl}(\Sigma)$. On the other hand, we can apply the definition to an algebraic number field k, and associate a divisor class group to k. A fundamental theorem asserts that this is a finite group: its order is the *class number*. The basic fact is that the ring of algebraic integers in k is a UFD if and only if the class number is 1.

12.2   Jacobians of Riemann surfaces

12.2.1   The Abel–Jacobi Theorem

Let $= p_1 + \cdots + p_d$ be a divisor in a compact Riemann surface, with p_i distinct. In Chapter 8 we saw that there is a meromorphic function with poles only at the p_i if and only if there are tangent vectors v_i at p_i—not all zero—such that for all holomorphic 1-forms θ on Σ we have $\sum \langle \theta, v_i \rangle = 0$. Now we want to answer another question. Given another such divisor $q_1 + \cdots + q_d$, when is there a meromorphic function with simple poles at p_i and simple zeros at q_i? The answer to this question is again expressed in terms of holomorphic 1-forms.

Theorem 21. *Such a meromorphic function exists if and only if there are paths γ_i from p_i to q_i such that, for all holomorphic 1-forms θ, we have $\sum_i \int_{\gamma_i} \theta = 0$.*

This can be seen as a 'multiplicative' or 'integrated' version of the proof of the previous result. We begin by proving the sufficiency of the condition. Suppose that we have paths γ_i as stated. Of course, by Cauchy's theorem we can always deform γ_i slightly, with fixed end points. We first choose *smooth* maps F_i from Σ to $\mathbf{CP}^1$ with the desired local behaviour. Thus we arrange (after choosing γ_i suitably) that:

- F_i is a meromorphic function in small discs around p_i and q_i, with a zero at q_i, a pole at p_i and no other poles or zeros;
- F_i is not equal to 0 or ∞ except at p_i, q_i, respectively;
- $F_i^{-1}[-\infty, 0] = \gamma_i$, so F_i is real and negative on γ_i.

For example, we can do this by choosing F_i to be equal to 1 outside a neighbourhood of γ_i. Now set $F = \prod F_i$, with the obvious interpretation regarding the point at ∞. We seek the desired solution in the form $f = e^u F$ for a smooth function u on Σ. The equation we need to solve is $\bar{\partial}(e^u F) = 0$, which, away from the poles and zeros, can be rearranged as

$$\bar{\partial} u = - \sum \chi_i F_i^{-1} \bar{\partial} F_i, \tag{12.2}$$

where $\chi_i = F_i^{-1} \bar{\partial} F_i$. Now $\bar{\partial} F_i$ has been to chosen to vanish in neighbourhoods of the pole and zero of F_i, so χ_i extends to a smooth 0,1-form on Σ, and it is clear that a smooth solution u will give the desired solution to our problem. By our main result, we can solve this equation (12.2) if and only if, for all holomorphic 1-forms, we have

$$\sum_i \int_\Sigma \chi_i \wedge \theta = 0.$$

So what we need to establish are identities

$$\int_\Sigma \chi_i \wedge \theta = 2\pi i \int_{\gamma_i} \theta. \tag{12.3}$$

To see this, cut the Riemann surface along the path γ_i. On this cut surface, we can define a branch of $\log F_i$ and write $\chi_i = \bar{\partial} \log F_i$. Now we apply Stokes' Theorem in the usual way. The function $\log F_i$ jumps by $2\pi i$ across the cut, and hence the result.

Another approach to proving the identities (12.3) is to show first that the left-hand side is independent of the choice of F_i. If the path γ_i is parametrised by the interval $[0, 1]$, we can define a function $I(t)$ given by replacing p_i by $\gamma_i(t)$ in the left-hand side of equation (12.3). Then we show that $I(0) = 0$ and that the derivative dI/dt coincides with the corresponding derivative for the right hand side.

Conversely, we need to show that if a meromorphic function f with these prescribed poles and zeros exists, then the identity involving path integrals holds, for a suitable choice of paths. Regard f as a map to the Riemann sphere. Multiplying by a scalar, we can suppose that there are no branch points of f on the real axis. Then the pre-image of $[-\infty, 0]$ gives smooth paths γ_i in Σ from p_i to q_i, after possibly reordering the q_i, say. It follows that we can write $f = e^u F$ for some smooth function u, where F is as constructed above, and the result, follows by the corresponding statement in our Main Theorem. Finally, it is is easy to see that the statement is independent of the ordering of the q_i.

We will now formulate this result in a slightly different way, involving the *Jacobian* of the Riemann surface. As a first step, we consider the dependence on the choice of paths. If γ is any path, we have a map

$$\theta \mapsto \int_\gamma \theta,$$

from $H^{1,0}$ to the complex numbers. So the path defines an element I_γ of the dual space $(H^{1,0})^*$. In particular, taking closed loops, we get a map from $H_1(\Sigma; \mathbf{Z})$ to $(H^{1,0})^*$, which is obviously a group homomorphism. Let $\Lambda \subset (H^{1,0})^*$ be the image of this map. Now, given end points p, q, we can choose different paths between them but the integrals will differ only by elements of Λ. For convenience, fix a base point p_0 in Σ. Then, choosing paths from p_0 to p_i and integrating holomorphic forms, we get a well-defined element $j(p_1 + p_2 + \ldots p_d) \in (H^{1,0})^*/\Lambda$. Then the statement is that there is a meromorphic function with poles p_i and zeros q_i if any only if $j(p_1 + \ldots p_d) = j(q_1 + \ldots q_d)$.

Now let us examine the nature of the quotient group $(H^{1,0})^*/\Lambda$. We saw in Chapter 8 that $H^{1,0}$ can be identified with the *real* cohomology group $H^1(\Sigma, \mathbf{R})$, under the map which takes a holomorphic 1-form α to the de Rham cohomology class of the real form $\alpha + \overline{\alpha}$. If we have a complex linear map $l : H^{1,0} \to \mathbf{C}$, its real part gives an $\mathbf{R}$-linear map from $H^{1,0}$ to $\mathbf{R}$. In this way, we get an identification between $(H^{1,0})^*$ and the real *homology* group $H_1(\Sigma; \mathbf{R})$. So, making this identification, what we are considering above is a group homomorphism from $H_1(\Sigma; \mathbf{Z})$ to $H_1(\Sigma; \mathbf{R})$. Unwinding definitions, one finds that this is just the standard inclusion induced by $\mathbf{Z} \subset \mathbf{R}$. So Λ is a lattice, and the quotient space is a $2g$ (real)-dimensional torus. On the other hand, the description as a quotient of $(H^{1,0})^*$ shows that it is naturally a complex torus. This complex torus is the *Jacobian* $\mathrm{Jac}(\Sigma)$ of Σ.

We can now formulate our result as follows (extending it in an obvious way to the case of multiple poles and zeros).

- Integration along paths defines a map $\underline{j} : \Sigma \to \mathrm{Jac}(\Sigma)$, which can also be characterised by the fact that $\underline{j}(p_0) = 0$ and the derivative of $\underline{j}$ at point $p \in \Sigma$ is equal to the evaluation map of holomorphic 1-forms at p.
- $\underline{j}$ extends uniquely to a *group homomorphism* from the divisor group of Σ to $\mathrm{Jac}(\Sigma)$.
- Two divisors are linearly equivalent if and only they have the same degree and they map under $\underline{j}$ to the same point in $\mathrm{Jac}(\Sigma)$.

12.2.2   Abstract theory

Here we take a different point of view, using the language of cohomology and line bundles. However, the real content is essentially identical to that of the previous subsection. Recall that for any divisor D, we have defined a holomorphic line bundle $\mathcal{L}_D$. Recall also that there is a holomorphic function with poles p_i and zeros q_i if and only if the line bundles $\mathcal{L}_D$ and $\mathcal{L}_{D'}$ are isomorphic, where $D = p_1 + \cdots + p_d$, $D' = q_1 + \cdots + q_d$. So what we want to do is to understand the set of isomorphism classes of holomorphic line bundles, which, as we have seen, is in a natural way a group $\mathrm{Pic}(\Sigma)$.

The next step is the following.

Lemma 35. *$\mathrm{Pic}(\Sigma)$ is isomorphic to $H^1(\Sigma; \mathcal{O}^*)$, where $\mathcal{O}^*$ is the sheaf of non-vanishing holomorphic functions with a group structure given by multiplication.*

This can be seen using Čech cohomology. A line bundle can be described by a system of transition functions $g_{\alpha\beta} : U_\alpha \cap U_\beta \to \mathbf{C}^*$ with the consistency condition $g_{\alpha\gamma} = g_{\alpha\beta} g_{\beta\gamma}$ on triple intersections. This is the cocycle condition written multiplicatively. Changing the local trivialisations over U_α changes the transition functions by a coboundary.

Next we consider the exact sequence of sheaves

$$0 \to \mathbf{Z} \to \mathcal{O} \to \mathcal{O}^* \to 0,$$

which just asserts that a non-vanishing function has a logarithm, locally. Now we get a long exact sequence

$$H^1(\Sigma; \mathbf{Z}) \to H^1(\Sigma; \mathcal{O}) \to \mathrm{Pic}(\Sigma) \to \mathbf{Z} \to 0,$$

using the fact that $H^2(\Sigma; \mathbf{Z}) = \mathbf{Z}$. If we choose an element of $\mathrm{Pic}(\Sigma)$ mapping to 1 in $\mathbf{Z}$, we get an isomorphism $\mathrm{Pic}(\Sigma) = \mathrm{Pic}_0(\Sigma) \times \mathbf{Z}$, where $\mathrm{Pic}_0(\Sigma)$ is the quotient $H^1(\Sigma; \mathcal{O})/H^1(\Sigma; \mathbf{Z})$.

To match this up with the previous section, we take a point p_0 in Σ, which gives a line bundle $\mathcal{L}_{p_0}$ and thus a class in $H^1(\Sigma; \mathcal{O}^*)$. It is a fact that the image of this under the boundary map above is $1 \in H^2(\Sigma; \mathbf{Z}) = \mathbf{Z}$. More generally, for a line bundle defined by a divisor, the boundary map yields the degree of the divisor. Our previous result asserts that the group of divisor classes modulo linear equivalence is $\mathbf{Z} \times \mathrm{Jac}(\Sigma)$, with the isomorphism defined by (degree, j). So what we need to establish is an isomorphism between the two quotients

$$\mathrm{Jac}(\Sigma) = (H^{1,0})^*/H_1(\Sigma; \mathbf{Z}), \quad \mathrm{Pic}_0(\Sigma) = H^{0,1}(\Sigma)/H^1(\Sigma; \mathbf{Z}).$$

But this is just the isomorphism induced by the duality between $H^{1,0}$ and $H^{0,1}$, which one can check is compatible with the two integral lattices.

12.2.3   Geometry of symmetric products

For any space X and $d \geq 0$, we can form the *symmetric product* $\mathrm{Sym}^d(X)$: the quotient of the product of d copies of X by the action of the permutation group. If X is a manifold, then, generally, the symmetric products will not be, they will have singularities corresponding to the 'diagonals' in the product. It is remarkable that, in the case of Riemann surfaces, these singularities do not appear.

Proposition 40. *If Σ is a Riemann surface, then $\mathrm{Sym}^d(\Sigma)$ is a complex manifold of complex dimension d.*

To see this, it is clear that one can reduce the proposition to case when Σ is the **C**, since the question is really local. Now identify $\mathrm{Sym}^d(\mathbf{C})$ with $\mathbf{C}^d$ by the map which takes $\lambda_1, \ldots, \lambda_d$ to the elementary symmetric functions, the co-efficients of the polynomial

$$(z - \lambda_1)(z - \lambda_2) \cdots (z - \lambda_d).$$

A projective version of this construction shows that $\mathrm{Sym}^d(\mathbf{CP}^1) = \mathbf{CP}^d$.

The tangent space of the symmetric product can be described as follows. Regarding $D \in \mathrm{Sym}^d(\Sigma)$ as a divisor, we form the sheaf $\mathcal{O}(D)$ of meromorphic functions f with $(f) \leq D$. Then $\mathcal{O} \subset \mathcal{O}(D)$, and the quotient is a sheaf supported at the points of the divisor. The space of global sections $H^0(\Sigma; \mathcal{O}(D)/\mathcal{O})$ is the tangent space of $\mathrm{Sym}^d(\Sigma)$ at D. Away from the diagonals, when $D = p_1 + \cdots + p_d$ with the p_i distinct, this is just the assertion that the residue of a meromorphic function at p_i is naturally an element of $T\Sigma_{p_i}$.

Now let Σ be a compact Riemann surface of genus $g \geq 1$ with a fixed base point. Regarding elements of the symmetric products as divisors, we get maps

$$j_d : \mathrm{Sym}^d(\Sigma) \rightarrow \mathrm{Jac}(\Sigma),$$

which are clearly holomorphic. If z is a point in the image of j_d, the 'fibre' $j_d^{-1}(z)$ consists of positive divisors in a fixed linear equivalence class. In other language, z corresponds to a line bundle L_z over Σ and the fibre consists of the divisors of zeros of holomorphic sections of L_z. Thus the points of the fibre can be identified with the projective space $\mathbf{P}(H^0(\Sigma; L_z))$. So the map j_d has the remarkable property that all its fibres are complex projective spaces. The derivative of j_d at a point $D \in \mathrm{Sym}^d(\Sigma)$ can be described as follows. We stated that the tangent space of $\mathrm{Sym}^d(\Sigma)$ at D is $H^0(\mathcal{O}(D)/\mathcal{O})$. The tangent space of the Jacobian is naturally isomorphic to $H^1(\Sigma; \mathcal{O})$. The derivative of j_d is the boundary map in the long exact sequence

$$0 \rightarrow \mathbf{C} = H^0(\mathcal{O}) \rightarrow H^0(\mathcal{O}(D)) \rightarrow H^0(\mathcal{O}(D)/\mathcal{O}) \rightarrow H^1(\mathcal{O}) \rightarrow H^1(\mathcal{O}(D))$$

$$\rightarrow 0.$$

The geometry of the maps j_d contains all the information about the line bundles and divisors on Σ. For $d = 1$, the map j_1 is injective (since there is no meromorphic function with just one pole). When d is 'large', in fact $d > 2g - 2$, the fibres all have the same dimension, $d - g$. For the Riemann–Roch formula gives

$$\dim H^0(L) - \dim H^0(K \otimes L^*) = d - g + 1,$$

and if $d > 2g - 2$, the line bundle $K \otimes L^*$ has negative degree and so cannot have holomorphic sections. Likewise, it follows from the exact sequence above that the derivative of j_d is surjective at each point. It follows easily that j_d is surjective (since the image is a compact open subset of the Jacobian). (In fact, a similar argument shows that j_d is surjective once $d \geq g$.) As a consequence, we have the following corollary.

Corollary 13. *Any holomorphic line bundle is isomorphic to $\mathcal{L}_D$ for some divisor D.*

In the range $d > 2g - 2$, the map j_d is a *holomorphic fibration*. In the intermediate region $1 < d < 2g - 2$, the structure of the maps j_d is more complicated and may vary depending on which Riemann surface of a given genus we take. For example:

- $d = 2$. If Σ is not a double cover of the Riemann sphere, then j_2 is injective.
- $d = 2g - 2$. If L has degree $2g - 2$, then $K \otimes L^*$ has degree 0 and so has a holomorphic section if and only if it is trivial, which is the case when $L = K$. So there is just one fibre of j_{2g-2} which is of dimension $g - 1$, and all the others have dimension $g - 2$.

12.2.4   Remarks in the direction of algebraic topology

If X is any space (say a finite CW complex), we can form various associated spaces:

- the symmetric products $\mathrm{Sym}^d(X)$, as above;
- the 'Jacobian torus' $\mathrm{Jac}(X) = H_1(X, \mathbf{R})/H_1(X, \mathbf{Z})$;
- the dual torus $\mathrm{Pic}_0(X) = H^1(X, \mathbf{R})/H^1(X, \mathbf{Z})$, which can be interpreted as the space of representations $\pi_1(X) \to S^1$.

Some of the constructions we have discussed above can be extended to this more general situation. For example, if X is a manifold, we can define maps $j_d(\mathrm{Sym}^d(X)) \to \mathrm{Jac}(X)$ by choosing a set of closed 1-forms representing the cohomology group $H^1(X, \mathbf{R})$. Up to homotopy, the map we get is independent of choices (and a similar construction can be performed when X is not a manifold). It is worth noticing, however, that in general the tori Jac and Pic should be distinguished. Although they are both tori of the same dimension they are not naturally isomorphic. The fact that they are isomorphic in the case of surfaces is the Poincaré duality.

There is a general theorem of Dold and Thom (1958) that for any X and any index p, the *homotopy group* $\pi_p(\mathrm{Sym}^d(X))$ is naturally isomorphic to the *homology group* $H_p(X)$, once d is sufficiently large. In the case of a Riemann surface, this follows from the description of the symmetric products, for large d, as bundles over the Jacobian. In general, the map j_d induces the stated isomorphism on π_1.

The statement that the symmetric products $\text{Sym}^d(\Sigma)$, for large d, are *homeomorphic* to projective-space bundles over the Jacobian is a purely topological statement, but it seems not to be very easy to see it by purely topological methods. What one can easily do is check the homology or cohomology groups. It is a general fact that if a finite group G acts on a space Y, then the *rational* cohomology of the quotient Y/G is isomorphic to the G-invariant part of the rational cohomology of Y. (This is *not* generally the case for cohomology with integral co-efficients.) We can use this to calculate the rational cohomology of $\text{Sym}^d(X)$, taking Y to be the product of d copies of X and G to be the permutation group. In particular, we can do this when $X = \Sigma$ is a compact Riemann surface. Let u be the standard generator of $H^2(\Sigma)$, and write U for the two-dimensional space spanned by u and $1 \in H^0(\Sigma)$. Thus $H^*(\Sigma) = U \oplus H^1(\Sigma)$. By the Künneth formula, the cohomology of the product of d copies of Σ is isomorphic to the tensor product of d copies of $U \oplus H^1(\Sigma)$. Taking invariants under the permutation group (and due account of signs) gives the formula

$$H^*(\text{Sym}^d(\Sigma)) = \bigoplus_{0 \le j \le d} s^{d-j}(U) \otimes \Lambda^j,$$

where Λ^j is the exterior power $\Lambda^j H^1(\Sigma)$ and $s^i(U)$ is the i'th symmetric power. (We set $\Lambda^j = 0$ if $j > 2g$ or $j < 0$.) Taking account of grading, we have, if $i \le d$,

$$H^i(\text{Sym}^d(\Sigma)) = \Lambda^i \oplus \Lambda^{i-2} \oplus \cdots . \qquad (12.4)$$

It is a fact that the cohomology of a bundle with a fibre that is a complex projective space is the same (additively) as that for a product. So if E is a bundle over the Jacobian with fibre $\mathbf{CP}^{d-g}$, we have

$$H^*(E) = H^*(\mathbf{CP}^{d-g}) \otimes H^*(\text{Jac}(\Sigma)).$$

Since the cohomology of the torus is given by the exterior powers, we obtain

$$H^i(E) = \bigoplus_{p \le d-q} \Lambda^{i-2p}. \qquad (12.5)$$

Now, the condition that $d > 2g - 2$ is exactly the condition that the two sums (12.4), (12.5) are the same, when $i \le d$. Thus, when this condition is met, we have $H^i(\text{Sym}^d(\Sigma)) = H^i(E)$ for $i \le d$ and, by Poincaré duality, the same is true for $i > d$.

When $d \leq 2g - 2$, the sums (12.4), (12.5) are not the same, and one can extract some topological information about the map j_d by comparing them.

12.2.5   Digression into projective geometry

We will now examine in some detail the special case when Σ has genus 2, so that $\text{Jac}(\Sigma)$ is a complex surface and we have a surjective map $j : \text{Sym}^2(\Sigma) \to \text{Jac}$. We would like to 'see' this complex torus from the algebraic point of view. It is a fact that the Jacobians of Riemann surfaces are complex projective varieties, so can be defined by polynomial equations, but for high genus it is hard to write these down explicitly. For genus 2, one can do this up to a double covering map, and the relevant theory is that of *Kummer surfaces* in 3-space, for which we follow Hudson (1990).

If L is a line bundle of degree 2, the line bundle $L^* \otimes K$ has degree 0 so can only have holomorphic sections if it is trivial, in which case $L = K$. There is a two-dimensional space of sections of K, which define a projective line E in $\text{Sym}^2(\Sigma)$. This line is collapsed to a point by j_2, but otherwise j_2 is bijective. Explicitly, we know that there is a double branched cover $f : \Sigma \to \mathbf{CP}^1$, which corresponds to an involution $\sigma : \Sigma \to \Sigma$. The set E consists of the pairs of the form $(x, \sigma(x))$ in $\text{Sym}^2(\Sigma)$, which can be parametrised by $f(x) \in \mathbf{CP}^1$. In fact, for those readers familiar with the concept, $\text{Sym}^2(\Sigma)$ is the 'blow-up' of the Jacobian at a single point and E is the exceptional curve.

Now f induces a map $f^{(2)} : \text{Sym}^2(\Sigma) \to \text{Sym}^2(\mathbf{CP}^1) = \mathbf{CP}^2$, and this is generically 4–1: for any distinct points p, p' in Σ which are not ramification points of f, and with p' not equal to σp, the four points $(p, p'), (\sigma p, p'), (p, \sigma p'), (\sigma p, \sigma p')$ in $\Sigma \times \Sigma$ map to four distinct points in $\Sigma^{(2)}$, each with the same image in $\mathbf{CP}^2$. Let $\hat{X}$ be the space obtained from $\Sigma^{(2)}$ by identifying (p, p') with $(\sigma p, \sigma p')$ for all p, p'. Then $f^{(2)}$ factors through natural maps $\Sigma^{(2)} \to \hat{X} \to \mathbf{CP}^2$, each of which is generically 2–1. In particular, $\hat{X}$ is a 'double branched cover' of $\mathbf{CP}^2$.

It is straightforward to find the branch locus of the map $\hat{X} \to \mathbf{CP}^2$. Under the identification $\text{Sym}^2(\mathbf{CP}^1) = \mathbf{CP}^2$, the diagonal maps to a conic curve in $\mathbf{CP}^2$. To simplify the formulae, we work in affine co-ordinates. Then, if we make the identification by writing $x = u + v, y = uv$, the conic is given by the equation $x^2 - 4y = 0$. There are six branch points of f in $\mathbf{CP}^1$ and, taking the tangents to the conic at these points, we get six lines in $\mathbf{CP}^1$. Let B be the union of these lines: a singular curve of degree 6 in $\mathbf{CP}^2$. Then it is easy to check from the definitions that the pre-image in $\hat{X}$ of a point $q \in \mathbf{CP}^2$ consists of two points if $q \notin B$ and one point if $q \in B$. The curve E in Sym^2 maps to a curve, E' say, in $\hat{X}$.

Now we want to exhibit a complex surface X in $\mathbf{CP}^3$, defined by a homogeneous polynomial of degree 4, which is 'almost the same' as $\hat{X}$. More precisely,

X is obtained from $\hat{X}$ by collapsing the curve E' to a point. To keep formulae compact, we will work in affine co-ordinates. We have co-ordinates (x, y) on the plane and six lines tangent to the conic $\{x^2 - 4y\} = 0$, as above. Let $\Pi_6(x, y)$ be the polynomial defining the union of these lines,

$$\Pi_6(x, y) = \prod_{i=1}^{6}(y - \lambda_i x - \mu_i),$$

say. We seek polynomials $a(x, y), b(x, y), c(x, y)$, of degrees $2, 3, 4$, respectively, such that if X_0 is the affine surface in 3-space defined by the equation

$$a(x, y)z^2 - 2b(x, y)z + c(x, y) = 0, \tag{12.6}$$

then projection to the x, y plane yields a double cover branched over these six lines. This occurs if the discriminant $\Delta = b^2 - ac$ is a multiple of Π_6. So, the question is how to find polynomials a, b, c satisfying this equation.

Take two copies of $\mathbf{C}$ with co-ordinates u, v and let $k_1, \ldots, k_6$ be the critical values of f. Write $u_i = u - k_i$, $v_i = v - k_i$. The expression

$$\phi_3 = u_1 u_2 u_3 v_4 v_5 v_6 + v_1 v_2 v_3 u_4 u_5 u_6$$

is symmetric under the interchange of u and v co-ordinates. This means that it can be written as a polynomial $b(u, v)$ in the elementary symmetric functions $x = u + v$, $y = uv$. The expression $\psi_3 = u_1 u_2 u_3 v_4 v_5 v_6 - v_1 v_2 v_3 u_4 u_5 u_6$ is antisymmetric under interchange of u and v, and this means that it can be written as

$$\psi_3 = (u - v)\phi_2,$$

where ϕ_2 is symmetric. Then ϕ_2 can be expressed as a polynomial in u, v. Write $\phi_2^2 = c(u, v)$.

Now

$$\phi_3^2 - \psi_3^2 = (u_1 u_2 u_3 v_4 v_5 v_6 + v_1 v_2 v_3 u_4 u_5 u_6)^2 - (u_1 u_2 u_2 v_4 v_5 v_6 - v_1 v_2 v_3 u_4 u_5 u_6)^2,$$

which is equal to

$$4 u_1 u_2 u_3 u_4 u_5 u_6 v_1 v_2 v_3 v_4 v_5 v_6.$$

A product $u_i v_i$ is symmetric and can be written as

$$u_i v_i = (u - k_i)(v - k_i) = (uv - k_i(u + v) + k_i^2) = y - k_i x + k_i^2,$$

and this is precisely the linear polynomial defining the line tangent to the conic at the given point. So we have $\phi_3^2 - \psi_3^2 = 4\Pi_6(x, y)$ when we substitute for x, y. Thus, by our definitions,

$$b(x, y)^2 - (u - v)^2 c(x, y) = 4\Pi_6(x, y).$$

If we take $a(x, y) = (u - v)^2 = x^2 - 4y$, we have found polynomials a, b, c satisfying the desired condition, and one can check that they have the correct degrees.

To see the connection of this construction with $\mathrm{Sym}^2(\Sigma)$, we start with two copies of Σ on which we have well-defined square roots

$$U = \sqrt{u_1 u_2 u_3 u_4 u_5 u_6}, \quad V = \sqrt{v_1 v_2 v_3 v_4 v_5 v_6}.$$

For given x, y, the 'solution' of equation (12.6) for z is, by the quadratic formula,

$$z = \frac{1}{2a}(b \pm 2\sqrt{\Pi_6}).$$

If we write $x = u + v$, $y = uv$, then we can set $\sqrt{\Pi_6} = UV$ to solve our equation. That is, we have a map from an open set in $\Sigma \times \Sigma$ to X given by

$$x = u + v, \quad y = uv, \quad z = \frac{1}{2(u - v)^2}(\phi^3 + UV).$$

This is symmetric under interchange of u and v and so defines a map from an open set in $\mathrm{Sym}^2(\Sigma)$. It is also symmetric under changing the sign of both U and V, so it defines a map on $\hat{X}$. On the face of it, this map is not defined when $u = v$, but we get a well-defined map to the corresponding projective surface X in $\mathbb{CP}^3$ (given by a homogeneous equation of degree 4). What happens, as one can check, is that the curve E' is collapsed to a single point in the plane at infinity (the point defined by $x = y = 0$) but otherwise $\hat{X}$ is mapped bijectively to X.

We will illustrate this construction with a particular example. Let Σ be the Riemann surface associated to the function $\sqrt{z(z^4 - 1)}$, so the branch points k_i are $0, \pm 1, \pm i, \infty$. In the discussion above, we tacitly assumed that the k_i were finite, but the construction extends in a straightforward way. We take $\{k_1, k_2, k_3\} = \{0, 1, -1\}$ and then

$$\phi_3 = u(u^2 + 1)(v^2 - 1) + v(v^2 + 1)(u^2 - 1),$$

$$\psi_3 = u(u^2 + 1)(v^2 - 1) - v(v^2 + 1)(u^2 - 1),$$

so $\phi_3^2 - \psi_3^2 = 4uv(u^4 - 1)(v^4 - 1)$. We find that

$$\phi_3 = (u + v)\left(u^2 v^2 - 1 - (u - v)^2\right),$$

so $b(x, y) = x(y^2 - 1 - x^2 + 4y)$. Similarly,

$$\psi_3 = (u - v)\left(u^2 v^2 - 1 - (u + v)^2\right)$$

and $c(x, y) = (y^2 - x^2 - 1)^2$. Our identity is

$$x^2(y^2 - x^2 - 1 + 4y)^2 - (x^2 - 4y)(y^2 - x^2 - 1)^2 = 4\pi(x, y),$$

where

$$\pi(x, y) = y(y - x - 1)(y + x - 1)(y - ix + 1)(y + ix + 1).$$

In the affine plane, the equation $\pi = 0$ defines the union of five lines: the other line is the line at infinity. Our surface is given by the equation

$$(x^2 - 4y)z^2 - 2x(y^2 - x^2 - 1 + 4y)z + (y^2 - x^2 - 1)^2 = 0.$$

We will now make some projective transformations to put this equation in to a tidier form. First, replace z by $z - x$, which gives a transformed equation

$$(x^2 - 4y)z^2 - 2x(y^2 - 1)z + (y^2 - 1)^2 + 4yx^2 = 0.$$

Now write the equation in homogeneous co-ordinates as

$$(x^2 - 4yt)z^2 - 2x(y^2 - t^2)z + (y^2 - t^2)^2 + 4ytx^2 = 0$$

and rearrange to give

$$(zx - y^2 + t^2)^2 + 4yt(x^2 - z^2) = 0.$$

Write $x = X + iZ, z = X - iZ, y = iY, t = T$ to arrive at the equation

$$(X^2 + Y^2 + Z^2 + T^2)^2 - 16XYZT = 0.$$

The surface defined by this equation has 16 singular points. In general, if F is a homogeneous polynomial in four variables, the singularities of the surface $F = 0$ correspond to common zeros of all the first partial derivatives of F. In this case

$$\frac{\partial F}{\partial X} = 4XQ - 16YZT,$$

where we write $Q = X^2 + Y^2 + Z^2 + T^2$. If the partial derivative vanishes we have $X^2 Q = XYZT$, and symmetrically in the other variables. So if Q is not zero we have $X^2 = Y^2 = Z^2 = T^2$. There are eight possibilities $(X, Y, Z, T) = (\pm 1, \pm 1, \pm 1, \pm 1)$, taking account of an overall multiplication by -1. Then $Q = 4$, and we need $X = YZT$ etc. This cuts us down to just four solutions. On the other hand, if $Q = 0$, we must have $YZT = XZT = XYT = XYZ = 0$ and two of the co-ordinates must vanish. There are six ways to choose such a pair of co-ordinates. If the pair is X, Y, say, then the equation $Q = 0$ is $Z^2 + T^2 = 0$, which has two solutions, projectively. So we get 12 singular points of this second kind, giving 16 in total.

These singular points are readily understood in terms of our earlier branched-cover description. The six lines in the plane have a total of 15 intersection points. These give singularities in the double cover. The basic model is to take the plane curve $xy = 0$ as a branch curve, which gives a branched cover $z^2 = xy$ of the plane, as a surface S in 3-space. The singularity is a 'node'. The surface S can also be viewed as the quotient $\mathbf{C}^2 / \pm 1$, via the map $(p, q, r) \mapsto (p^2, q^2, pq)$. In this way, we see 15 of the singularities in our surface X. The other singularity is the point at infinity, in our previous affine co-ordinates, which is the image of the curve $E' \subset \hat{X}$.

Now consider more generally a surface defined by an equation

$$Q_\mu^2 + \lambda\, XYZT = 0,$$

where

$$Q_\mu(X, Y, Z, T) = X^2 + Y^2 + Z^2 + T^2 + \mu(XY + XZ + XT + YZ + YT + ZT),$$

for some fixed $\lambda, \mu \in \mathbf{C}$. For a typical value of μ, straightforward calculation shows that this surface has a singular point only if $\lambda = \lambda(\mu) = 4(\mu + 1) (\mu - 2)^2$. With this choice of λ there are exactly 16 singular points, as in the case $\mu = 0$ above. In fact, for each $\mu \neq -1, 2$, the resulting surface is projectively equivalent to one constructed as above from a suitable curve Σ_μ of genus 2.

We want to use this family to 'see' the relation between the symmetric product $\mathrm{Sym}^2(\Sigma)$ and a torus explicitly. For this, we take the limit as $\mu \to -1$, so $\lambda(\mu) \to 0$. If we just take the naive limit, we get the equation $Q_{-1}^2 = 0$. The *set* of solutions is obviously the quadric surface $A = \{(X, Y, Z, T) : Q_{-1}(X, Y, Z, T) = 0\}$. This is a non-singular quadric and, as such, is isomorphic to the product $\mathbf{CP}^1 \times \mathbf{CP}^1$, with two 'rulings' by lines in $\mathbf{CP}^3$. If we set $T = 0$ in the equation $Q_{-1} = 0$, we get

$$X^2 + Y^2 + Z^2 - (XY + YZ + XZ) = 0,$$

which factorises as $(X + \rho Y + \rho^2 Z)(X + \rho^2 Y + \rho Z) = 0$, where ρ is a cube root of unity. Thus the plane $T = 0$ cuts the quadric surface A in two lines, one from each ruling. Similarly for the other planes $X = 0, Y = 0, T = 0$. So we get eight lines in A, four from each ruling.

The situation here is very similar to what we encountered in Section 12.1.3 for curves of genus 3 mapping to $\mathbf{CP}^2$. If μ is close to but not equal to -1, we have a symmetric product $\mathrm{Sym}^2(\Sigma_\mu)$ with a map $f_\lambda : \mathrm{Sym}^2(\Sigma_\epsilon) \to X_\lambda$ which is generically two-to-one but collapses a $\mathbf{CP}^1$ to one of the singular points. This is the same $\mathbf{CP}^1$ which is collapsed to a point in the Jacobian $\mathrm{Jac}(\Sigma_\mu)$. As μ approaches -1, the image surface approaches the quadric A but there are two 'sheets' of the image lying close to A. Just as in the lower-dimensional case, there is a natural 'limiting surface' X_{-1}, but it is a double cover of A, branched over the eight lines where $XYZT$ vanishes. (In more precise, technical language, there is a 'flat family' of surfaces whose fibre over -1 is this double cover.)

Now we identify A with $\mathbf{CP}^1 \times \mathbf{CP}^1$. The four lines in one family give four points p_1, p_2, p_3, p_4 in the first $\mathbf{CP}^1$ factor, and the four lines in the other give four points q_1, q_2, q_3, q_4 in the second factor. We get a pair S_1, S_2 of Riemann surfaces of genus 1 as double branched covers of $\mathbf{CP}^1$ with these branch points. Thus we have a generically 4–1 map from $S_1 \times S_2$ to $A = \mathbf{CP}^1 \times \mathbf{CP}^1$ with a branch set that is the union of the eight lines. It is not hard to convince oneself that that this is the limit, in the above sense, of the Jacobians $\mathrm{Jac}(\Sigma_\mu)$ regarded as double covers of the X_μ, that is, as the symmetric products with the special curve collapsed. In particular, one sees in this way that, on collapsing this special curve, the space we get is *homeomorphic* to the 4-torus $S_1 \times S_2$.

From another point of view, we start with the Jacobian $\mathrm{Jac}(\Sigma)$ of a curve of genus 2, defined as a complex torus—a quotient of $H^{0,1}(\Sigma)$—and with the origin corresponding to the canonical divisor. The map $j : \mathrm{Sym}^2(\Sigma) \to \mathrm{Jac}(\Sigma)$ intertwines the involution $\sigma^{(2)}$ of Sym^2 with the map $\tau : \mathrm{Jac}(\Sigma) \to \mathrm{Jac}(\Sigma)$ induced by multiplication by -1 on $H^{0,1}(\Sigma)$. Our surface X is the quotient of $\mathrm{Jac}(\Sigma)$ by τ, and this has 16 singular points corresponding to the 16 points of order 2 in $\mathrm{Jac}(\Sigma)$, which are fixed points of τ.

Exercises

1. Let Σ be a compact Riemann surface of genus ≥ 2 and let $\mathrm{Aut}(\Sigma)$ be its group of holomorphic automorphisms. Show that the image of $\mathrm{Aut}(\Sigma)$ in the general linear group of $H^{1,0}(\Sigma)$ is finite. [*Hint*: It preserves a Hermitian metric. In fact, we will see that $\mathrm{Aut}(\Sigma)$ is itself finite. See Exercise 3 in Chapter 14.]
2. Prove Lemma 31.
3. Suppose we know that the cohomology $H^1(L)$ of a line bundle over a compact Riemann surface is finite-dimensional. Deduce directly from this that L is defined by a divisor.
4. Let T_0, T_1, T_2 be 'generic' $n \times n$ complex matrices. Show that the equation $\det(Z_0 T_0 + Z_1 T_1 + Z_2 T_2) = 0$ defines a complex curve C in the projective plane of degree n and that the kernel of $Z_0 T_0 + Z_1 T_1 + Z_2 T_2$ defines a holomorphic line bundle over C. (You may choose to find exactly what is required of 'generic' matrices here, or treat the matter informally.)
5. Study the map to $\mathbf{CP}^3$ defined by the canonical line bundle of a Riemann surface of genus 4. How is this related to the genus formula for curves in $\mathbf{CP}^1 \times \mathbf{CP}^1$ in Exercise 1 in Chapter 7?
6. Suppose that the discrete group $\Gamma \subset PSL(2, \mathbf{R})$ acts freely on H. Let f be a holomorphic function on H which satisfies the transformation law

$$f\left(\frac{az+b}{cz+d}\right) = (cz+d)^{2r} f(z),$$

for the Möbius maps in Γ. Show that f defines a holomorphic section of the rth tensor power of the cotangent bundle over the quotient surface H/Γ. Can you relate this to the sums G_p considered in Exercise 6 in Chapter 6?

7. Verify the description of the derivative of j_d away from the diagonals in the symmetric product.

8. (a) By considering the symmetries, or otherwise, identify the points in the Klein quartic $x^3 y + y^3 z + z^3 x = 0$ which correspond to the cusps in the hyperbolic-geometry picture.

(b) Find linear maps inducing automorphisms of order 2 of the Klein quartic.

9. For fixed $\alpha \in \mathbf{C}$ with $\alpha \neq 0, \alpha^4 \neq -1$, find the quartic surface in 3-space associated to the function $\sqrt{z(z^2 - \alpha^2)(z^2 + \alpha^{-2})}$.

13 Moduli and deformations

A striking thing about Riemann surfaces is that they often arise in families, depending upon continuous parameters. For example, we can continuously vary the coefficients in a polynomial defining a Riemann surface. In the case of genus 1, we have studied this systematically in Chapter 6, giving a complete classification of isomorphism classes. In this chapter and the next, we give an introduction to a large circle of ideas, discussing connections between

- topological properties of surface diffeomorphisms;
- the *moduli space* parametrising isomorphism classes of surfaces of fixed genus;
- the compactification of the moduli space and Dehn twists in 'vanishing cycles'.

13.1 Almost-complex structures, Beltrami differentials and the integrability theorem

In Chapter 3, we defined the two notions of smooth surfaces and Riemann surfaces. A Riemann surface is a smooth surface with an atlas of charts which are related by holomorphic maps. There is another way of proceeding, which turns out to be equivalent. Let S be an oriented smooth surface. An *almost-complex structure* on S is a collection of linear maps $J_x : TS_x \to TS_x$ with $J_x^2 = -1$, varying smoothly with $x \in S$ and compatible with the given orientation in that for any non-zero $v \in TS_x$, the pair $v, J_x v$ form an oriented basis. (Here the meaning of 'varying smoothly with x' is the obvious one, defined in the same manner as for differential forms etc.) Now any Riemann surface has a natural almost-complex structure, as we discussed in Chapter 3. Conversely, we have the following theorem.

Theorem 22. *If (S, J) is a smooth surface with an almost-complex structure, then there is a unique compatible Riemann surface structure.*

Here, the uniqueness is clear. We can define a holomorphic function on an almost-complex surface as a smooth map from S to $\mathbf{C}$ whose derivative is complex linear with respect to the almost-complex structure, and the Riemann surface charts are just those given by local holomorphic functions. What is not clear is that such charts exist. Thus the content of Theorem 22 is to show that for each point x of S, there is a holomorphic function defined on a neighbourhood of x and with a non-vanishing derivative at x. We shall postpone the proof of Theorem 22.

There are two useful ways of encoding an almost-complex structure. In fact, although this is perhaps something of a coincidence, the issue is really the same as that of the different descriptions of the hyperbolic plane. Let V be a two-dimensional oriented real vector space. We want to describe the set J_V of $\mathbf{R}$-linear maps $J : V \to V$ with $J^2 = -1$, compatible with the orientation. The group $SL(V)$ acts transitively on J_V by conjugation. If we fix some standard structure $J_0 \in J_V$, which is the same as fixing an identification of V with $\mathbf{C}$, the stabiliser of J_0 is, by definition, the set of complex linear maps $\cos\theta \, 1 + i\sin\theta J_0$, which we can identify with the standard S^1. So J_V can be expressed as the quotient space $SL(V)/S^1$ or, fixing co-ordinates, $SL(2,\mathbf{R})/S^1$, and we know that this is a description of the hyperbolic plane. Now we get different points of view corresponding to our different models for the hyperbolic plane. In one point of view, we can regard a point of J_V as a Euclidean metric, defined up to a positive scale factor or, equivalently, normalised to have determinant 1 (fixing an area element on V). Passing to a smooth oriented surface S, we apply this construction in each tangent space to see that an almost-complex structure can be identified with a 'conformal class' of Riemannian metrics on S or, equivalently, Riemannian metrics with a fixed area form. Then, for each unit tangent vector v in TS_x, there are exactly two unit vectors $\pm v'$ orthogonal to v, and just one of these, v' say, has the property that (v, v') form an oriented basis. We define an almost-complex structure by decreeing that $J(v) = v'$, $J(v') = -v$. This is essentially the same as the relation between holomorphic functions and conformal mappings.

These ideas are important in two-dimensional differential geometry. Suppose, for example, we have an oriented surface S in $\mathbf{R}^3$. Then the induced Riemannian metric (or 'first fundamental form', in classical terminology) makes S into a Riemann surface, which is perhaps surprising since complex numbers do not appear in the definitions. Many differential-geometric concepts have a Riemann surface interpretation; see, for example, Exercise 3 in Chapter 4.

The other important way of describing almost-complex structures corresponds to the disc model of the hyperbolic plane. Consider $\mathbf{C}^2$ with the anti-linear map $\sigma(z_1, z_2) = (\bar{z}_2, \bar{z}_1)$. Given $\mu \in \mathbf{C}$ with $|\mu| < 1$, let Γ_μ be the graph

$$\Gamma_\mu = \{(z, \mu z) : z \in \mathbf{C}\}.$$

An intersection point of Γ_μ and $\sigma(\Gamma_\mu)$ is given by a solution of the equations $z = \overline{\mu w}, \mu z = \overline{w}$, and since $\mu\overline{\mu} \neq 1$, this can occur only when $z = w = 0$. Thus Γ_μ and $\sigma(\Gamma_\mu)$ are complementary subspaces, and

$$\mathbf{C}^2 = \Gamma_\mu \oplus \sigma(\Gamma_\mu).$$

Now we can define $J_\mu^{\mathbf{C}} : \mathbf{C}^2 \to \mathbf{C}^2$ to act as multiplication by i on Γ_μ and by $-i$ on $\sigma(\Gamma_\mu)$. Then $J_\mu^{\mathbf{C}}$ anti-commutes with σ and, obviously, $\left(J_\mu^{\mathbf{C}}\right)^2 = -1$. Let V be the real subspace of $\mathbf{C}^2$ consisting of the fixed points of σ. This is the set of points of the form $(z, \overline{z})$. Then $J_\mu^{\mathbf{C}}$ preserves V, and its restriction gives $J_\mu : V \to V$ with $J_\mu^2 = -1$, that is, a point of J_V. Of course, this construction sets up a 1–1 correspondence between points μ in the disc and points of J_V, which, as we have said, just amounts to the correspondence between the disc and Klein models of the hyperbolic plane. In terms of explicit formulae, if $\mu = re^{i\alpha}$, then in a standard basis J_μ has a matrix

$$J_\mu = \begin{pmatrix} -2r\sin\alpha & 1 + r^2 - 2r\cos\alpha \\ -(1 + r^2 + 2r\cos\alpha) & 2r\sin\alpha \end{pmatrix}.$$

A good exercise is to show that the 'circles' with respect to the complex structure J_μ are ellipses with eccentricity $(1 + r)/(1 - r)$ and with major axis making an angle $\alpha/2$ with the first co-ordinate axis.

To repeat the above in more invariant language, let U be a one-dimensional complex vector space and μ a complex linear map from U to $\overline{U}$. Then the modulus $|\mu|$ is invariantly defined, even though μ is not strictly a complex number. The set of complex structures on U compatible with the given orientation is identified with the set of μ with $|\mu| < 1$. Now we go to a point p in a Riemann surface Σ and apply this with $U = T^*\Sigma_p$. We see that the almost-complex structures on Σ can be identified with sections $\underline{\mu}$ of the bundle $T\Sigma \otimes \overline{T}^*\Sigma$ which are the same as $(0, 1)$-forms with values in the bundle T. In a local co-ordinate z, we can write

$$\underline{\mu} = \mu(z)d\overline{z} \otimes \frac{\partial}{\partial z}$$

for a smooth function μ, with $|\mu| < 1$ pointwise. These sections are called *Beltrami differentials*. We write $J_{\underline{\mu}}$ for the almost-complex structure on Σ defined by $\underline{\mu}$.

A section $\underline{\mu}$ defines a bundle map from $T^*\Sigma$ to $\overline{T^*\Sigma}$, which we denote by the same symbol.

Lemma 36. *A function f on Σ is holomorphic with respect to the almost-complex structure $J_{\underline{\mu}}$ if and only if*

$$(\overline{\partial} + \underline{\mu}\partial)f = 0.$$

The proof is a matter of following through definitions, which we leave as a valuable exercise. In local co-ordinates, the equation appearing above is

$$\frac{\partial f}{\partial \bar{z}} + \mu \frac{\partial f}{\partial z} = 0. \tag{13.1}$$

This is known as the *Beltrami equation* defined by μ.

Now we can return to Theorem 22, stated at the beginning of this section. The actual proof will be given later (in the appendix to this chapter), but we can see that the real content is the following.

Theorem 23. *Let μ be a smooth function defined in a neighbourhood of $0 \in$ $\mathbf{C}$, with $|\mu| < 1$. Then there is a solution f to the partial differential equation (13.1), defined on a possibly smaller neighbourhood, and with a non-vanishing derivative ∂f at 0.*

The condition of the non-vanishing of the derivative is just what is required to give a local co-ordinate for the deformed structure. For the Beltrami equation implies that the derivative at the origin, seen as a 2×2 matrix, is invertible, and we can apply the inverse function theorem.

We fix a compact oriented smooth surface S and define the *moduli set* $\mathcal{M}_S$ to be the set of equivalence classes of Riemann surface structures on S, compatible with the given orientation. Thus, by definition,

$$\mathcal{M}_S = \mathcal{J}/\text{Diff},$$

where $\mathcal{J}$ is the set of almost-complex structures, or conformal classes, and Diff is the group of orientation-preserving diffeomorphisms. We can give $\mathcal{M}_S$ a topology, just the quotient topology, say, and so hereafter we use the standard terminology of 'moduli space'. (We will not go into the details of defining the topology on these infinite-dimensional spaces, but take an informal point of view. These details contain no surprises, so there is essentially no ambiguity in the meaning of phrases such as 'a continuously varying family of Riemann surfaces'. See, however, the discussion in Section 14.4 below, where this issue arises.)

It is clear from either point of view—metrics or Beltrami differentials—that the space of almost-complex structures on a Riemann surface is *contractible*. In terms of Riemannian metrics, any two metrics g_0, g_1 can be joined by a linear path $g_t = (1 - t)g_0 + tg_1$. So, in particular, any two compact Riemann surfaces of the same genus can be joined by a continuous path of Riemann surfaces, and the moduli space is connected. A formal consequence of the definition is that if S' is another surface diffeomorphic to S, then there is a canonical identification between $\mathcal{M}_S$ and $\mathcal{M}_{S'}$. So we get one moduli space for each genus g, which we also denote by $\mathcal{M}_g$.

13.2   Deformations and cohomology

The description of deformations of the complex structure on a Riemann surface gives a good illustration of the value of the machinery of cohomology which we have sketched in Chapter 12. We will give an informal discussion of this here. The analysis of small deformations can be thought of as describing a neighbourhood of a point in the relevant moduli space, and cohomology theory identifies 'infinitesimal deformations', which correspond to the tangent vectors of the moduli space at this point. Later, in Section 14.3, we give a direct construction of the moduli space using hyperbolic geometry, and this means that we can bypass many of the foundational issues which need to be addressed with care in other related problems. (For example 'moduli spaces', defined in a naive way, may fail to be smooth or Hausdorff, and the study of non-compact manifolds yields essentially new phenomena.) But for compact Riemann surfaces, the picture that finally emerges is much what one would expect from a naive approach.

The appearance of cohomology, of various kinds, in describing infinitesimal deformations of structures is a very general phenomenon. Making a digression, consider for example finite-dimensional associative algebras over $\mathbf{R}$. Thus if we fix a dimension n, an object of our theory is a bilinear map $M : \mathbf{R}^n \times \mathbf{R}^n \to \mathbf{R}^n$ satisfying the associative law

$$M(M(x, y), z) = M(x, M(y, z)). \tag{13.2}$$

If we describe M by n^3 structure constants M_{ijk}, this is just a system of quadratic equations. Two structures M_1, M_2 are equivalent if there is a $g \in GL(n, \mathbf{R})$ such that $M_2(gx, gy) = g(M_1(x, y))$. So the 'moduli space of algebras' of dimension n is the quotient

$$\mathcal{M} = \mathcal{X} / GL(n, \mathbf{R}),$$

where $\mathcal{X}$ is the set of solutions of equation (13.2). Let M_0 be a point in $\mathcal{X}$, and write $M_0(x, y)$ simply as the product xy. Consider a small deformation $M_0 + \epsilon$. The quadratic equation (13.2) becomes

$$\epsilon(x, y)z + \epsilon(xy, z) - \epsilon(x, yz) - x\epsilon(y, z) + Q(\epsilon) = 0,$$

where Q is quadratic in ϵ. Similarly, the action of a $g = 1 + \delta \in GL(n, \mathbf{R})$ takes M_0 to M_1 with

$$M_1(x, y) = xy + \delta(x)y + x\delta(y) - \delta(xy) + O(\delta^2),$$

using the fact that $(1 + \delta)^{-1} = 1 - \delta + O(\delta^2)$. Thus, ignoring technicalities, we expect that the infinitesimal version of the classification problem—which should correspond to the tangent space of $\mathcal{M}$ at $[M_0]$—is given by taking the solutions ϵ of the *linear* equation

$$\epsilon(x, y)z + \epsilon(xy, z) - \epsilon(x, yz) - x\epsilon(y, z) = 0$$

modulo the linear subspace of solutions of the form

$$\epsilon(x, y) = \delta(x)y + x\delta(y) - \delta(xy).$$

Now write W_p for the vector space of $(p + 1)$-linear maps from the product of $(p + 1)$ copies of $\mathbf{R}^n$ to $\mathbf{R}^n$. Thus, in the above, we are considering $\epsilon \in W_1$ and $\delta \in W_0$. Define $B_0 : W_0 \to W_1$ by

$$B_0(\delta)(x, y) = \delta(x)y + x\delta(y) - \delta(xy),$$

and $B_1 : W_1 \to W_2$ by

$$B_1(\epsilon)(x, y, z) = \epsilon(x, y)z + \epsilon(xy, z) - \epsilon(x, yz) - x\epsilon(y, z).$$

Then, as one can check directly, $B_1 \circ B_0 = 0$, and our candidate for the tangent space of $\mathcal{M}$—the 'infinitesimal deformations of the product modulo equivalence'—is the quotient ker $B_1/\text{Im } B_0$. In any event, we have been led to this definite mathematical object associated to the algebra M_0. As the reader can check, there is a natural way to extend the definitions to define $B_p : W_p \to W_{p+1}$ for all p, making a cochain complex. This gives cohomology groups H^p associated to the algebra, and our conclusion is that the elements of H^1 should be interpreted as infinitesimal deformations. (In the same way, elements of H^0 can be interpreted as infinitesimal automorphisms of the algebra.)

Leaving this digression, we return to our real concern: the deformations of a compact Riemann surface Σ. We approach this in three ways.

Approach through almost-complex structures

We have seen that the moduli space $\mathcal{M}$ is the quotient of the infinite-dimensional space of almost-complex structures $\mathcal{J}$ by the infinite-dimensional diffeomorphism group $\text{Diff}(\Sigma)$. Our description via Beltrami differentials identifies $\mathcal{J}$ with an open subset of $\Omega^{0,1}(T\Sigma)$. To understand the tangent space of the moduli space, we want to understand how 'infinitesimal diffeomorphisms' act on almost-complex structures. The obvious candidate for the infinitesimal diffeomorphisms is provided by the space of vector fields on Σ, smooth sections of the tangent bundle. For example, if we have a smooth family of diffeomorphisms $f_t : \Sigma \to \Sigma$ parametrised by $t \in \mathbf{R}$ and with f_0 the identity, then the derivative with respect to t of f_t at $t = 0$ is a vector field v on Σ. Now these diffeomorphisms act on the almost-complex structures, and what we want to identify is

$$L_v = \frac{d}{dt}(f_t^*(J_0))|_{t=0}, \tag{13.3}$$

where J_0 is a given almost-complex structure on Σ. Further, we want to identify L_v as an element of $\Omega^{0,1}(T)$. Now, in Chapter 12 we saw that, as for any holomorphic line bundle, there is an intrinsically defined operator

$$\bar{\partial} : \Omega^0(T) \to \Omega^{0,1}(T),$$

and the fact we need is the following.

Lemma 37. *In the above situation, $L_v = \bar{\partial}(v)$.*

Given this, we expect that the tangent space to the moduli space should be the quotient

$$\Omega^{0,1}(T)/(\operatorname{Im} \bar{\partial}),$$

which is just the Dolbeault description of $H^1(T\Sigma)$.

To establish equation (13.3), we assume that the reader is acquainted with the notion of the *Lie derivative*. If v is a smooth vector field and τ is a tensor of any kind (i.e. a section of a bundle on which a diffeomorphism acts naturally), we have a Lie derivative $\mathcal{L}_v\tau$ which is a tensor of the same kind. To be more precise, τ is a section of a bundle R, say, and for each point p there is an action of the general linear group of the tangent space at p on the fibre of R. The derivative of this action is a bundle map p from End $TS \otimes R$ to R. We need the general formula, for a function f,

$$\mathcal{L}_{fv}\tau = f\mathcal{L}_v(\tau) + p(df \otimes v \otimes \tau).$$

By definition, L_v is exactly the Lie derivative $\mathcal{L}_v J_0$. We work in a local complex co-ordinate $z = x_1 + ix_2$ and write $\partial_i = \partial/\partial x_i$. Clearly, $\mathcal{L}_{\partial_i}(J_0) = 0$, so if $v = f_1\partial_1 + f_2\partial_2$ our formula gives

$$\mathcal{L}_v(J_0) = p(A \otimes J_0),$$

where A is the endomorphism $\sum_{ij} f_{i,j}\, dx_j \otimes \partial_i$, and we write $f_{i,j}$ for the partial derivatives. The map p is given by the bracket: $p(A \otimes J_0) = [A, J_0]$. This is

$$\begin{pmatrix} f_{1,1} & f_{2,1} \\ f_{1,2} & f_{2,2} \end{pmatrix} \begin{pmatrix} 0 & 1 \\ -1 & 0 \end{pmatrix} - \begin{pmatrix} 0 & 1 \\ -1 & 0 \end{pmatrix} \begin{pmatrix} f_{1,1} & f_{2,1} \\ f_{1,2} & f_{2,2} \end{pmatrix},$$

which is

$$\begin{pmatrix} f_{1,2} - f_{2,1} & f_{1,1} - f_{2,2} \\ f_{1,1} - f_{2,2} & f_{1,2} - f_{2,1} \end{pmatrix}.$$

Now, looking at the formula (13.3), we see that for small $\mu = \mu_1 + i\mu_2$,

$$J_\mu = J_0 + 2 \begin{pmatrix} \mu_2 & -\mu_1 \\ -\mu_1 & -\mu_2 \end{pmatrix} + O(\mu^2).$$

So, to interpret $\mathcal{L}_v(J_0)$ as a Beltrami differential, we need to take

$$2\mu_1 = f_{1,1} - f_{2,2}, \quad 2\mu_2 = f_{1,2} - f_{2,1}.$$

On the other hand, shifting to complex notation, the vector field v is the real part of $F(\partial/\partial z)$ where $F = f_1 + if_2$. Then

$$\bar{\partial}(v) = \frac{\partial F}{\partial \bar{z}} d\bar{z} \otimes \frac{\partial}{\partial z}$$

Now the assertion is just the formula that $\partial F/\partial \bar{z} = \frac{1}{2}(\partial F/\partial x_1 + i\,\partial F/\partial x_2)$.

Of course, there are more invariant ways of presenting the calculation. Note too that it is clear that if $\bar{\partial}v = 0$ then $L_v = 0$, for in that case v is a holomorphic vector field and it generates (locally) a holomorphic flow on Σ which preserves J_0. This observation can be developed into a proof of the formula. The main point of going through the proof above, in local co-ordinates, is to illustrate explicitly how the definitions work.

Approach through coordinate charts

We go back to the definition of the Riemann surface Σ by an atlas of co-ordinate charts $\chi_\alpha : U_\alpha \to \tilde{U}_\alpha$, where $U_\alpha \subset \Sigma$. The structure is encoded in the collection of open sets $\tilde{U}_\alpha \subset \mathbf{C}$ and compatible holomorphic maps $f_{\alpha\beta} : \tilde{U}_{\alpha\beta} \to \tilde{U}_{\beta\alpha}$ for the appropriate $\tilde{U}_{\alpha\beta} \subset \tilde{U}_\alpha$. We expect that small deformations of Σ can be realised by retaining the same sets $\tilde{U}_\alpha$ and deforming the $f_{\alpha\beta}$. The exact domain of the $f_{\alpha\beta}$ may also change slightly, but we will not try to keep track of this in the notation. The key point is that if $f_{\alpha\beta}^{(t)}$ is a deformation, parametrised by $t \in \mathbf{R}$ and with $f_{\alpha\beta}^{(0)} = f_{\alpha\beta}$, then the derivative of $f_{\alpha\beta}^{(t)}$ with respect to t, evaluated at $t = 0$, is naturally a holomorphic vector field on $\tilde{U}_{\beta\alpha}$ and this can also be considered as a vector field on $U_\alpha \cap U_\beta \subset \Sigma$. Thus what we get is a Cech 1-cochain $\underline{v} = (v_\alpha)$ for the sheaf of holomorphic vector fields, with respect to the covering $\{U_\alpha\}$.

Not all deformations $f_{\alpha\beta}^{(t)}$ of the transition maps correspond to Riemann surfaces. Those that do are just the ones that satisfy the compatibility conditions

$$f_{\alpha\beta}^{(t)} \circ f_{\beta\gamma}^{(t)} = f_{\alpha\gamma}^{(t)} \tag{13.4}$$

on triple intersections. On the other hand, we get some deformations $f_{\alpha\beta}^{(t)}$ by deforming the charts on the same Riemann surface. The derivative of such a deformation of the charts yields a collection of holomorphic vector fields on the U_α; that is to say, a Čech 0-cochain. If we differentiate the compatibility identity (13.4) with respect to t and then set $t = 0$, the equation which results is just the assertion that $\underline{v}$ is a *cocycle* in the Čech complex. The deformations $\underline{v}$ which arise from trivial deformations are just those which are coboundaries in this complex. In sum, this approach also indicates that the space of infinitesimal deformations of Σ is the cohomology $H^1(T)$, but now appearing through the Čech description.

Abstract approach

Here we think of a Riemann surface structure as being a sheaf of rings $\mathcal{O}_\Sigma$. Suppose R is a commutative ring and t is an element of R with $t^2 = 0$. Then t maps to zero in the quotient ring R/tR of R by the ideal tR. We define an infinitesimal deformation of Σ to consist of

- a sheaf S of rings on Σ,
- a global section $t \in \Gamma(\Sigma, S)$ with $t^2 = 0$,
- an isomorphism of sheaves of rings $\alpha : S/(tS) \to \mathcal{O}_\Sigma$.

We say that two infinitesimal deformations S, S' are equivalent if there is an isomorphism of sheaves of rings from S to S' which takes t to t' and is compatible with the isomorphisms α, α'.

Now there is a trivial deformation $\mathcal{O}_\Sigma \oplus \mathcal{O}_\Sigma t$, with the obvious ring structure. We make it part of our definition that the deformations we consider are locally equivalent to this one.

To motivate this definition, consider a genuine deformation of Σ in the form of a complex surface $\mathcal{X}$ and a holomorphic map $\pi : \mathcal{X} \to \Delta$, where Δ is the unit disc in C such that π is a differentiable fibration and $\Sigma = \pi^{-1}(0)$. Then we have a sheaf of rings $\hat{S}$ on Σ given by the restriction of the holomorphic functions on $\mathcal{X}$. This contains an element t, lifting the standard co-ordinate function on C. Locally, around a point of Σ, we can trivialise the fibration holomorphically and we can represent an element of $\hat{S}$ in the form $\sum_{i=0}^{\infty} f_i t^i$, where the f_i are local holomorphic functions on Σ. But, of course, this representation depends on the trivialisation and is not canonical. Now, if we take the quotient $\hat{S}/t^2\hat{S}$, we get a sheaf representing an infinitesimal deformation as defined above. What this amounts to is considering functions on $\mathcal{X}$ up to first order in the transverse co-ordinate t.

Now we have an abstract sheaf-theoretic fact, expressed in the following proposition.

Proposition 41. *The equivalence classes of infinitesimal deformations of Σ are in natural 1–1 correspondence with classes in $H^1(T\Sigma)$.*

To make the correspondence, given an infinitesimal deformation S, one considers the sheaf $\mathcal{D}$ of *derivations* on S and shows that this is a rank-2 locally free sheaf of $\mathcal{O}_\Sigma$ modules over Σ. For a genuine deformation, this sheaf is the sheaf of sections of the restriction of the tangent bundle of $\mathcal{X}$ to Σ. Moreover, there is an exact sequence of sheaves of $\mathcal{O}_\Sigma$-modules

$$0 \to T\Sigma \to \mathcal{D} \to \mathcal{O}_\Sigma \to 0,$$

corresponding, for a genuine deformation, to the maps induced by inclusion of the fibre in $\mathcal{X}$ and the projection to C. Now, take the long exact sequence in cohomology to get a boundary map

$$H^0(\mathcal{O}_\Sigma) \to H^1(T\Sigma),$$

and the image of the element $1 \in H^0(\mathcal{O}_\Sigma)$ gives our deformation class in $H^1(T\Sigma)$.

Studying deformations without the language of cohomology is more awkward. By Serre duality, the dual of the deformation space can be identified with $H^0(T^* \otimes T^*)$—the space of holomorphic quadratic differentials—which can thus be be viewed (roughly) as the *cotangent* space of the moduli space. To give an illustration of how this deformation theory works, suppose that Σ admits a branched cover $f : \Sigma \to \mathbf{CP}^1$ with simple branch points mapping to points $p_i \in \mathbf{CP}^1$. Suppose we have tangent vectors v_i of $\mathbf{CP}^1$ at p_i. We can deform Σ, along with the branch covering, by moving the branch points in the directions v_i. So this should give us a deformation class $b \in H^1(T\Sigma)$. Now let α be a holomorphic quadratic differential on Σ. We pair b with α to get a number $\langle b, \alpha \rangle$. The question is how to express this in terms of the input v_i.

Let $q_i \in \Sigma$ be the branch point lying over p_i. The derivative of f at q_i vanishes by hypothesis, but (since the branch point is simple) the second derivative induces an isomorphism from $(T\Sigma \otimes T\Sigma)_{q_i}$ to the tangent space of $\mathbf{CP}^1$ at p_i. Now, using this isomorphism, we can regard v_i as an element in $(T\Sigma \otimes T\Sigma)_{q_i}$. This is the dual of the space in which the value of α at q_i lies, so there is a pairing $\langle v_i, \alpha \rangle \in$ C. Of course, the desired formula is

$$\langle b, \alpha \rangle = \sum_i \langle v_i, \alpha \rangle.$$

The proof of this, using one of the approaches above, is a good exercise for the reader.

We should clarify one point before finishing this chapter. We have seen that $H^1(T\Sigma)$ ought to be the tangent space to the moduli space at $[\Sigma]$, which would tacitly imply that the moduli space is modelled on a neighbourhood of 0 in $H^1(T\Sigma)$. This turns out to be true with one reservation. Suppose the automorphism group G of Σ is non-trivial. Then G acts linearly on $H^1(T\Sigma)$, and the correct statement, which we will essentially see in Section 14.3 below, is that a neighbourhood in the moduli space is modelled on a neighbourhood of 0 in the quotient of $H^1(T\Sigma)$ by G. Again this is a general phenomenon; for example, similar remarks apply in the case of associative algebras discussed above.

13.3 Appendix

Recall the problem. We have a smooth function μ defined on a neighbourhood B of 0 in $\mathbf{C}$, with $|\mu| < 1$ at each point of B, and we want to find a solution of the equation $f_{\bar{z}} + \mu f_z = 0$ in a possibly smaller neighbourhood of 0 with $f_z(0)$ not equal to 0.

We begin with some elementary observations which will simplify the problem.

Observation 1. *We can suppose that $\mu(0) = 0$.*

For this, we just make an **R**-linear change of co-ordinates.

Observation 2. *We can suppose that B is the unit disc, and for any ϵ we can suppose that μ and its derivatives $\mu_z, \mu_{\bar{z}}$ are bounded in modulus by ϵ over the disc.*

For this, we just make a dilation $z = \epsilon\tilde{z}$ which transforms μ to $\tilde{\mu}(\tilde{z}) = \mu(\epsilon\tilde{z})$.

Observation 3. *In addition, we can suppose that μ has compact support.*

For this, we multiply by a cut-off function equal to 1 on the disc of radius $1/2$. We do not change the problem, because it suffices to find a solution on a subset of this smaller disc. With adjustment of constants, we can still suppose that μ and its derivatives are as small as we please.

So now we have a smooth μ defined on all of $\mathbf{C}$ but supported in the unit disc, and we can suppose that μ and its derivatives are very small. We seek a solution f to the equation $f_{\bar{z}} + \mu f_z = 0$ on the whole of $\mathbf{C}$. We write $f = z + \phi$, so we need

$$\phi_{\bar{z}} + \mu\phi_z = -\mu. \tag{13.5}$$

Thus ϕ will be holomorphic outside the unit disc. If we can achieve

$$|\phi_z(0)| < 1, \tag{13.6}$$

then $f_z(0)$ will not vanish, and we will be done.

To solve the partial differential equation (13.5), we introduce an integral operator

$$(Tu)(z) = \frac{1}{2\pi} \int_{\mathbf{C}} \frac{u(w)}{z - w} d\mu_w. \tag{13.7}$$

Here the notation is something of a mix of real and complex variables: $d\mu_w$ is supposed to indicate the ordinary Lebesgue measure on the plane but with respect to w. The integrand is singular, but there is no difficulty in defining the integral if u is in C_c^∞, say—the space of smooth functions of compact support. Likewise, one can show that in such a situation Tu is a smooth function. The crucial property is that, like the Newton potential for the Poisson equation, this integral operator inverts the differential operator $\partial/\partial\bar{z}$.

Proposition 42. *For $u \in C_c^\infty$, we have $\partial Tu/\partial\bar{z} = u$.*

This was Exercise 1 in Chapter 9. See also the proof of Lemma 38 below.

Given this, we seek our solution in the form $\phi = Tv$, and the equation becomes

$$v - \mu S v = -\mu,$$

where we define

$$Sv = -\frac{\partial Tv}{\partial z}.$$

The basic idea now is that if μS is a small operator in a suitable sense, we should be able to regard $1 - \mu S$ as a small perturbation of the identity and, when μ is small, find a small solution v. More explicitly, we try to construct a solution as a Neumann series

$$v = v_0 + v_1 + v_2 + \cdots, \tag{13.8}$$

where $v_0 = -\mu$ and $v_i = \mu S(v_{i-1})$ for $i > 0$. Notice that this recipe does define a sequence of smooth functions v_i of compact support (since μ has compact support), so the issue is the convergence of the sum (13.8).

The foundation for this analysis is an estimate for the operator S on Hölder spaces. Fix $\alpha \in (0, 1)$ and, for a function ψ on $\mathbf{C}$, define the Hölder (semi)-norm

$$[\psi]_\alpha = \sup_{x \neq y} \frac{|\psi(x) - \psi(y)|}{|x - y|^\alpha},$$

allowing the value $+\infty$ in the obvious way. Certainly, $[\psi]_\alpha$ is finite for $\psi \in C_c^\infty$ and, indeed, $[\,]_\alpha$ is a norm on this space.

Theorem 24. *There is a constant K_α such that for any $u \in C_c^\infty$ we have*

$$[Su]_\alpha \leq K_\alpha [u]_\alpha.$$

This result, which is a particular case of the 'Schauder estimates', is almost saying that S is a bounded operator with respect to the norm $[\,]_\alpha$, but not quite, because Su need not have compact support and so does not (in an obvious way) lie in a space on which $[\,]_\alpha$ is a norm.

If we differentiate equation (13.7) formally inside the integral sign, we are led to write down an expression

$$(Su)(z) = \frac{1}{2\pi i} \int_C \frac{u(w)}{(w-z)^2} d\mu_w. \tag{13.9}$$

But this formula does *not* make sense on the face of it, because, unlike equation (13.7), the expression inside the integral will not generally be an integrable function. One can make sense of this as a *singular integral operator* Stein (1970), but we will not need to make explicit use of that theory. Instead, we begin with the following lemma.

Lemma 38. *Suppose $u \in C_c^\infty$ and $u(z) = 0$ for some given $z \in \mathbf{C}$. Then $u(w)/(w-z)^2$ is an integrable function of w, and the formula (13.9) is true at this point z.*

Without loss of generality, we can suppose $z = 0$. Since $u(0) = 0$ and u is smooth, we have $|u(w)| \le C|w|$ for some C, and this means that $u(w)/w^2$ is integrable near zero. Since u has compact support, this function is integrable on the whole of $\mathbf{C}$.

Now we can write

$$(Tu)(z) - (Tu)(0) = \frac{1}{2\pi i} \int_C \frac{u(w)}{w-z} - \frac{u(w)}{w} \, d\mu_w.$$

Making a change of variable in one term, this is the same as

$$(Tu)(z) - (Tu)(0) = \frac{1}{2\pi i} \int_C \frac{u(w+z) - u(w)}{w} \, d\mu_w.$$

It follows from this that

$$\left(\frac{\partial Tu}{\partial z}\right)(0) = \frac{1}{2\pi i} \int_C \frac{\partial u}{\partial w} \frac{1}{w} d\mu_w. \tag{13.10}$$

Now we write I_ϵ for the integral of the function appearing on the right-hand side of equation (13.10) over the complement in $\mathbf{C}$ of the disc of radius ϵ, so the right-hand side of equation (13.10) is the limit of I_ϵ as ϵ tends to 0. We can apply Stokes' Theorem to write

$$I_\epsilon = \frac{1}{2\pi i} \left(\int_{|w| \ge \epsilon} \frac{u(w)}{w^2} d\mu_w + \int_{|w| = \epsilon} \frac{u}{w} dw \right).$$

The fact that $u(0) = 0$ implies that the second term tends to zero with ϵ, and we obtain our result.

Now we can begin the proof of Theorem 24. A crucial point, which the reader should verify, is that the operator S is 'scale-invariant'—it commutes with the action on functions arising from maps $z \mapsto \lambda z$ of $\mathbf{C}$. This means that it suffices to establish the bound for $|(Su)(z_1) - (Su)(z_2)|$ when $|z_1 - z_2| = 1$. Since the problem is obviously invariant under rotation and translations, we can suppose that $z_1 = 1, z_2 = 0$. So what we need to show is that

$$|(Su)(1) - (Su)(0)| \le K_\alpha [u]_\alpha \qquad (13.11)$$

for some fixed K_α and all $u \in C_c^\infty$.

Now we first show the following lemma.

Lemma 39. *Equation (13.11) is true in the case when $u(0) = u(1) = 0$.*

Let Δ_0, Δ_1 be the discs of radii $1/2$ centred on $0,1$, respectively, and let $\Omega = \mathbb{C} \setminus (\Delta_0 \cup \Delta_1)$. Then, applying the previous result,

$$2\pi i (Su)(1) = \int_{\Delta_0} \frac{u}{(w-1)^2} d\mu_w + \int_{\Delta_1} \frac{u}{(w-1)^2} d\mu_w + \int_\Omega \frac{u}{(w-1)^2} d\mu_w$$

$$= I(1, \Delta_0) + I(1, \Delta_1) + I(1, \Omega),$$

say. Similarly,

$$2\pi i (Su)(0) = \int_{\Delta_0} \frac{u}{w)^2} d\mu_w + \int_{\Delta_1} \frac{u}{w}^2 d\mu_w + \int_\Omega \frac{u}{w^2} d\mu_w$$

$$= I(0, \Delta_0) + I(0, \Delta_1) + I(0, \Omega).$$

Now, since $u(0) = 0$, we have $|u(w)| \le [u]_\alpha |w|^\alpha$. This means that

$$|I(0, \Delta_0)| \le 2\pi [u]_\alpha \int_0^{1/2} r^{\alpha-1}\, dr = 2\pi [u]_\alpha \alpha^{-1} 2^{-\alpha}.$$

By symmetry, we have the same estimate for $I(1, \Delta_1)$ and even more obvious estimates for $I(0, \Delta_1), I(1, \Delta_0)$. So it suffices to prove that $|I(1, \Omega) - I(2, \Omega)| \le K'[u]_\alpha$. But

$$|I(1, \Omega) - I(2, \Omega)| \le \int_\Omega |u(w)| \left| \frac{1}{w^2} - \frac{1}{(w-1)^2} \right| d\mu_w.$$

On Ω, we have a $|w - 1| \ge (1/3)|w|$ and $|w| \ge 1/2$, and these give

$$\left| \frac{1}{w^2} - \frac{1}{(w-1)^2} \right| = \left| \frac{2w-1}{w(w-1)} \right| \le 12\, |w|^{-3},$$

It follows that

$$|I(1, \Omega) - I(2, \Omega)| \le 12 \times 2\pi [u]_\alpha \int_{1/2}^\infty r^{\alpha-2}\, dr = 24\pi [u]_\alpha (1 - \alpha)^{-1} 2^{\alpha-1}.$$

To finish the proof of Theorem 24, we need to give an argument to extend the above result to functions which do not vanish at 0 and 1. Obviously, it suffices to prove that

$$|Su(1) - Su(0)| \le K$$

for $u \in C_c^\infty$ with $[u]_\alpha = 1$. If $g \in C_c^\infty$, an argument like that in the proof of Lemma 38 above shows that

$$T\left(\frac{\partial g}{\partial \bar{z}}\right) = g.$$

Choose a function $g_0 \in C_c^\infty$ supported in the unit disc and equal to $\bar{z}$ on the 1/2-disc Δ_0. Set $u_0 = \partial g_0/\partial \bar{z}$. Then $u_0 \in C_c^\infty$ is equal to 1 on Δ_1 and is supported in the unit disc and we have $Su_0 = \partial g_0/\partial z$, so Su_0 is in C_c^∞ and $[Su_0]_\alpha$ is finite. Let

$$C = \max\left([u_0]_\alpha, [Su_0]_\alpha\right).$$

Introduce a parameter $\lambda \geq 0$ and define

$$u_{0,\lambda}(z) = u_0(\lambda^{-1}z).$$

Then, from the scaling behaviour, we have

$$[u_{0,\lambda}]_\alpha, [Su_{0,\lambda}]_\alpha \leq C\lambda^{-\alpha},$$

and $u_{0,\lambda}$ is equal to 1 on the disc of radius $\lambda/2$ about 0. Let $u_1(z) = u_0(z-1)$, so, by translation invariance,

$$[u_1]_\alpha, [Su_1]_\alpha \leq C,$$

and $u_1(1) = 1, u_1(0) = 0$. Now write

$$u = \tilde{u} + u(0)u_{0,\lambda} + (u(1) - u(0))u_1.$$

Choose $\lambda > 2$ so that $u_{0,\lambda}(1) = 1$. This implies that $\tilde{u}$ vanishes at $0, 1$. We have

$$|Su(1) - Su(0)| \leq |S\tilde{u}(0)\ -S\tilde{u}| + |u(0)|\,[Su_{0,\lambda}]_\alpha + |u(1) - u(0)|\,[Su_1]_\alpha$$

$$\leq |S\tilde{u}(0) - S\tilde{u}(1)| + C\,|u(0)|\,\lambda^{-\alpha} + C\,|u(1) - u(0)|.$$

Applying equation (13.11) to $\tilde{u}$ and having in mind that we are assuming $[u]_\alpha = 1$, we get

$$|Su(1) - Su(0)| \leq K'[\tilde{u}]_\alpha + C\,|u(0)|\,\lambda^{-\alpha} + C.$$

Now $\tilde{u} = u - u(0)u_{0,\lambda} - (u(1) - u(0))u_1$, and so

$$[\tilde{u}]_\alpha \leq [u]_\alpha + |u(0)|\,[u_{0,\lambda}]_\alpha + [u_1]_\alpha \leq 1 + C\lambda^{-\alpha}\,|u(0)| + C.$$

Putting this together, we get

$$|Su(0) - Su(1)| \leq K' + |u(0)|\,(K' + C)\lambda^{-\alpha} + C + 1.$$

The point is now that although we have no control of $|u(0)|$, *we can make λ as large as we please.* So we get $|Su(1) - Su(0)| \leq K$ for any $K > K' + C + 1$. This completes the proof of Theorem 24.

Armed with this result, we go back to our main problem. Recall that the v_i are supported in the unit disc D and $v_{i+1} = \mu S v_i$. Since v_i vanishes at the point $2 \in \mathbf{C}$, we can apply the formula (13.9),

$$(S v_i)(2) = \frac{1}{2\pi i} \int_{\mathbf{C}} \frac{v_i(w)}{(w-2)^2} d\mu_w,$$

and $|w - 2| \geq 1$ on the support of v_i, so clearly

$$|S(v_i)(2)| \leq c \, \|v_i\|_\infty,$$

where $\| \ \|_\infty$ is the usual L^∞ norm. On the other hand, since $|w - 2| \leq 3$ for w in D, the definition of $[\ ]_\alpha$ gives

$$\sup_D |S v_i| \leq c \, \|v_i\|_\infty + 3^\alpha [S v_i]_\alpha \leq c' \|v_i\|_\infty + K[v_i]_\alpha,$$

by equation (13.11). We can suppose that $\|\mu\|_\infty, [\mu]_\alpha \leq \epsilon$, where ϵ is as small as we please. We get

$$\|v_{i+1}\|_\infty \leq \epsilon(c' \, \|v_i\|_\infty + K[v_i]_\alpha).$$

Similarly,

$$[v_{i+1}]_\alpha \leq [\mu]_\alpha \left(\sup_D |S v_i| \right) + \|\mu\|_\infty [S v_i]_\alpha \leq \epsilon(c' \, \|v_i\|_\infty + 2K[v_i]_\alpha).$$

So, in sum, if we define $\|v_i\|_{,\alpha} = \|v_i\|_\infty + [v_i]_\alpha$, we have

$$\|v_{i+1}\|_{,\alpha} \leq \epsilon c'' \, \|v_i\|_{,\alpha}$$

for a computable constant c'', depending on α. When $\epsilon < 1/c''$, we see that the sum $\sum v_i$ converges in the Hölder norm $\| \ \|_{,\alpha}$. It follows that $\sum T v_i$ and its derivatives converge in $\| \ \|_{,\alpha}$ (for $\partial T v_i / \partial \bar{z} = v_i$ and $\partial T v_i / \partial z = S v_i$). Thus we get a solution ϕ of our equation (13.5) which is a differentiable function with Hölder continuous derivatives. Moreover, we have an estimate $|\phi_{\bar{z}}(0)| \leq c''' \epsilon$, so if ϵ is small we can arrange that $|\phi_{\bar{z}}| < 1$.

This solves our problem, except for the proof of the fact that ϕ is actually a *smooth* function. We will only sketch this proof. Let us prove first that we can choose a ϕ such that the *second* partial derivatives of ϕ are in $C^{,\alpha}$. To do this, we want to consider the action of μS on $C^{1,\alpha}$, the functions supported on the disc with norm $\|v\|_{1,\alpha} = \|v\|_\infty + [v]_\alpha + [\nabla v]_\alpha$. Since the derivative ∇ commutes with S, we find that $\mu S : C^{1,\alpha} \to C^{1,\alpha}$ has a small operator norm if the derivatives of μ are suitably small. Then we can perform the whole construction in this more restricted space to obtain a $C^{2,\alpha}$ solution. The same applies if we differentiate any number of times. So we conclude that for each $k \geq 1$ there is a $C^{k,\alpha}$ solution f_k, say, to our problem, defined on a small disc about 0. Now the composite map $f_k \circ f_1^{-1}$ is a holomorphic map, in a suitable neighbourhood of the origin with respect to the standard complex structure.

So we know that this is actually smooth. It follows that f_1 is actually in C^k for all k.

One can also establish smoothness by standard partial-differential-equation arguments. In fact, for many of our purposes we do not need this smoothness. Once we have these $C^{1,\alpha}$ solutions to equation (13.5), we can show easily that μ defines a Riemann surface structure, using the fact that $C^{1,\alpha}$ solutions of the Cauchy–Riemann equations are smooth. Establishing the smoothness of solutions to equation (13.5) is the same as showing that the smooth structure underlying this Riemann surface is the same as the smooth structure we started with.

Exercises

1. Use the Riemann–Roch Theorem to show that for a compact Riemann surface Σ of genus $g \geq 2$, the dimension of $H^1(\Sigma; T\Sigma)$ is $3g - 3$.
2. Show that the L^∞ norm on Beltrami differentials and the L^1 norm on quadratic differentials are invariantly defined.
3. Let $\underline{\mu}$ be a Beltrami differential on a compact Riemann surface Σ. Define a sheaf over Σ by pairs (f_0, f_1), where $\bar{\partial} f_0 = 0$ and $\bar{\partial} f_1 = \mu\, \partial f_0$. Show that this defines an infinitesimal deformation of Σ, as considered in the 'abstract approach' above.
4. Suppose $\underline{\mu}$ is a *real analytic* Beltrami differential. Give another proof of the existence of local solutions of the equation $(\bar{\partial} + \mu\partial) f = 0$. (It is easiest to write μ locally as a power series in $z, \bar{z}$.)
5. Formulate, and prove, a 'Schauder estimate' for the Laplace operator on $\mathbf{R}^n$.
6. Let T be the integral operator considered in equation (13.7) and let $p > 2$. Show that there is a constant C_p such that for all compactly supported u,

$$[Tu]_\alpha \leq C_p \,\|u\|_{L^p},$$

where $\alpha = (p - 2)/2$.

14 Mappings and moduli

14.1 Diffeomorphisms of the plane

We apply the ideas of the previous chapter to prove a topological result here. A diffeomorphism $f : \mathbf{R}^2 \to \mathbf{R}^2$ is said to have *compact support* if it is equal to the identity outside some compact set. Such diffeomorphisms form a group $\mathrm{Diff}_c(\mathbf{R}^2)$ which has a natural topology (a sequence of diffeomorphisms converges if they converge in C^r on compact subsets for all r).

Theorem 25. $\mathrm{Diff}_c(\mathbf{R}^2)$ *is weakly contractible.*

Recall that a space X is called weakly contractible if all its homotopy groups vanish, i.e. for all k, any map of a k-sphere to X can be extended to the $(k + 1)$-ball.

To prove this theorem, we identify $\mathbf{R}^2$ with $\mathbf{C}$ and introduce another group Diff^* of diffeomorphisms $f : \mathbf{C} \to \mathbf{C}$ which have the form $f(z) = z + \psi(z)$, where $\psi(z)$ is holomorphic outside a compact set and $\psi(z) = O(z^{-1})$; that is to say, the Laurent expansion of ψ is $\sum_{r \geq 1} a_r z^{-r}$. Now we will establish the following proposition.

Proposition 43. Diff^* *is contractible.*

Suppose $f \in \mathrm{Diff}^*$. Then, by the discussion in the previous chapter, we have

$$|f_{\bar{z}}| < |f_z|$$

at each point. We define $\mu = -f_{\bar{z}}/f_z$, so $|\mu| < 1$ and μ has compact support (it vanishes outside a compact set in $\mathbf{C}$). By construction, f is a solution of the Beltrami equation

$$\frac{\partial f}{\partial \bar{z}} + \mu \frac{\partial f}{\partial z} = 0. \tag{14.1}$$

Let $\mathcal{B}$ be the set of smooth complex-valued functions μ on $\mathbf{C}$ of compact suppport with $|\mu| < 1$ everywhere. Clearly $\mathcal{B}$ is contractible. The above construction defines a map $\Phi : \mathrm{Diff}^* \to \mathcal{B}$. What we want to show is that Φ is a homeomorphism: this will prove the proposition. Suppose now that we have any $\mu \in \mathcal{B}$. This defines an almost-complex structure on $\mathbf{C}$, equal to the standard one outside a compact set. By Theorem 22, we know that this yields a new Riemann surface structure on $\mathbf{C}$, which clearly extends to a structure, $(\mathbf{C} \cup \{\infty\})_\mu$ say, on the sphere. From Chapter 8, we know that this 'new' Riemann surface structure is equivalent to the standard one, so there is a holomorphic diffeomorphism $f : (\mathbf{C} \cup \{\infty\})_\mu \to \mathbf{C} \cup \{\infty\}$ from the new structure to the standard one. By composing with a Möbius map, we can suppose that $f(\infty) = \infty$. Since μ has compact support, f is holomorphic in the ordinary sense outside a compact set, and the condition at ∞ is equivalent to the statement that $f(z) = az + b + \psi(z)$, where ψ is holomorphic outside a compact set, $\psi(z) = O(z^{-1})$ and a is not zero. Replacing f by $a^{-1}(f - b)$, we can suppose that $a = 1$ and $b = 0$. So we conclude that for any $\mu \in \mathcal{B}$, there is a solution f of equation (14.1) with $f = z + \psi(z)$, ψ holomorphic outside a compact set and $\psi = O(z^{-1})$. That is to say, f lies in Diff*. Moreover, the same argument (the classification of automorphisms of the Riemann sphere) shows that this f is unique, so we see that $\Phi : \text{Diff}^* \to \mathcal{B}$ is a bijection. We leave aside the issue of showing that Φ and its inverse are continuous with respect to appropriate topologies on Diff*, $\mathcal{B}$.

Now we need to relate the topologies of the groups Diff* and Diff_c. Let $\beta(t)$ be a standard cut-off function, equal to 1 when $t < 1$ and vanishing when $t > 2$, and introduce a real parameter $R > 0$. Given $f = z + \psi(z) \in \text{Diff}^*$, we define a complex-valued function by

$$T_R(f)(z) = z + \beta(R^{-1}|z|)\psi(z).$$

Then $T_R(f)$ equals f inside the R-ball and $T_R(f)$ is the identity outside the $2R$-ball. For a given f, we choose R large enough that ψ is holomorphic in the R-ball. We have $\psi = O(R^{-1})$, $\nabla\psi = O(R^{-2})$ on the annulus $\{R \leq |z| \leq 2R\}$. It follows easily that on this annulus we have

$$\left| \frac{\partial T_R(f)}{\partial \bar{z}} \right| = O(R^{-2})$$

and

$$\left| \frac{\partial T_R(f)}{\partial z} - 1 \right| = O(R^{-2}).$$

Thus, for fixed f, we can choose R large enough so that $T_R(f)$ lies in $\text{Diff}_c(\mathbf{R}^2)$. The same holds if f ranges over any given compact subset of Diff*. Now suppose that K is a compact space and we have a map $\alpha : K \to \text{Diff}^*$. We choose R large enough that the composite $T_R \circ \alpha$ maps K to $\text{Diff}_c(\mathbf{R}^2)$. If $J \subset K$ is a closed subset such that α maps J to $\text{Diff}_c(\mathbf{R}^2) \subset \text{Diff}^*$, we can also choose R large enough so that $T_R \circ \alpha = \alpha$ on J. Taking $K = B^{k+1}$ and $J = S^k$, we see that the contractibility of Diff* implies the weak contractibility of $\text{Diff}_c(\mathbf{R}^2)$.

It is certainly possible to prove this theorem without using complex analysis. But it cannot be proved using merely elementary formal properties of the situation—more precisely, the corresponding statement is *false* if we replace $\mathbf{R}^2$ by $\mathbf{R}^n$ for arbitrary n. In general, $\mathrm{Diff}_c(\mathbf{R}^n)$ is weak homotopy equivalent to the quotient $\mathrm{Diff}(S^n)/SO(n+1)$, where $\mathrm{Diff}(S^n)$ is the group of orientation-preserving diffeomorphisms of the sphere. For certain values of n (for example $n = 6$), there are diffeomorphisms $g : S^n \to S^n$ which cannot be deformed through diffeomorphisms to the identity. Hence, in this case, $\mathrm{Diff}_c(\mathbf{R}^n)$ has more than one component. Such diffeomorphisms g can be used to construct 'exotic $(n+1)$-spheres'—manifolds homeomorphic but not diffeomorphic to S^{n+1}. To do this, one glues two copies of the $(n+1)$-ball along their boundaries using the diffeomorphism g. The existence of these exotic spheres was a famous discovery of Milnor (1956).

14.2   Braids, Dehn twists and quadratic singularities

14.2.1   Classification of branched covers

Let $\Delta \subset \mathbf{C}$ be a finite set. Fix a base point $x_0 \in \mathbf{C} \setminus \Delta$. In Chapter 4, we saw how a representation $\rho : \pi_1(\mathbf{C} \setminus \Delta, x_0) \to S_d$ on the permutation group of d objects $\{1, 2, \ldots, d\}$ leads to a Riemann surface with a holomorphic map to $\mathbf{C}$ branched over the points of Δ. We consider the case where the monodromy around a large circle is trivial, which means that we get a compact Riemann surface $X(\Delta, \rho)$ mapping to the Riemann sphere, with no branch points at infinity. The d objects $\{1, \ldots, d\}$ are canonically identified with the pre-image $f^{-1}(x_0)$.

Now suppose that we have another system Δ', x_0', ρ' of such data, with the same value of d and the same number of branch points. Suppose that f is a compactly supported diffeomorphism of $\mathbf{C}$ mapping Δ to Δ' and x_0 to x_0'. Suppose also that near to points of Δ the map is holomorphic. We have an induced map $f_* : \pi_1(\mathbf{C} \setminus \Delta) \to \pi_1(\mathbf{C} \setminus \Delta', x_0')$. Suppose that there is a $\lambda \in S_d$ such that

$$\rho' \circ f_*(\gamma) = \lambda \rho(\gamma) \lambda^{-1},$$

for all $\gamma \in \pi_1(\mathbf{C} \setminus \Delta)$. Then we have the following proposition.

Proposition 44. *In the above situation, there is an induced diffeomorphism from $X(\Delta, \rho)$ to $X(\Delta', \rho')$ which is uniquely determined by the condition that it covers f on $\mathbf{C}$ and acts as λ on the pre-images of the base points.*

This is straightforward from the construction of the branched covers. The condition that f be holomorphic near the points of Δ will not be important, it is a technicality somewhat similar to that encountered in Section 14.1 (with the introduction of the group Diff*). Any diffeomorphism f can be deformed slightly in small discs around points of Δ to be holomorphic in interior discs, and the same holds for compact families of diffeomorphisms.

A first consequence of this proposition is that we may fix attention on any given standard set Δ_0, since it is clear that the diffeomorphisms of the above type act transitively on these sets. We take as a standard model a set of points on the real axis, and we take the base point x_0 to be $-Ri$ for large R. The fundamental group of the punctured plane is a free group on standard generators γ_i, as indicated in Figure 14.1.

From now on we restrict our attention to the case when the monodromy around each of the branch points is a transposition, so our holomorphic map has simple branch points. This implies that the number of points in Δ_0 is an even number, $2b$ say. A representation ρ is specified by giving the monodromies $\delta_i \in S_d$ around γ_i, and the condition on the monodromy at infinity is the requirement

$$\delta_1\delta_2\cdots\delta_{2b} = 1 \tag{14.2}$$

in S_d.

Now we take an algebraic viewpoint. Suppose we have any $2b$ transpositions $\delta_1, \delta_2, \ldots, \delta_{2b}$ in S_d satisfying the identity (14.2). We call this data a *factorisation* and denote it by $\underline{\delta}$. The δ_i generate a subgroup of S_d, which we call the monodromy group of the factorisation. If this acts transitively on $\{1, \ldots, d\}$, we say the factorisation is transitive. For each transitive factorisation $\underline{\delta}$, we

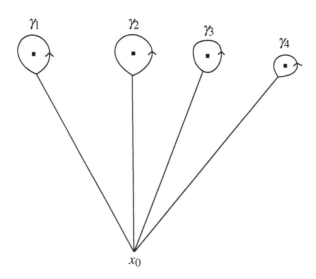

Figure 14.1 *Standard generators of* π_1

construct a branched cover as above, with a compact connected Riemann surface which we denote $X(\underline{\delta})$. Suppose that $b = (d - 1) + g$ for $g \geq 0$. Then there is a factorisation

$$(12)^{2g+1}(13)^2(14)^2 \cdots (1d)^2 = 1.$$

(Here we use notation in an obvious way, so we mean, for example, $\delta_i = (12)$ for $i \leq 2g + 1$.) It is easy to verify that this is transitive: we call it the *standard transitive factorisation* for given b, d.

Next we introduce an important object: the *braid group* Br_{2b} on $2b$ 'strands'. There are a number of different approaches: for us, the simplest is to define it to be the group of connected components of compactly supported diffeomorphisms of $\mathbf{C}$ which map Δ_0 to Δ_0. We can choose a representative in each connected component which fixes the base point (thinking of R as very large) and which is holomorphic near to Δ_0. Then, on the one hand, we get an induced map on $\pi_1(\mathbf{C} \setminus \Delta_0)$ which is independent of the representative. This gives an action of the braid group as automorphisms of $\pi_1(\mathbf{C} \setminus \Delta_0)$ and hence an action on representations, and so on factorisations. For any $g \in \mathrm{Br}$ and transitive factorisation $\underline{\delta}$, we have another $g(\underline{\delta})$. We can also conjugate a factorisation by an element of S_d. We say that two transitive factorisations $\underline{\delta}, \delta'$ are *equivalent* if δ' is conjugate to $g(\underline{\delta})$ for some $g \in \mathrm{Br}_{2b}$.

On the other hand, this representative defines a diffeomorphism of branched covers, showing in particular that if $\underline{\delta}, \delta'$ are equivalent, then $X(\underline{\delta}), X(\underline{\delta}')$ are diffeomorphic.

Now we will prove the following theorem.

Theorem 26. *If $\underline{\delta}$ is any transitive factorisation, we have $b \geq d - 1$, and $\underline{\delta}$ is equivalent to the standard transitive factorisation.*

This result, and the method of proof by 'Hurwitz moves', goes back to Hurwitz (1891). As a corollary of this, we get another proof of the classification of compact oriented surfaces, by the following route.

- It is easy to show that any compact surface S has a Riemannian metric.
- By Theorem 22, if S is oriented, we get an induced Riemann surface structure.
- From Chapter 8, there is a non-trivial meromorphic function on S representing it as a branched cover of the Riemann sphere.
- Using the Riemann–Roch formula, it is easy to show that this branched cover can be chosen to have simple branch points, so it corresponds to a factorisation as above (see Exercise 2 in Chapter 8).
- Theorem 26 shows that S is diffeomorphic to the surface constructed from the standard factorisation, and it is easy to check that this is (of course!) diffeomorphic to a standard surface of genus g.

This proof of the classification theorem is not completely different from the 'Morse theory' proof sketched in Chapter 2. Rather than choosing a real-valued Morse function and studying the changes in the topology of the fibres (the level sets) as we cross a critical value, we choose a meromorphic function and study the monodromy of the fibres as we move around branch points. However, we should emphasise that the content of Theorem 26 definitely goes beyond the topological classification of surfaces. *A priori*, there could be 'topologically different' ways of representing a given smooth surface as a branched cover of the Riemann sphere of a given degree, with simple branch points, and Theorem 26 asserts that this is in fact not the case.

Before proceeding with the proof, we shall say a bit more about the braid group. The definition we have given is rather abstract, but we will see that actually we only need to use very simple, explicit diffeomorphisms. We shall also describe other ways of thinking about the braid group. Let C_{2b} be the 'configuration space', the open subset of $\mathbf{C}^{2b}$ defined by the condition that all the components are distinct. Our standard set Δ_0 can be regarded as a base point in C_{2b}, and we have the following proposition.

Proposition 45. *The braid group Br_{2b} is naturally isomorphic to the fundamental group $\pi_1(C_{2b}, \Delta_0)$.*

This follows from the fact, proved in Theorem 25, that the space of compactly supported diffeomorphisms $\mathrm{Diff}_c(\mathbf{R}^2)$ is weakly contractible. We can identify C_{2b} with the quotient space $\mathrm{Diff}_c(\mathbf{R}^2)/\mathrm{Diff}_c(\mathbf{R}^2, \Delta_0)$, where $\mathrm{Diff}_c(\mathbf{R}^2, \Delta_0) \subset \mathrm{Diff}_c(\mathbf{R}^2)$ is the subgroup of maps preserving Δ_0. Now the statement follows from the exact sequence of homotopy groups

$$1 \to \pi_1(C_{2b}) \to \pi_0(\mathrm{Diff}_c(\mathbf{R}^2, \Delta_0)) \to 1.$$

More explicitly, suppose we are given a compactly supported diffeomorphism f mapping Δ_0 to Δ_0. We join f to the identity by a path f_t of diffeomorphisms and get a loop $f_t(\Delta_0)$ in the configuration space. This construction sets up an isomorphism between the groups, in fact the inverse of the connecting homomorphism above.

Now suppose we have an embedded curve α in the plane joining two points p, q of Δ_0 but otherwise disjoint from Δ_0. We can associate an element of the braid group—a 'half-twist'—to α (Figure 14.2). Thinking of the braid group as the fundamental group of the configuration space, this half-twist fixes all points of Δ_0 except p, q and interchanges p, q by an 'anticlockwise rotation around α'. Thinking of the braid group as the components of the diffeomorphism group, we choose a neighbourhood N of α disjoint from all other elements of Δ_0 such that there is a diffeomorphism ψ_N taking N to the standard disc D_4 of radius 4 and α to the interval $[-1, 1] \subset D_4$. Now we define a diffeomorphism $\tau : D_2 \to D_2$ by

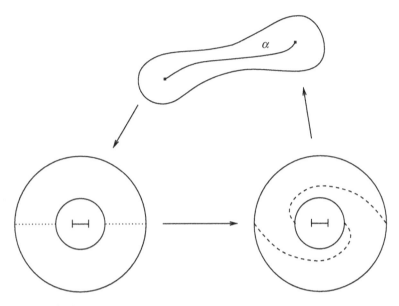

Figure 14.2 *A half-twist*

$$\tau(z) = e^{i\chi(|z|)}z,$$

where χ is a real-valued function with $\chi(r) = 0$ if $r \geq 3$ and $\chi(r) = \pi$ if $r \leq 2$.

(Notice the similarity between this definition and that of a Dehn twist.) Now our half-twist map $\tau_\alpha : \mathbf{C} \to \mathbf{C}$ is defined to be the composite $\psi_N^{-1} \circ \tau \circ \psi_N$. This is a diffeomorphism of $\mathbf{C}$ which is supported in N and interchanges p, q. Of course, one should really check that the corresponding element of the braid group is independent of the choices of N, ψ_N, χ. But, for our purposes, this is not really necessary. In fact, we can just consider $2b - 1$ standard arcs $\alpha_1, \ldots, \alpha_{2b-1}$ joining consecutive points of B_0 on the real axis and we can then write down corresponding standard elements $\sigma_1, \ldots, \sigma_{2b-1}$ of Br_{2b} (Figure 14.3).

It can be shown that these elements generate Br_{2b}, but we do not need this, because to prove Theorem 26 it clearly suffices to show that any transitive factorisation is equivalent to the standard one by the action of the part of the braid group generated by the σ_i.

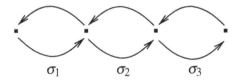

Figure 14.3 *Standard generators for the braid group*

Tracing through the definitions, we see that the action of σ_i on a factorisation $\underline{\delta} = (\delta_1, \ldots, \delta_{2b})$ is given by

$$\sigma_i(\underline{\delta}) = (\ldots, \delta_{i+1}, \delta_{i+1}^{-1} \delta_i \delta_{i+1}, \ldots). \tag{14.3}$$

Notice that since the δ_i are transpositions, they are equal to their inverses, but this way of writing the formula applies in more general situations. The inverse operation is

$$\sigma_i^{-1}(\underline{\delta}) = (\ldots, \delta_i \delta_{i+1} \delta_i^{-1}, \delta_i, \ldots). \tag{14.4}$$

So what we will now show is that any transitive factorisation can be transformed into the standard one by conjugation in S_d (which just amounts to a change of the labels $1, \ldots, d$) and a sequence of 'moves' of the type (14.3), (14.4).

Lemma 40. *Any transitive factorisation is equivalent to one of the form*

$$(12)^2 \delta_3 \ldots \delta_{2b}.$$

We use the convention that products in the permutation group are read from the left. Let $\underline{\delta}$ be a transitive factorisation, and consider the image of the object 1 in $\{1, \ldots, d\}$ under the action of the successive elements $\delta_1, \delta_1 \delta_2, \delta_1 \delta_2 \delta_3, \ldots$. This generates a list of objects in $\{1, \ldots, d\}$, where we omit successive repeats. For example, if the factorisation was

$$(12)(34)(45)(25)(15)\,W,$$

where W does not involve 1, our list would be $1, 2, 5, 1$. Since the factorisation is transitive, the list contains more than one item, and since the product of the δ_i is the identity, the list begins and ends with 1. It might be that this list contains repeats, for example it might be $1, 2, 5, 2, 1$, but it is clear that we can find some symbol $p_0 \in \{1, \ldots, d\}$ so that our list has the form $A p_0, p_1, \ldots, p_{k-1} p_0 B$, where the p_i are all distinct. This means that the transpositions $(p_j p_{j+1})$ appear, in sequence, in our factorisation. Now we can use the move (14.3) repeatedly to move these transpositions to the left. So we see that our factorisation is equivalent to one of the form

$$(p_0 p_1)(p_1 p_2)(p_2 p_3) \ldots (p_{k-1} p_0)\,W$$

and, without loss of generality, we may suppose this is

$$(12)(23) \ldots (k\ k-1)(k\ 1)\,W.$$

Now the move (14.3) shows that $U(k\ k-1)(k\ 1)V$ is equivalent to $U(k-1\ 1)(k\ 1)V$. Applying this move repeatedly, we see that our factorisation is equivalent to one of the form $(12)^2 V$, as required.

Lemma 41. *A factorisation of the form $U\delta^2\delta_*^2 V$ can be transformed by the moves (14.3), (14.4) to $U\delta_*^2\delta^2 V$.*

The proof is just

$$U\delta\delta\delta_*\delta_* V \mapsto U\delta\delta_*(\delta_*\delta\delta_*)\delta_* V \mapsto U\delta_*(\delta_*\delta\delta_*)(\delta_*\delta\delta_*)\delta_* V \mapsto$$
$$U\delta_*(\delta_*\delta\delta\delta_*)\delta_*\delta V \mapsto U\delta_*\delta_*\delta\delta V.$$

Now we prove our theorem by induction. We can suppose the result holds in any case where d is smaller or where d is the same but b is smaller. We use Lemma 40 to see that our transitive factorisation is equivalent to $(12)^2 W$. Since $(12)^2 = 1$, we have another factorisation defined by W, with a smaller value of b. If this is transitive, we know by induction that it is equivalent to $(12)^{2g+1}(13)^2 \ldots (1d)^2$. But we should be careful that the conjugation involved in the notion of equivalence will act on $1, 2$, so what we immediately see is that our factorisation is equivalent to

$$(pq)^2(12)^{2g+1}(13)^2 \ldots (1d)^2 \tag{14.5}$$

for some p, q. Suppose that the factorisation defined by W is not transitive. There must be exactly two orbits of the subgroup it generates in S_d in $\{1, \ldots, d\}$, and $1, 2$ are in different orbits. We change notation to write one orbit as $1', 2', \ldots, d'$ and the other as $1'', 2'', \ldots, d''$, so $d = d' + d''$ and our set of d objects is $\{1', \ldots, d', 1'', \ldots, d''\}$. Then W defines transitive factorisations for the two smaller sets and, by induction, we see that our factorisation is equivalent to

$$(p'q'')^2(1'2')^{2g'+1}(1'3')^2(1'd')^2 \ldots (1''2'')^{2g''+1}(1''3'')^2 \ldots (1''d'')^2, \tag{14.6}$$

where $p' \in \{1, \ldots, d'\}, q'' \in \{1'', \ldots, d''\}$. We leave it as an exercise for the reader to show that the factorisations (14.5), (14.6) are equivalent to the standard one, using Lemma 41 and similar applications of the moves.

Remark

1. Although we have been concerned with the braid group on an even number of strands, this is completely irrelevant as far as most of the definitions go.
2. Elements of the braid group can be represented as 'braids' in $\mathbf{R}^3$, hence the name. These are given by representing a path in the configuration space by a union of arcs in $\mathbf{R}^2 \times [0, 1] \subset \mathbf{R}^3$, as indicated in Figure 14.4.
3. It can be shown that the braid group is generated by the σ_i with a complete set of relations

$$\sigma_i\sigma_j = \sigma_j\sigma_i \ \text{ if } |i - j| > 1, \tag{14.7}$$

$$\sigma_i\sigma_j\sigma_i = \sigma_j\sigma_i\sigma_j \ \text{ if } |i - j| = 1. \tag{14.8}$$

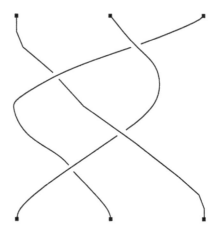

Figure 14.4 *Braid on three strands*

4. Let G be any group and let C be a conjugacy class in G. We may consider factorisations $\delta_1\delta_2\ldots\delta_a = 1$ with $\delta_i \in C$, and the equivalence relation on such factorisations modulo conjugation in G and the moves (14.3), (14.4). The problem is then to describe the equivalence classes, which is essentially what we have achieved above in the case of the conjugacy class of a transposition in the symmetric group. The general problem seems very difficult, but a solution for certain G, C would have important applications. For example, if $G = SL(2,\mathbf{Z})$ and C is the conjugacy class of $\begin{pmatrix} 1 & 1 \\ 0 & 1 \end{pmatrix}$, this is related to the classification of 'elliptic surfaces'.

14.2.2 Monodromy and Dehn twists

We begin by setting up some general facts and definitions. Let F be a compact manifold. Then we have a group $\mathrm{Diff}(F)$ of diffeomorphisms $\phi : F \to F$. An *isotopy* on F is a smooth map $\Phi : F \times [0,1] \to F$ such that for each $t \in [0,1]$, the map $\phi_t : F \to F$ defined by $\phi_t(x) = \Phi(x,t)$ is a diffeomorphism. When such a Φ exists, we say that ϕ_0, ϕ_1 are isotopic. This is an equivalence relation, and the equivalence class of the identity is a normal subgroup $\mathrm{Diff}_0(F) \subset \mathrm{Diff}(F)$. This subgroup Diff_0 can also be viewed as either the connected component or the path-connected component of the identity. We define the *mapping class group* of F to be the quotient $\mathrm{Diff}(F)/\mathrm{Diff}_0(F)$. Notice that if F is a 0-dimensional manifold—a discrete set—then the mapping class group is the permutation group of this set.

Now suppose we have a diffeomorphism $\phi : F \to F$. The *mapping torus* $\mathcal{X}_\phi$ is the manifold obtained from $F \times \mathbf{R}$ by identifying (x,t) with $\phi(x,t+1)$. Thus we can also construct $\mathcal{X}_\phi$ by starting with the product $F \times [0,1]$ and gluing the ends $F \times \{0\}, F \times \{1\}$ using ϕ. Clearly, $\mathcal{X}_\phi$ comes with a natural fibration $p : \mathcal{X}_\phi \to S^1$ over the circle $S^1 = \mathbf{R}/\mathbf{Z}$. A basic fact is that this sets up a 1–1 correspondence between fibrations over the circle, with fibre F, and elements of the mapping class group of F. Now consider any smooth fibre bundle $p : \mathcal{X} \to B$ with a connected base manifold B, base point $b_0 \in B$ and fibre $F = p^{-1}(b_0)$. For any smooth map $\gamma : S^1 \to B$ which maps a base point on S^1 to b_0, we may consider the 'pull-back'

$$\gamma^*(\mathcal{X}) = \{(x, t) \in \mathcal{X} \times S^1 : p(x) = \gamma(t)\}.$$

The projection to the S^1 factor makes $\gamma^*(\mathcal{X})$ a bundle over S^1 with fibre F, so this corresponds to an element μ_γ, say, of the mapping class group. Another basic fact is that μ_γ depends only on the (based) homotopy class of γ and the construction defines a group homomorphism from $\pi_1(B)$ to the mapping class group of the fibre. This homomorphism is the *monodromy* of the fibration $p : \mathcal{X} \to B$. In the case when the fibre is discrete, so p is a covering, this recovers the earlier notion discussed in Chapter 5.

We now go back to the set-up discussed in Section 14.2.1 with a Riemann surface $X(\Delta, \rho)$ associated to a finite subset $\Delta \subset \mathbf{C}$ and a representation ρ. Suppose we have a diffeomorphism $f \in \mathrm{Diff}_c(\mathbf{R}^2)$ which maps Δ to Δ and which is holomorphic near to the points of Δ and suppose that, for all $\gamma \in \pi_1(\mathbf{C} \setminus \Delta)$,

$$\rho(f_*(\gamma)) = \lambda \rho(\gamma) \lambda^{-1}.$$

Then we get an induced diffeomorphism $\tilde{f}$ of $X(\Delta, \rho)$, and the isotopy class of this depends only on that of f. In other words, what we are considering is the *stabiliser* of the factorisation corresponding to ρ for the action of $\mathrm{Br}_{2b} \times S_d$ on the factorisations. From another point of view, if we regard the braid as a loop Δ_t in the configuration space, we can apply our construction in this family to construct a three-dimensional manifold $\mathcal{X}$ fibring over the circle with monodromy $\tilde{f}$. We choose a path of diffeomorphism f_t for $t \in [0, 1]$ with $f_t(\Delta_t = B_0)$ and define $\rho_t = \rho \circ (f_t)_*$. Then we construct a space fibring over the interval where the fibre over t is $X(\Delta_t, \rho_t)$. The hypothesis gives us a canonical way to identify the fibres over the end points.

We will restrict our attention to the simplest case, in which λ is the identity and the monodromies around a pair of branch points give the same transposition. So, without loss of generality, our factorisation has the form $\delta^2 W$. We consider the diffeomorphism $\tilde{f}$ corresponding to the arc α_1. To make things explicit, consider for the moment the model case when $d = 2$ and there are two branch points ± 2. The Riemann surface X in this case is the Riemann sphere, but we work over the subset $\mathbf{C} \setminus \{0\}$. Our branched covering is given by the map $\phi(\tau) = \tau + \tau^{-1}$, so $\phi(\tau = \phi(\tau^{-1})$ and the branch points correspond to $\tau = \pm 1$. We consider our induced diffeomorphism $\tilde{f}$ as a compactly supported diffeomorphism of $\mathbf{C} \setminus \{0\}$. Thus the diffeomorphism is the identity when $|\tau|$ is either very large or very small. As usual, the diffeomorphism depends upon choices, but its isotopy class, among isotopies through compactly supported diffeomorphisms, is independent of these choices.

Now recall that, in Chapter 2, we defined *Dehn twists* about embedded curves in surfaces. In particular, we have a Dehn twist about the unit circle $|\tau| = 1$ in $\mathbf{C} \setminus \{0\}$. We have the following proposition.

Proposition 46. *The induced diffeomorphism $\tilde{f}$ is isotopic to the Dehn twist about the unit circle in $\mathbf{C} \setminus \{0\}$.*

In a nutshell, the proof is that the 'half-twist' downstairs, which interchanges the two points, lifts on the double cover to two half-twists, which are the same as one full twist (see Figure 14.5).

But we will now express this in formulae. Consider the image $z = x + iy$ of $\tau = Re^{i\theta}$ under $z = \phi(\tau) = \tau + \tau^{-1}$, where $R > 1$. So,

$$x = (R + R^{-1})\cos\theta, \quad y = (R - R^{-1})\sin\theta$$

and the point lies on the ellipse

$$\left(\frac{x}{R + R^{-1}}\right)^2 + \left(\frac{y}{R - R^{-1}}\right)^2 = 1.$$

It is elementary to see that any point in $\mathbf{C}$ which is not in the interval $[-2, 2] \subset \mathbf{R} \subset \mathbf{C}$ lies on exactly one ellipse of this kind, for a particular value $R(x, y) > 1$. In fact, if we set $U = R^2 + R^{-2}$, then algebraic manipulation gives

$$2U = (x^2 + y^2) \pm \sqrt{(x^2 - 4)^2 + y^4 + 2(x^2 + 4)y^2},$$

and one can check that only the $+$ root gives a value of U greater than 2. Fixing this sign, we can then solve the equation $R^2 + R^{-2} = U$ to find $R = R(x, y)$ explicitly. Now we define a circle-valued function $\theta(x, y)$ which gives the argument of a point τ lying over $x + iy$. We fix values $R_2 > R_1 > 1$. We can choose our diffeomorphism f to be the identity in the region $\{(x + iy) : R(x, y) \geq R_2\}$ and to be multiplication by -1 inside the inner region $\{x + iy : R(x, y) \leq R_1\}$. We choose a monotone function β with $\beta(r) = 0$ if $r \geq R_2$ and $\beta_r = 1$ if $r \leq R_1$, and define f in the annular region between the ellipses by decreeing that it preserves the ellipses and maps the ellipse $R(x, y) = r$ by increasing θ to $\theta + \beta(R)\pi$.

When $R < 1$, the circle $|\tau| = R$ maps to the same ellipse as the circle $|\tau| = R^{-1}$. However, the orientation is different: as τ moves clockwise around a circle of radius greater than 1, the point $\phi(\tau)$ moves clockwise around the ellipse, while if τ moves around a circle of radius less than 1, we move anticlockwise around the ellipse. It follows that the induced diffeomorphism is given by

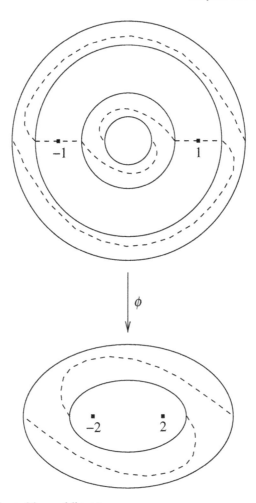

Figure 14.5 *A half-twist lifts to a full twist*

$$\tilde{f}(Re^{i\theta}) = Re^{i(\theta + \chi(R))},$$

where

- $\chi(r) = 0$ if $r \geq R_2$,
- $\chi(r) = \pi\beta(r)$ if $R_1 \leq r \leq R_2$,
- $\chi(r) = \pi$ if $R_1^{-1} \leq r \leq R_1$,
- $\chi(r) = 2\pi - \pi\beta(r^{-1})$ if $R_2^{-1} \leq r \leq R_1^{-1}$,
- $\chi(r) = 2\pi$ if $r < R_2^{-1}$.

Now it is clear that this map is a Dehn twist. Indeed, we get the same map, up to isotopy, if we use any function $\chi(r)$ which vanishes when r is large and is equal to 2π when r is small.

The same discussion applies to any factorisation $\delta^2 W$. The pre-image of the arc α_1 in our Riemann surface is an embedded circle, and the induced diffeomorphism is the corresponding Dehn twist. This is just because we can choose a diffeomorphism from a neighbourhood of this circle to a neighbourhood of the unit circle in the model case above, compatible with the branched coverings, and arrange that our maps are the identity outside this neighbourhood. Saying the same thing slightly differently, if we have any branch set Δ and an arc α joining two points of Δ such that the monodromy around the points is the same when compared by transporting along α, then the pre-image of α is an embedded circle in the Riemann surface and the diffeomorphism of the surface associated to the half-twist about α is the Dehn twist about this circle.

A particularly simple situation is when the degree d of the covering is 2, so we do not have to worry about the representation. We get a homomorphism from the braid group Br_{2b} to the mapping class group of a surface of genus $b - 1$ which takes the basic 'twist maps' in the braid group to Dehn twists in the mapping class group. It can be shown that this homomorphism is injective.

We should say that there are many ways of proceeding in this discussion. For example, in Section 15.2.3 we will write down a slightly different model, which would do just as well. In fact, the group of isotopy classes of compactly supported diffeomorphisms of the punctured plane is isomorphic to $\mathbf{Z}$, with the Dehn twist as a generator; so if one establishes that fact, the calculation is just a matter of pinning down a single integer.

14.2.3   Plane curves

We consider one of the most obvious ways in which we meet a 'continuously varying' collection of Riemann surfaces. Suppose that f, g are homogeneous polynomials in three variables, of degree d. For each complex number ϵ, we have an algebraic curve $C_\epsilon \subset \mathbf{CP}^2$ defined by the equation $f - \epsilon g = 0$. We can put all these together into a 'family', defining

$$\mathcal{C} = \{(x, \epsilon) \in \mathbf{CP}^2 \times \mathbf{C} : (f - \epsilon g)(x) = 0\}.$$

We will focus on the restriction of this family to small values of ϵ, restricting to a disc $D_r \subset \mathbf{C}$. First we consider the case when C_0 is smooth.

Proposition 47. *Suppose that C_0 is a smooth curve and that f vanishes with multiplicity 1. Then, for small enough r,*

- *the curves C_ϵ are smooth when $|\epsilon| < r$;*
- *there is a smooth map*

$$\tau : C_0 \times D_r \to \mathbf{CP}^2,$$

with $\tau(x, 0) = 0$ and such that for each $\epsilon \in D_r$, the map $\tau(\ , \epsilon)$ is a diffeomorphism from C_0 to C_ϵ.

We shall only sketch a proof. The first statement is elementary: if not, we would have a sequence of points $y_i \in \mathbf{CP}^2$ which were singular points of C_{ϵ_i}, where $\epsilon_i \to 0$. By compactness of $\mathbf{CP}^2$, we can suppose that the y_i converge, and the limit would be a singular point of C_0, a contradiction.

To prove the second statement, we make use of a Hermitian metric on the underlying three-dimensional complex vector space $\mathbf{C}^3$. At each point $x \in C_0$, there is a unique complex projective line tangent to C_0 at x. This line corresponds to a two-dimensional vector subspace $\Pi \subset \mathbf{C}^3$. The point x yields a one-dimensional subspace $\langle x \rangle$ of Π. The orthogonal complement of Π is a one-dimensional subspace, and the span of $\langle x \rangle$ is a two-dimensional subspace which yields another projective line L_x through x. (If the reader is happy with the notion of a Riemannian metric on $\mathbf{CP}^2$, then this line L_x can also be viewed as the unique line through x whose tangent space at x is orthogonal to that of C_0.) Now one can show without difficulty that when ϵ is small, there is a unique point of $L_x \cap C_\epsilon$ close to x. We use this to define a map from C_0 to C_ϵ and verify the stated properties.

The map given by Proposition 47 means that when C_0 is non-singular, the restriction of the family $\mathcal{C}$ to a small disc D_r has a smooth product structure. In any case, we define $E \subset \mathbf{C}$ to be the set of parameters ϵ such that C_ϵ is singular (including the case of a non-singular curve with multiplicity). It follows from Proposition 47 (applying translations in the ϵ variable) that E is closed and that the restriction of $\mathcal{C}$ to the complement $\mathbf{C} \setminus E$ is a smooth fibre bundle. (In fact, E is a finite set, the set of roots of a discriminant polynomial, but we do not need to know this.) We take some base point ϵ_0 in the complement $\mathbf{C} \setminus E$, so we have a monodromy homomorphism from $\pi_1(\mathbf{C} \setminus E)$ to the mapping class group of the curve C_{ϵ_0}.

We want just to consider the simplest case. Suppose now that C_0 is not smooth but has just one singular point, at the origin in $\mathbf{C}^2 \subset \mathbf{CP}^2$. We suppose that this an 'ordinary double point', which is to say that in affine co-ordinates z_1, z_2, the matrix of second derivatives of $\partial^2 f / \partial z_1 \partial z_2$, (evaluated at the origin) is non-degenerate. There is no loss in supposing that

$$f(z_1, z_2) = z_1^2 - z_2^2 + \text{higher-order terms.}$$

We also suppose that $g(0,0)$ does not vanish. We claim that, in this situation, the following proposition is true.

Proposition 48. *The curve C_ϵ is smooth for all small but non-zero ϵ, and the monodromy around a small loop $|\epsilon| =$ constant is a Dehn twist.*

Set $F(z_1, z_2) = f/g$. Since $g(0,0) \neq 0$, this is a holomorphic function in a neighbourhood of the origin, and the curve C_ϵ near the origin is given by the equation $F = \epsilon$.

Lemma 42. *There are local holomorphic coordinates $\tilde{z}_1, \tilde{z}_2$ in a neighbourhood of zero such that*

$$F = \tilde{z}_1^2 - \tilde{z}_2^2.$$

Moreover, we can suppose that $\tilde{z}_1$ is a function of z_1 and that the matrix of partial derivatives $\partial \tilde{z}_i / \partial z_j$ is equal to the identity at the origin.

Throughout the proof, we restrict our attention to a sufficiently small neighbourhood of the origin. We know that $F = z_1^2 - z_2^2 +$ higher-order terms. Write $F_2 = \partial F / \partial z_2$. Then $\partial F_2 / \partial z_2 \neq 0$ and $\partial F_2 / \partial z_1 = 0$ at the origin, so by our basic result, Theorem 1, the curve $F_2 = 0$ is, near the origin, a graph of the form $z_2 = a(z_1)$, where $a(0) = a'(0) = 0$. Set $z_2' = z_2 - a(z_1)$ and consider F as a function of z_1, z_2'. Then we have arranged that $\partial F / \partial z_2'$ vanishes when $z_2' = 0$ and so has the form $z_2' P(\tilde{z}_1, \tilde{z}_2)$, say, where $P(0,0) = 2$. By examining the power series in two variables, one sees that this implies that $F = b(z_1) - (z_2')^2 Q(z_1, z_2')$, where $Q(0,0) = 1$ and $b(z_1) = z_1^2 +$ [higher-order terms]. Thus there are local holomorphic square roots of Q and b. We set

$$\tilde{z}_1 = \sqrt{b(z_1)}, \quad \tilde{z}_2 = z_2' \sqrt{Q},$$

so $F = \tilde{z}_1^2 - \tilde{z}_2^2$, as desired.

Remark This result is a holomorphic version of the 'Morse Lemma', which gives a similar description of a smooth function around a non-degenerate critical point.

We see immediately from the lemma that for $\epsilon \neq 0$ there are no singular points of C_ϵ near to the origin. On the other hand, the same argument as in Proposition 47 shows that there are no singular points far away from the origin.

Now we compute the monodromy by using branched covers. Projection to the z_1 coordinate in the affine piece extends to a holomorphic map

$$p_0 : C_0 \setminus \{(0,0)\} \rightarrow S^2.$$

There is no loss of generality in supposing that p_0 has only simple branch points, with distinct critical values $w_1, \ldots, w_N$, say. Now consider the same projection on C_ϵ, so we have

$$p_\epsilon : C_\epsilon \rightarrow S^2.$$

It follows easily from Proposition 47 that for small non-zero ϵ, the branch points of p_ϵ are either close to the branch points of p_0 or close to the origin in $\mathbf{C}^2$. Conversely, for each branch point of p_0 there is a corresponding branch point of p_ϵ, and we get critical values $w_1(\epsilon), \ldots, w_N(\epsilon)$, say, close to the respective w_i. Now examine the branch points of p_ϵ close to the origin. Since $\tilde{z}_1$ is a function of z_1, these are determined by the projection of the curve $\tilde{z}_1^2 - \tilde{z}_2^2 = \epsilon$ to the $\tilde{z}_1$ coordinate. Writing $\epsilon = \eta^2$ as before and using the parametrisation by τ, we see that these branch points occur at the points where $\tau = \pm 1$ and $\tilde{z}_1 = \pm\eta$. In sum, then, if we write $z_1 = \phi(\tilde{z}_1)$, the critical values of p_{η^2} are the $N + 2$ points

$$\phi(\eta), \phi(-\eta)w_1(\eta^2), \ldots, w_N(\eta^2).$$

Notice that $\phi(\eta) = \eta + O(\eta^2)$.

We now reach the crucial point. Let η move over a small semicircle in the upper half-plane, from δ to $-\delta$, say. Then $\epsilon = \eta^2$ completes a whole circuit around the origin but the two branch points $\phi(\eta), \phi(-\eta)$ are *interchanged*. When δ is small, the points $\phi(\eta)$ and $\phi(-\eta)$ trace paths which are very close to semicircles in the upper and lower half-planes, respectively. So we are in just the situation considered above, except that now, rather than being fixed, the other branch points are allowed to move slightly. But it is clear that this does not affect the analysis of the monodromy. This completes the proof of Proposition 48.

The same considerations apply much more generally, not just to plane curves. Roughly, if we have any family of curves parametrised by a disc π : $\mathcal{X} \to \Delta$ where the fibre $\pi^{-1}(0)$ has an ordinary double-point singularity (modelled on the singular curve $z_1 z_2 = 0$), then the monodromy around the boundary of the disc is a Dehn twist in a 'vanishing cycle'—a small embedded circle in the smooth fibre. More precisely, we should really speak of an isotopy class of embedded circles. We get a vivid picture of this if we consider a slightly different basic model: plane curves defined by affine equations $F(z_1, z_2) = t$, where

$$F(z_1, z_2) = z_1^2 + z_2^2 + \text{higher-order terms.}$$

Thus we have merely changed the sign in our discussion above, which of course makes no difference when working with complex numbers. But suppose now that F is a polynomial with *real* co-efficients, and consider real values of t. Then we can consider the real solutions of our equation. In the model case when $F = z_1^2 + z_2^2$, we get a circle of radius $\sqrt{t}$ when t is positive and the empty set when t is negative. The same holds true in general: as t decreases through 0, there is a small component of the real curve which shrinks to a point and disappears. This component, thought of as a circle in the complex curve, is a representative of the 'vanishing-cycle' isotopy class.

14.3 Hyperbolic geometry

Fix a genus $g \geq 2$. Recall that the *moduli space* $\mathcal{M}_g$ is, as a set, the set of equivalence classes of compact Riemann surfaces of this genus. In this section, we give a hyperbolic-geometry description of this moduli space, as the quotient of the *Teichmüller space* T_g by the action of the mapping class group MC_g. We take a slightly informal point of view, skimming over some subtleties and details. These can all be filled in, of course, but as well as taking space, the precision required for that might obscure the simplicity and geometric nature of the basic ideas which we seek to convey.

Given real numbers $l_1, l_2, l_3 > 0$, we know from Lemma 27 that we have a unique right-angled hexagon $ABCDEF$, up to isometry, with side lengths $AB = l_1, CD = l_2, EF = l_3$. The other three sides have lengths

$$BC = l_1^*, \quad DE = l_2^*, \quad FA = l_3^*,$$

where l_i^* is a certain (computable) function of l_j. (See Exercise 11 in Chapter 11.)

Take two copies of this hexagon and glue them along the sides of lengths l_i^*. The result is a surface-with-boundary Y_l which has a hyperbolic metric in which each component of the boundary is a closed geodesic. This surface is diffeomorphic to a sphere with three discs removed. We have 'marked points' on each boundary component, corresponding to A, C, E. The lengths of the three boundary components are $2l_1, 2l_2, 2l_3$. The crucial thing is that if we have two of these objects, say $Y_l, Y_{l'}$, and one boundary component of Y_l has the same length as a boundary component of $Y_{l'}$, then we can glue Y_l and $Y_{l'}$ along the relevant boundaries to produce a new hyperbolic surface with a geodesic boundary, just as in the discussion in Section 11.2.5. To organise this we need some combinatorial data, which can be taken to be a connected graph Γ with $2(g-1)$ vertices and three edges emanating from each vertex, but we allow a pair of edges from the same vertex to be joined, forming a loop. For example, if $g = 2$ there are two essentially different possibilities Γ_1 and Γ_2 (Figure 14.6).

Fix such a graph Γ. A 'decoration' of the graph is the assignment of a pair of real numbers (l_E, θ_E) to each edge E, with $l_E > 0$. Label the vertices by $1, \ldots, 2(g-1)$ and fix an ordering of the three edges emanating from each vertex. Then the set of decorations can be identified with $(\mathbf{R} \times \mathbf{R}^+)^{3(g-1)}$. For each vertex α, we take a copy Y_l^α of Y_l with edge lengths defined in the obvious way. The graph Γ supplies exactly the information required to specify

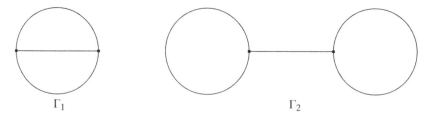

Figure 14.6 Γ_1 *and* Γ_2

a matching in pairs of the boundary components of the disjoint union $\sqcup_\alpha Y_l^\alpha$. The parameters θ_E specify the gluing: that is, when $\theta_E = 0$, we glue the marked points on the corresponding boundaries, and for general θ_E we rotate by a distance $\theta_E l_E$. Thus, if $\theta_E \in \mathbf{Z}$, we construct the *same surface* as we did with $\theta_E = 0$, but it is important, as we shall see, to keep θ_E as a real parameter for the time being. (Notice that there is a question of orientation here in specifying the rotation: one might guess that the direction of rotation depend on a choice of an 'arrow' on the edge of the graph, but a little thought shows that in fact it does not.)

The upshot is that for each choice of length and 'rotation' parameters, we get a compact hyperbolic surface. Let $\mathcal{T}_\Gamma = (\mathbf{R}^+ \times \mathbf{R})^{3(g-1)}$, so a choice of parameters is a point $\tau \in \mathcal{T}_\Gamma$ and we have a hyperbolic surface S_τ.

The striking fact is that we have actually constructed *all* compact hyperbolic surfaces in this way.

Theorem 27. *Any compact hyperbolic surface of genus g is isometric to one of the S_τ.*

Later we will want a more detailed and precise statement, but to give the main ideas we will indicate the very simple proof of Theorem 27. The starting point is the following proposition.

Proposition 49. *Any hyperbolic metric on the three-holed sphere, with geodesic boundaries, is isometric to one of the Y_l.*

To see this, one shows that, given such a hyperbolic metric and a pair of boundary components, there is a unique geodesic of minimal length running between the boundary components and meeting them at right angles. This can be proved by an analogue of the argument for Proposition 36, or by applying Proposition 36 to the appropriate free-homotopy class in the 'double' (the compact surface corresponding to the graph Γ_2 above). Now fix a base point τ_0 in $\mathcal{T}_\Gamma$, and so a surface $S_0 = S_{\tau_0}$. Write μ_0 for the metric on S_0. (The symbol g is in use for the genus!) Let S be any compact oriented hyperbolic surface of genus g, and choose a diffeomorphism $f : S_0 \to S$. Pulling back by f, we get another hyperbolic metric μ, isometric to the original one, on S_0. We join μ_0 to μ by a path in the space of metrics, say a linear path. Each of the metrics on this path is conformal to a hyperbolic metric, so we get a path of *hyperbolic* metrics μ_s on Σ_0, for $s \in [0, 1]$ with $\mu_1 = \mu$. We fix our attention on one of the three-holed spheres $Y \subset \Sigma_0$ given by our construction. The boundary circles are disjoint embedded representatives in their free-homotopy classes, so for each metric μ_s there are unique disjoint embedded geodesic representatives. We hope the reader will agree that it is then fairly clear that there is a one-parameter family of diffeomorphisms $\psi_s : \Sigma_0 \to \Sigma_0$ with ψ_0 equal to the identity and such that, for each s, the diffeomorphism ψ_s maps the boundary components of Y to these geodesics. Further, we can choose ψ_s to have this property for each of the $2(g-1)$ three-holed spheres used in constructing Σ_0. Now let $\mu' = \psi_1^*(\mu)$. Then μ' is a hyperbolic metric on Σ_0, isometric to the original one and such that each three-holed sphere $Y \subset \Sigma_0$ has a geodesic boundary with respect to μ'. Now the statement follows immediately from Proposition 49.

To sum up so far, we have constructed in this simple way 'all' hyperbolic surfaces of genus g in a family parametrised by T_Γ. Hence we have constructed 'all' compact Riemann surfaces of this genus. The subtlety of the theory of moduli spaces comes from the fact that the 'same' hyperbolic surface, or Riemann surface, is represented by infinitely many points of T_Γ. To understand this, we take a rather more abstract and sophisticated point of view.

A *family* of hyperbolic surfaces parametrised by a manifold B is a C^∞ fibration $\pi : \mathcal{C} \to B$ and a Euclidean metric on the vertical tangent bundle $\ker d\pi \subset T\mathcal{C}$ such that each fibre $\pi^{-1}(b)$ is a smooth surface on which the given metric is hyperbolic.

Proposition 50. *The product $T_\Gamma \times S_0$ and the fibration $\pi : T_\Gamma \times S_0 \to T_\Gamma$ can be endowed with the structure of a family of hyperbolic surfaces in such a way that $\pi^{-1}(\tau)$ is isometric to Σ_τ for each $\tau \in T_\Gamma$.*

What one should keep in mind here is that the family $\mathcal{C} \to T_\Gamma$ that we construct is a canonical object but the identification of $\mathcal{C}$ with the product $T_\Gamma \times \Sigma_0$ is not—it corresponds exactly to a trivialisation of the fibre bundle and depends on arbitrary choices. Said in a different way, what we want to do is to construct hyperbolic metrics μ_τ on the fixed surface S_0 varying smoothly with τ and with (S_0, μ_τ) isometric to S_τ.

To explain the main point, and simplify the notation, let us just construct this family over the part of T_Γ where the length parameters are fixed, equal to those of τ_0. Let Y be one of the three-holed spheres with hyperbolic metrics appearing in the construction. Fix attention on one boundary component $C \subset Y$ of length l, say. For each $\theta \in \mathbf{R}$, we have a diffeomorphism $\psi_\theta : C \to C$ which rotates the boundary by a distance $l\theta/2$. So ψ_θ is the identity if $\theta \in 2\mathbf{Z}$. Extend ψ_θ to an isometry of a small neighbourhood of S in Y. Then we can extend it further, over the whole of Y, by a diffeomorphism equal to the identity outside a slightly larger neighbourhood. We can make the whole construction vary continuously with the parameter $\theta \in \mathbf{R}$. For example, we could fix an identification of a neighbourhood of S with $S \times [0, \delta)$ and, for $d < \delta$, rotate $S \times \{d\}$ by $\beta(d)\theta l/2$, where β is a smooth function equal to 1 when d is very small and to 0 when $d \geq \delta/2$. Now we do this for each of the three boundary circles, so we have a continuous family of diffeomorphisms

$$\psi_{\underline{\theta}} : Y \to Y,$$

equal to the identity outside a small neighbourhood of ∂Y and acting as isometries of the hyperbolic metric near the boundary.

Now consider our surface S_τ obtained from $\sqcup_\alpha Y^\alpha$ by gluing the boundaries. For each α, we have a diffeomorphism $\psi^\alpha : Y^\alpha \to Y^\alpha$ defined as above, using the θ parameters of τ; by construction, these glue together to induce a diffeomorphism from S_0 to S_τ. (Notice that for each gluing circle C, corresponding to an edge E of Γ, we rotate by $\theta_E l_E/2$ on 'each side' of C so we change the overall identification by $\theta_E l_E$, as required.) Now pull back the hyperbolic metric on S_τ to get a hyperbolic metric μ_τ on the *fixed* surface S_0, as required.

Consider pairs S, ϕ where S is a compact oriented hyperbolic surface of genus g and $\phi : S_0 \to S$ is an oriented diffeomorphism. We say two such pairs $(S, \phi), (S', \phi')$ are *equivalent* if there is a diffeomorphism $f : S_0 \to S_0$ *isotopic to the identity* such that the pull-back of the metric on S' by $\phi' \circ f$ is the same as the pull-back of the metric on S by ϕ. This is an equivalence relation, and we define the Teichmüller space $\mathcal{T}_g$ (initially as a set) to be the set of equivalence classes. The mapping class group MC_g acts on $\mathcal{T}_g$ as follows. If $f : S_0 \to S_0$ is a diffeomorphism, then, given one pair (S, ϕ), we get another one $(S, \phi \circ f^{-1})$: by definition, $(S, \phi \circ f^{-1})$ is equivalent to (S, ϕ) if f is isotopic to the identity, so we get an induced action of MC_g—the quotient of the diffeomorphisms modulo those isotopic to the identity. It also follows from the definition that there is a canonical identification

$$\mathcal{M}_g = \mathcal{T}_g / MC_g.$$

It is not hard to define a natural topology on $\mathcal{T}_g$.

For each point $\tau \in \mathcal{T}_\gamma$, we have a metric μ_τ on S_0. The hyperbolic surface (S_0, μ_τ) together with the identity map from S_0 to S_0 is a pair as above and defines a point of $\mathcal{T}_g$. While the construction of μ_τ was *not* canonical, any two choices differ by the action of diffeomorphisms isotopic to the identity, so this point in $\mathcal{T}_g$ is independent of choices. By varying τ, we get a map from $\mathcal{T}_\Gamma$ to $\mathcal{T}_g$. The more precise version of Theorem 27 is the statement that this map is a *bijection* (and of course a homeomorphism). From this point of view, the choice of combinatorial data Γ amounts to a choice of coordinate system on the canonical object $\mathcal{T}_g$. From another point of view, we can consider our family $\mathcal{C} \to \mathcal{T}_\Gamma$ as a family parametrised by $\mathcal{T}_g$. We have an action of MC_g on $\mathcal{C}$ covering the action on $\mathcal{T}_g$.

There are no fundamental difficulties in the proof of all these statements, given the following fact.

Proposition 51. *The set of diffeomorphisms of the three-holed sphere which map each boundary component to itself is connected.*

One way of proving this is to use the fact that $\mathrm{Diff}_c(\mathbf{R}^2)$ is connected, by extending diffeomorphisms over the holes.

The mapping class group does not act freely on $\mathcal{T}_g$. In fact, the stabiliser of a point τ is isomorphic to the group of oriented isometries of the hyperbolic surface S_τ—or, equivalently, the automorphism group of the corresponding Riemann surface. It is a general fact that these stabilisers are finite groups (see Exercise 3 at the end of this chapter). If $g > 2$, then a typical surface has no non-trivial isometries, so there is a dense open set $\mathcal{T}_g^0 \subset \mathcal{T}_g$ on which MC_g acts freely. In fact, the complement of $\mathcal{T}_g^0$ has codimension 2 or, more so, $\mathcal{T}_g^0$ is connected. If $\mathcal{C}^0$ is the restriction of our family to this open set, we can take the quotient of $\mathcal{C}^0$ by MC_g to get a family of hyperbolic surfaces

$$\mathcal{C}^0/\mathrm{MC}_g \to \mathcal{M}_g^0,$$

for an open subset $\mathcal{M}_g^0 \subset \mathcal{M}_g$. We choose the base point τ_0 to be in $\mathcal{T}_g^0$. For each $\gamma \in \mathrm{MC}_g$, we can join τ_0 to $\gamma(\tau_0)$ by a path in $\mathcal{T}_g^0$, and this maps down to a loop in $\mathcal{M}_g^0$. Now, being a fibration, any family of hyperbolic surfaces leads to a monodromy homomorphism from $\pi_1(B)$ to the mapping class group of the fibre. Of course, the monodromy around the loop we have constructed is exactly γ. This follows immediately from the definitions.

Most of the action of the mapping class group is hard to see from the point of view of our original description as $\mathcal{T}_\Gamma$, but we can see the action of some elements. Let E be an edge of Γ corresponding to a circle C. As we observed before, increasing θ_E by 1 does not change the isometry class of the surface we construct. Tracing through the constructions, we see that the relevant diffeomorphism of S_0 is given by applying our maps ψ_1 on the two sides of C and, of course, this yields the Dehn twist about C. In other words, we can factorise the map to $\mathcal{M}_g$ into

$$\mathcal{T}_g = (\mathbf{R} \times \mathbf{R}^+)^{3(g-1)} \to (S^1 \times \mathbf{R}^+)^{3(g-1)} \to \mathcal{M}_g, \qquad (14.9)$$

where, for the first map, we take the quotient by the subgroup $\mathbf{Z}^{3(g-1)} \subset \mathrm{MC}_g$ generated by Dehn twists about the $3(g-1)$ circles.

We can compare this whole picture with that for the torus. The only difference is that a torus, unlike a surface of higher genus, has automorphisms which are isotopic to the identity (by translations). To fix this, we consider instead tori with a marked point (a choice of origin). Then the analogue of the Teichmüller space $\mathcal{T}_g$ is the upper half-plane, the mapping class group is $SL(2,\mathbf{Z})$ and the 'moduli space' is the quotient $H/SL(2,\mathbf{Z})$. The group ± 1 acts trivially, so the action factors through $PSL(2,\mathbf{Z})$. The complex tori which have stabilisers larger than ± 1 correspond to the points in H with a non-trivial stabiliser in $PSL(2,\mathbf{Z})$, which are the orbits of $i, e^{i\pi/3}$.

14.4 Compactification of the moduli space

Recall that in Section 11.2 we defined the injectivity radius $i(S)$ of a compact hyperbolic surface S and showed that it was equal to half the length of the shortest closed geodesic. Given $r > 0$, let $\mathcal{M}_{g,r}$ be the subset of $\mathcal{M}_g$ corresponding to surfaces with $i(S) \geq r$.

Theorem 28. *For each $r > 0$, $\mathcal{M}_{g,r}$ is compact.*

This may be surprising from the point of view of our description of $\mathcal{T}_g = \mathcal{T}_\Gamma$. Certainly, we can have a sequence of points in $\mathcal{T}_\Gamma$ which fails to have a convergent subsequence because some of the length parameters tend to zero. But we also have sequences which fail to converge because some of the length parameters tend to infinity. What the theorem says is that in such a case there are *other* free-homotopy classes—not one of the $3(g-1)$ which we have chosen for our description—where the length of the geodesic representative becomes small.

The main idea of the proof of the theorem is quite simple and takes us back to the definition of a surface by gluing together co-ordinate charts. For a hyperbolic surface, the transition functions can be taken to be defined by elements of the finite-dimensional group $PSL(2, \mathbf{R})$, so it is easy to arrange that a sequence of such elements has a convergent subsequence. More precisely, the basic fact that we need, but whose proof is completely obvious, is the following.

Lemma 43. *Let D be a hyperbolic disc in H and let g_n be a sequence in $PSL(2, \mathbf{R})$ such that $g_n(\overline{D}) \cap \overline{D}$ is non-empty. Then g_n has a convergent subsequence.*

Then we get a limiting surface by the taking the limits of the transition functions. However, there are a few complications and subtleties to take care of. We fix r and g throughout and consider a hyperbolic surface S of genus g with $i(S) \geq r$.

For each point p in S and for $\rho < r$, we have an embedded open disc $D(r, p) \subset S$ centred at p. First, choose a maximal set $p_1, \ldots, p_N$, such that the distances $d(p_\alpha, p_\beta)$ all exceed $r/10$, say. By maximal we mean that it is not possible to adjoin any other point q preserving the separation property. It is clear that such sets exist. By definition, for each point $q \in S$ there is some p_α with $d(q, p_\alpha) \leq r/10$, so the discs $D(r/5, p_\alpha)$ cover S. On the other hand, the separation property means that the discs $D(r/20, p_\alpha)$ are disjoint. Now the area of each disc $D(r/20, p_\alpha)$ is some fixed number determined by r and since, by the Gauss–Bonnet formula, the area of S is fixed, we deduce that

$N \le N(r, g)$ for some (computable) $N(r, g)$. Notice that this implies that the diameter of S is bounded by $d = (r/10)N(r, g)$.

Now let S_n be a sequence of such hyperbolic surfaces. For each n, we choose a cover by $N(r, g)$ discs of radii $r/5$, with centres $p_{\alpha,n}$ (ignoring now the separation condition). So we have numbers $d(\alpha, \beta, n) \le d$, the distances in S_n between $p_{\alpha,n}$ and $p_{\beta,n}$. Taking a subsequence, we can suppose that all these numbers converge to limits $d(\alpha, \beta)$ as $n \to \infty$. Then we can find ρ slightly bigger than $r/5$ and less than $r/4$ and $\delta > 0$ such that

$$|d(\alpha, \beta, n) - 2\rho| > 2\delta, \qquad (14.10)$$

for all α, β and all but a finite number of values of n. We may ignore this finite number and suppose that equation (14.10) holds for all n. Now we simplify the notation by writing $D_{\alpha,n} = D(p_\alpha, \rho) \subset S_n$. The point of the condition (14.10) is that two discs $D_{\alpha,n}, D_{\beta,n}$ either intersect 'by a definite amount'—their intersection contains a disc of radius δ—or are disjoint 'by a definite amount'—the distance between the discs exceeds 2δ. Furthermore, the fact that $\rho < r/4$ means that the intersections $D(\alpha, n) \cap D(\beta, n)$ are *connected* (exercise!).

We fix isometries

$$e_{\alpha,n} : D \to D_{\alpha,n}$$

where D is the standard ρ-disc in the hyperbolic plane. If $D_{\alpha,n}$ and $D_{\beta,n}$ intersect, we have an isometry $f_{\alpha,\beta,n}$ of the hyperbolic plane such that

$$\chi_{\beta,n} = \chi_{\alpha,n} \circ f_{\alpha,\beta,n}$$

on a non-empty connected set $\Delta_{\alpha,\beta,n} \subset \chi_{\alpha,i}^{-1}(D_{\alpha,i} \cap D)$. This implies that the intersection of D and $f_{\alpha,\beta,n}(D)$ is non-empty. In fact, by our choice of r, this intersection contains a disc of a fixed radius δ.

Let $R \subset \{1, \ldots, N(r, g)\}^2$ be the subset consisting of pairs (α, β) for which $d_{\alpha,\beta} < 2\rho$. As before, we can suppose that for all n, $d(\alpha, \beta, n) \ge 2\rho + 2\delta$ if $(\alpha, \beta) \notin R$ and $d(\alpha, \beta, n) \le 2\rho - 2\delta$ if $(\alpha, \beta) \in R$. In the first case, $D_{\alpha,n}, D_{\beta,n}$ are disjoint, and in the second they intersect.

By Lemma 43, we can suppose that for each (α, β) in R the maps $f_{n,\alpha,\beta}$ converge as $n \to \infty$ to some limit $f_{\alpha,\beta}$ and, clearly $U_{\alpha\beta} = f_{\alpha,\beta}(D) \cap D$ contains a disc of a fixed radius δ and, in particular, is non-empty. Now we define a set Σ, which will, of course, turn out to be our limiting hyperbolic surface. We start with the disjoint union $D \times \{1, \ldots, N\}$. For pairs $(\alpha, \beta) \in R$, we have a non-empty connected open set $U_{\alpha,\beta}$, and for $z \in U_{\alpha\beta}$ we identify (z, α) with $(f_{\alpha,\beta}^{-1}(z), \beta)$. By construction, we have 'charts'

$$\chi_\alpha : D \to \Sigma.$$

It is an exercise in the definitions to check that these make Σ into a compact hyperbolic surface. (Of course, the key point is that there is a consistency

condition which needs to be satisfied on triple intersections. This holds for
the $\chi_{\alpha,\beta,n}$ and so also in the limit. Our choice of ρ so that equation (14.10)
holds overcomes a more technical difficulty: it means that intersections cannot
'disappear' in the limit, which would yield a non-compact surface.)

To finish the proof, we must deal with another technical issue. We have
constructed a hyperbolic surface S which obviously ought to be the limit of our
sequence in $\mathcal{M}_{g,r}$. But, according to our definition of the topology on the moduli space, to establish this we should construct a sequence of diffeomorphisms
$\phi_n : S \to S_n$ such that the pull-backs of the hyperbolic metrics on S_n converge
to that on S. (This will also make it fairly clear that the injectivity radius of S
is at least r.) What we have, in obvious notation, is a cover of S by discs D_α
and hyperbolic isometries $\phi_{\alpha,n} : D_\alpha \to S_n$, but these do not agree exactly on the
intersections, although they are approximately equal. Suppose for the moment
that we were in a similar situation but the ϕ_n were *functions*, rather than maps
to surfaces. Then it would be clear what to do. We would choose a partition of
unity λ_α subordinate to the cover of S and set $\phi = \sum \lambda_\alpha \phi_\alpha$. So what we want
to do is to extend this idea to apply to maps into hyperbolic surfaces. We say
that a subset K of the hyperbolic plane is *convex* if, for each pair of points p, q
in K, the geodesic segment joining p, q also lies in K. Clearly, any finite set
of points $p_1, \ldots, p_m$ has a convex hull $K(p_1, \ldots, p_m)$. Given $\lambda_1, \ldots, \lambda_m \in [0,1]$
with $\sum \lambda_i = 1$, we want to define a 'weighted centre of mass'

$$Z(\lambda_1, \ldots, \lambda_m, p_1, \ldots, p_m) \in K(p_1, \ldots, p_m)$$

with certain natural properties, analogous to the centre of mass of particles in
$\mathbf{R}^n$. For example, a crucial property we want is hyperbolic invariance,

$$Z(\lambda_1, \ldots, \lambda_m, g(p_1) \ldots, g(p_m)) = g(Z(\lambda_1, \ldots, \lambda_m, p_1, \ldots, p_m)),$$

for isometries g of the hyperbolic plane. We will use the abbreviated notation
$Z(\lambda_i, p_i)$.

The Klein model gives an obvious way of doing this. First, it is clear that a
subset $K \subset Q$ is convex if and only if the corresponding cone in $\mathbf{R}^{2,1}$ is convex
in the ordinary sense. (And so, in the representation $\Delta \subset \mathbf{R}^2$, hyperbolic convexity is the same as standard convexity.) Second, we can define Z as follows.
Given points p_i in $Q \subset \mathbf{R}^{2,1}$, we consider

$$z = \sum \lambda_i p_i \in \mathbf{R}^{2,1}.$$

This will not lie in Q, but it is clear that $B(z,z) < 0$ and z has positive first
co-ordinate. Thus there is a positive multiple μ such that μz lies in Q, and we
can define $Z(\lambda_i, p_i) = \mu z$.

With this background, we can proceed as follows. Fix a partition of unity
λ_α on S subordinate to the cover D_α. Given a point $p \in S$, choose a chart
D_α containing it and let $p_n = \phi_{n,\alpha}(p) \in S_n$. Fix an exponential map $e : D \to S$

mapping 0 to p_n. Let $D_\beta \subset S$ be any other disc which has a non-empty intersection with D_α. There is a unique lift of $\phi_{n,\beta}$ to a map $\tilde{\phi}_{n,\alpha} : D_\beta \to D$ whose image is close to $0 \in D$. Now, define

$$\phi_n(p) = e(Z(\lambda_\beta(p), \tilde{\phi}_{n,\beta}(p))).$$

Then one can check that this independent of the choice of chart D_α. (This uses the fact that any finite intersection of hyperbolic discs is convex.) The construction provides us with a smooth map $\phi_n : S \to S_n$, and it is not hard to check further that for large n this map is a diffeomorphism and an approximate hyperbolic isometry, so the pull-backs by ϕ_n of the hyperbolic metrics on S_n converge to that on S.

14.4.1   Collars and cusps

Suppose S_n is a sequence of hyperbolic surfaces of genus g and the sequence of points $[S_n] \in \mathcal{M}_g$ does not have a convergent subsequence. We know that the injectivity radius of S_n tends to 0 and hence, for large enough n, the surface S_n contains at least one very short simple closed geodesic. But one can get a much more detailed picture of what can happen. Recall that we have defined the standard manifold $\Sigma_l = H/G_l$, where G_l is the group generated by $z \mapsto e^l z$, given a length l. This model contains a geodesic γ_0 of length l, corresponding to the imaginary axis. The set of points in H whose distance to the imaginary axis is less than D is the wedge

$$W_D = \{x + iy : y > 0, |x| < y \sinh D\}.$$

(See Exercise 8 in Chapter 11.) It follows that the set of points in Σ_l whose distance to γ_0 is less than D is the quotient $N_{D,l} = W_D/G_l \subset \Sigma_l$. This is the *standard collar* about a geodesic of length l. For any hyperbolic surface S which contains a simple closed geodesic γ of length l, we know that there is some D such that there is a neighbourhood of γ isometric to $N_{D,l}$. But now we want a much more precise result. We set

$$D(l) = \sinh^{-1}\left(\frac{2e^{-l}}{1 - e^{-2l}}\right)$$

and write $N(l) = N_{D(l),l}$.

Proposition 52. *If γ is a simple closed geodesic in a compact hyperbolic surface S, there is a neighbourhood of γ isometric to $N(l)$.*

To prove this, observe first that any simple closed curve on a surface can be included in a set of $3(g - 1)$ simple closed curves which give a standard decomposition of the surface, as considered in Section 14.3. Thus it suffices to prove that in any surface Y_l, the geodesic boundary of length l_1 has a neighbourhood

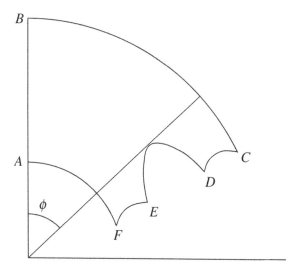

Figure 14.7 *The largest collar*

isometric to one-half of N_{D,l_1} for $D = D(l)$ (in the obvious sense). In turn, it suffices to prove the corresponding statement about hexagons. Take a right-angled hexagon $ABCDEF$ with one edge from $A = i$ to $B = e^l i$, as shown in Figure 14.7.

The points C and F lie respectively on the unit circle and the circle of radius e^l. We have to show that all the edges CD, DE, EF lie outside the wedge $W(D)$, where $D = D(l)$. Now, given the hexagon, there is some maximal angle ϕ such that CD, DE, EF lie outside an (open) sector of angle ϕ about the imaginary axis. Thus the ray $\{r(\sin\phi + i\cos\phi) : r > 0\}$ touches one of the three edges CD, DE, EF. A little thought will show the reader that the extremal (limiting) case is when $C = D = E$ and the hexagon degenerates into a quadrilateral (Figure 14.8).

For this case, elementary geometry gives the angle as ϕ_0, where

$$e^l = \frac{1 + \cos\phi_0}{\sin\phi_0},$$

and what we have shown is that $\phi > \phi_0$ for any other case, which is equivalent to the assertion in the proposition.

Now we want to give another description of $N(l)$. Recall that Σ_l can be thought of as being obtained by gluing the semicircle C_1 of (Euclidean) radius 1 to the semicircle C_2 of radius e^l by the map $z \mapsto e^l z$, that is, as the quotient of H by the group G_l. Consider the map $f(z) = \alpha(z-1)/(z+1)$, where $\alpha = \coth l/2$. This has $f(1) = 0$, $f(-1) = \infty$, $f(e^l) = 1$, $f(-e^l) = \alpha^2$, so it takes C_1 to the imaginary axis I and C_2 to the semicircle J of radius R centred at $1 + R$, where $1 + 2R = \alpha^2$. Let Ω_l be the region between I and J. Then Σ_l

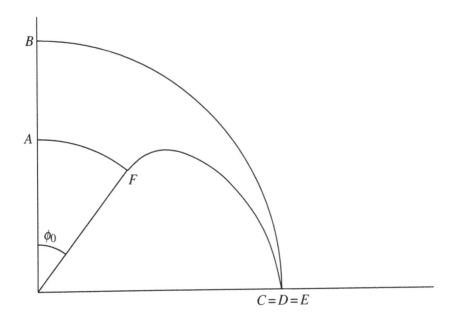

Figure 14.8 *The extremal case*

can be obtained from Ω_l by identifying the edges I, J. In other words, we are considering another cyclic subgroup in $PSL(2, \mathbf{R})$, conjugate by f to G_l, and a short calculation shows that this group is generated by the map

$$w \mapsto \frac{1 + w}{1 + \epsilon w},$$

where $\epsilon = \tanh^2(l/2)$. If we apply f to the configuration we had before, we obtain Figure 14.9.

Clearly, then, the 'limit' of these quotients as $l \to 0$ (so $R \to \infty$ and $\epsilon \to 0$) is the surface Z obtained by gluing I to the line $\mathrm{Re}\, w = 1$ by the map $w \mapsto w + 1$. That is, Z is the quotient of H by the group of maps $w \mapsto w + k$, $k \in \mathbf{Z}$. This hyperbolic surface Z is the 'standard cusp'.

To make things more precise, fix b_0 slightly bigger than 1. For $b > b_0$, let $Z_b \subset Z$ be the quotient of the strip $\{z : b_0 < \mathrm{Im}(z) < b\}$ by $\mathbf{Z}$. We also write Z_∞ for the quotient of $\{z : \mathrm{Im}\, z > 1\}$. Then we can certainly write down diffeomorphisms $\phi_{b,l}$ from Z_b to an open set in $N(l)$. It is easy to choose the $\phi_{b,l}$ so that for fixed (finite) b the pull-back by $\phi_{b,l}$ of the hyperbolic metric on $N(l)$ converges to the hyperbolic metric on Z_b. The point here is that in the diagram, the small circles have a radius of at most approximately 1 when l is small, so the lower boundary of the fundamental-domain hexagon is well below the line $\{\mathrm{Im}\, z = b_0\}$. Another explicit picture of these model

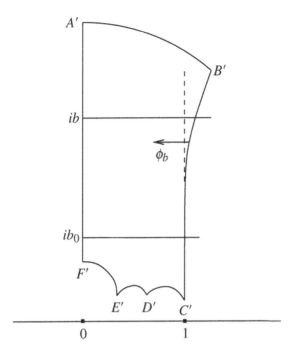

Figure 14.9 *Collar close to cusp*

surfaces can be obtained through surfaces of revolution in $\mathbf{R}^3$ (see Exercise 4 in Chapter 11).

We define a hyperbolic surface with cusps to be a complete non-compact hyperbolic surface which is the disjoint union of a compact set and a finite number of open subsets which are isometric to Z_∞. Since the complex structure of Z_∞ is clearly that of a punctured disc, we can compactify a hyperbolic surface S with cusps to obtain a compact Riemann surface $\overline{S}$, adding a finite number of points, one for each cusp. From that point of view, a hyperbolic surface with cusps is a metric on $\overline{S}$ with certain standard singularities at these points. Notice that we have already encountered examples of this picture in the 'modular curves', discussed in Section 7.2.4.

What our discussion above shows is that any sequence of hyperbolic surfaces of fixed genus has a subsequence which converges either to a compact hyperbolic surface or to a hyperbolic surface with cusps. Of course, one has to give a precise definition of 'convergence' in the latter case. But we know that if the injectivity radius tends to zero, we have a sequence of isometric copies $N(l_n)$ with $l_n \to 0$, and the discussion of the convergence around the cusp is reduced to the standard model. Of course, we may have more than one short simple closed geodesic. It is fairly clear that if γ is a simple closed geodesic of length l, and l is small, then no other short simple closed geodesic can intersect the isometric copy of $N(l)$ around γ. In fact, a more general version of the collaring theorem says that for any two simple closed geodesics γ_1, γ_2 of lengths less than l_1, l_2, there are *disjoint* neighbourhoods of γ_i comparable in size to the $N(l_1), N(l_2)$. Thus there is 'no interaction' between the short geodesics.

Suppose that a sequence of compact hyperbolic surfaces converges to a hyperbolic surface with cusps, and so we have a corresponding compact surface $\overline{S}$ (which may not be connected) and a finite set of marked points. We can record another piece of information from the converging sequence, which is the decomposition of this set of marked points into pairs. We can think of the resulting data algebro-geometrically as a singular curve with 'nodal singularities', modelled on the union of the two axes in C^2, and with the convergence near these singularities modelled on the family of smooth curves $z_1 z_2 = \epsilon$ for $\epsilon \to 0$.

Now we can extend the ideas further to consider a sequence of hyperbolic surfaces with cusps, including a pairing of points at infinity corresponding to the cusps, and—with some extra work—show that there is a subsequence which converges to a limit of the same kind. More formally, consider data consisting of a Riemann surface, possibly disconnected, with $2b$ marked points and a partition of these marked points into pairs. We require that the Euler characteristic of the surface be $2g - 2 - 2b$ and that there are at least three marked points on any component of genus 0 and at least one on any component of genus 1. This last is the condition that each component has a hyperbolic metric with cusps. Then we let $\overline{\mathcal{M}}_g$ be the set of equivalence classes of such data. Our arguments show that this space is *compact*, in the natural topology defined by the above notion of convergence. It is a compactification of the dense open subset $\mathcal{M}_g$.

There is a description of $\overline{\mathcal{M}}_g$ which is quite explicit, up to the action of appropriate mapping class groups. We start with any graph Γ as before and allow some of the lengths l_E to be zero, in which case we do not include the parameter θ_E. Now we do the same for all graphs Γ, and we have described all hyperbolic surfaces with cusps and pairings of points at infinity. As before, the complication is that the same geometric object may be described in many different ways. Recall that, given Γ, the action of the part of the mapping class group generated by the Dehn twists about the chosen $3g - 3$ disjoint loops is easy to see: it leads to the partial quotient $(\mathbf{R}^+ \times S^1)^{3g-3}$. In this picture, the compactification is modelled locally on $(\mathbf{R}^2)^{3g-3}$ when we make the standard identification of $\mathbf{R}^+ \times S^1$ with $\mathbf{R}^2 \setminus \{0\}$. The loops in $\mathcal{M}_g$ which correspond to the Dehn twists become contractible in $\overline{\mathcal{M}}_g$ and, in fact, $\overline{\mathcal{M}}_g$ is simply connected. (This statement is essentially equivalent to an important theorem of Lickorish (1962), that the Dehn twists generate the mapping class group.)

An easier problem than trying to describe the whole space $\overline{\mathcal{M}}_g$ is just to describe its 'strata'. These are labelled by the different topological types

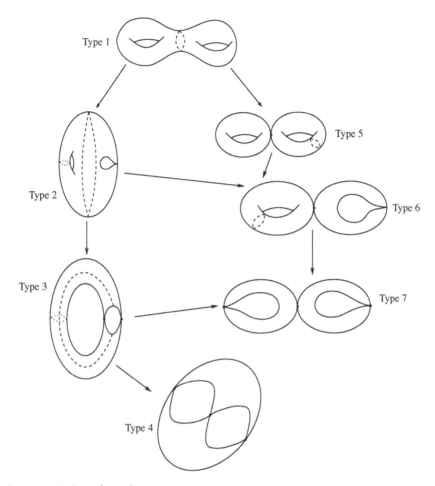

Type 1

Type 2

Type 3

Type 4

Type 5

Type 6

Type 7

Figure 14.10 *Strata for g = 2*

which arise. For example, when $g = 2$ there are seven strata, and the relation between them is described in Figure 14.10. Here an arrow means that one stratum contains the other in its closure—we can 'degenerate' one type to the other.

Some of these degenerations of Riemann surfaces of genus 2 are easy to see from the algebraic point of view. We know that we can describe these surfaces as double branched covers of a sphere, with six branch points $z_1, \ldots, z_6$, say. If we allow two of the points to come together, we get type 2. If we allow two pairs to come together, we get type 3, and for three pairs type 4. The other degenerations are a little less obvious. For example, suppose we take a pair of cubic curves in the plane given by equations $f_1 = 0$ and $f_2 = 0$, with nine intersection points $P_1, \ldots, P_9 \in \mathbf{CP}^2$. The product $f_1 f_2$ is a homogeneous polynomial of degree 6. One can show that there is a polynomial g of degree 6 which vanishes at $P_1, \ldots, P_8$ but not at P_9. Then, for small ϵ, the curve defined by the equation $f_1 f_2 + \epsilon g = 0$ has eight singular points close to $P_1, \ldots, P_8$. The normalisations of these curves give a family of Riemann surfaces of genus 2 degenerating to the union of the original elliptic curves identified at the point P_9, corresponding to type 5. We encountered this, at least implicitly, in Section 12.2.5 when we discussed Kummer surfaces and Jacobians. The discussion at the end of that subsection amounts to showing that the 'limit' of the Jacobians is the product of the two elliptic curves.

There are many algebraic degenerations which, at first sight, do not fit into this picture. For example, we could take a polynomial $f(z_1, z_2)$ of the form $f(z_1, z_2) = z_1^2 - z_2^3 +$ terms of order 4 or more. Then, for generic g, the plane curves C_ϵ defined by the equation $f - \epsilon g = 0$ will be smooth for small non-zero ϵ, but C_0 has a singularity at the origin. But this singularity is not of the nodal type considered above. The genus of the normalisation of C_0 is one less than the genus of C_ϵ. To fit this into the hyperbolic-geometry picture, we take the normalisation of C_0 and consider it as a compact Riemann surface with one marked point, corresponding to the singularity. We get a complete non-compact hyperbolic Riemann surface S with one cusp. Then one can show that, for generic f and g, the limit of the sequence C_ϵ in the compactified moduli space is the union of S and an elliptic curve E, with one marked point on each (Figure 14.11). For small but non-zero ϵ, we can describe the hyperbolic metric on C_ϵ by deforming the two cusps into a long collar. In the algebraic picture, the whole torus region is mapped down to a small neighbourhood of

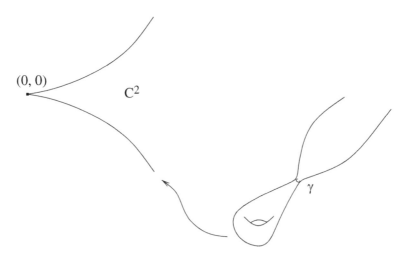

Figure 14.11   *Hyperbolic and complex pictures*

$(0, 0)$, collapsing into the origin as $\epsilon \to 0$. The monodromy of this family, as the parameter ϵ traces out a small circle around zero, is the Dehn twist around the loop γ separating the two regions.

Remark

1. One surprising thing is that when we discuss deformations of Riemann surfaces in the complex analytic framework, we get *complex* vector spaces of infinitesimal deformations and we expect the moduli space to be a *complex* manifold (or, more precisely, orbifold). The hyperbolic geometry picture presents Teichmüller space as a *real* manifold, and the moduli space as a real orbifold. This is similar to—and turns out to be a variant of—the fact we have encountered in Chapter 8 that the real first cohomology group of a Riemann surface acquires a complex structure through the identification with $H^{1,0}$.

2. The compactified moduli space can be constructed entirely algebro-geometrically as the 'Deligne–Mumford compactification'. However, there are subtleties which arise when discussing the 'smooth structure' of the compactification, matching up the algebro-geometric and hyperbolic-geometry points of view.

Exercises

1. Verify that the relations (14.7) and (14.8) hold in the braid group.
2. Let Y be a three-holed sphere with a hyperbolic metric for which the boundaries are geodesic. Show that the isometry group of Y is finite. What finite groups occur?
3. Use the preceding result to show that if Σ is a compact hyperbolic surface, then any isometry which is sufficiently close to the identity is equal to the identity. Deduce that the isometry group of Σ is finite.

15 Ordinary differential equations

In Chapter, 1 we discussed ordinary differential equations and, in particular, the hypergeometric equation

$$z(1-z)\frac{d^2 f}{dz^2} + (c - (1 + a + b)z)\frac{df}{dz} - abf = 0. \tag{15.1}$$

We will denote this equation by $HG_{a,b,c}(z)$. (This notation is not very logical, but will be useful in a moment.) In this chapter we take this discussion further, fitting it into some of the theory developed in the book.

15.1 Conformal mapping

Here we take the parameters a, b, c to be real. We write

$$\alpha = (1 - c), \quad \beta = (b - a), \quad \gamma = c - a - b.$$

For simplicity, suppose $0 < \alpha, \beta, \gamma < 1$. We define a *curvilinear triangle* (Figure 15.1) to be a domain in $\mathbf{CP}^1$ whose boundary consists of three arcs of circlines.

Theorem 29. *If F, G are two independent solutions of the hypergeometric equation in the upper half-plane H, then the ratio G/F maps H to a curvilinear triangle. The points $z = 0, \infty, 1$ map to the vertices of the triangle, and the angles at the vertices are $\pi\alpha, \pi\beta, \pi\gamma$, respectively. Conversely, any such triangle can be obtained in this way.*

We will give two proofs, the first elementary (and focused on the first statement in the theorem) and the second more sophisticated (and focused on the second, converse statement). To begin, notice that it suffices to prove the result for *some* pair of solutions F, G. For if $\tilde{F}, \tilde{G}$ are another pair, we can write

$$\tilde{F} = pF + qG, \quad \tilde{G} = rF + sG$$

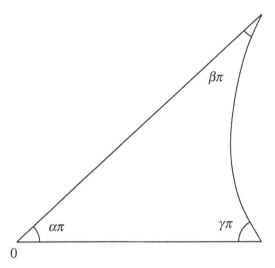

Figure 15.1 *The curvilinear triangle*

for constants p, q, r, s, and then $\tilde{G}/\tilde{F}$ is the composite of G/F and the Möbius map

$$w \mapsto \frac{pw + q}{rw + s}.$$

This uses the fact that Möbius maps preserve the class of curvilinear triangles, and the angles at vertices.

Back in Chapter 1, we mentioned that any second-order ordinary differential equation (ODE) with three regular singular points can be transformed to a hypergeometric equation by the action of Möbius maps and multiplication of the variable by a fixed function (Exercise 6 in Chapter 1). By the same token, the Möbius maps which permute the three points $0, 1, \infty$ transform one hypergeometric equation to another. Thus any property we can establish for solutions around 0, say, gives a corresponding property for solutions around 1 and around ∞.

To examine this symmetry, consider the Möbius map $T(z) = 1/(1 - z)$. This preserves the upper half-plane and permutes $0, 1, \infty$ cyclically. Write $w = T(z)$ and suppose that f satisfies the hypergeometric equation $HG_{a,b,c}(w)$, using notation in the obvious way. Rewrite this equation in terms of the variable z using the chain rule:

$$\frac{d}{dw} = (1 - z)^2 \frac{d}{dz}.$$

We get

$$
\frac{1}{1-z}\left(\frac{-z}{z-1}\right)(1-z)^2\frac{d}{dz}\left((1-z)^2\frac{df}{dz}\right)
$$

$$
+\left(c-\frac{(1+a+b)}{1-z}\right)(1-z)^2\frac{df}{dz}-abf=0.
$$

This simplifies to

$$
z(1-z)\frac{d^2f}{dz^2}+P\frac{df}{dz}+\frac{ab}{1-z}f=0,
$$

where $P = 1 + a + b - c(1-z) - 2z$. Now write $f = (1-z)^r\tilde{f}$, where r is a constant to be determined. We get a second-order equation for $\tilde{f}$ of the form

$$
z(1-z)\frac{d^2\tilde{f}}{dz^2}+\tilde{P}\frac{d\tilde{f}}{dz}+\tilde{Q}\tilde{f}=0. \tag{15.2}
$$

A little calculation gives

$$
\tilde{Q}=\frac{z(r^2+r-cr)+ab-(1+a+b)r+cr}{1-z}.
$$

To get an equation in hypergeometric form, we need the numerator to be a multiple of $1-z$, which is the condition $r^2 + r = (1+a+b)r - ab$ and hence $r^2 - (a+b)r + ab = 0$. Thus we need to take $r = a$ or b. It turns out to be convenient to choose $r = b$. With this choice,

$$
\tilde{P} = (1+a+b-c) - z(1+b-c), \quad \tilde{Q} = -b(1+b-c),
$$

so equation (15.2) is a hypergeometric equation with parameters a', b', c', where $c' = 1 + a + b - c, a' + b' = b + 1 - c, a'b' = b(1+b-c)$. We have another choice, of which parameter we call a' and which b'. The convenient choice is

$$
a' = 1 + b - c, \quad b' = b.
$$

So, to sum up, we have shown that a function $\tilde{f}(z)$ satisfies $HG_{a',b',c'}(z)$ if and only if $w^{-b}\tilde{f}((w-1)/w)$ satisfies $HG_{a,b,c}(w)$. If we take the equivalent choice of parameters α, β, γ and α', β', γ', the situation is much clearer, since we have

$$
\alpha' = 1 - c' = c - a - b = \gamma,
$$

$$
\beta' = a' - b' = 1 - c = \alpha,
$$

$$
\gamma' = c' - a' - b' = a - b = \beta.
$$

So, as T cyclically permutes the three points $0, 1, \infty$, we cyclically permute the parameters α, β, γ in the hypergeometric equation. (This is the reason for the particular choices made above—other choices would introduce signs.)

Recall the hypergeometric series

$$F(a, b, c; z) = 1 + \frac{ab}{1!c}z + \frac{a(a+1)b(b+1)}{2!c(c+1)}z^2 + \cdots . \tag{15.3}$$

This gives a holomorphic solution of equation (15.2) around 0. We denote this by F_0. By straightforward calculation, we find that a second solution is given by

$$G_0(z) = z^{1-c} F(a - c + 1, b - c + 1, 2 - c; z).$$

Here the power z^{1-c} is defined by cutting the plane along a ray in the lower half-plane. Now F_0 is real on the real axis. The second solution G_0 is real on the positive real axis but maps the negative real axis to the ray defined by the condition $\arg z = \pi\alpha$. So it is clear that G_0/F_0 maps the 'semi-disc' $\{z : |z| < 1, \operatorname{Im} z > 0\}$ to the wedge-shaped region $\{w : 0 < \arg w < \pi\alpha\}$. This map is illustrated in Figure 15.2. below. Now we apply our cyclic symmetry. We see that there are two solutions, F_1 and G_1 say, of the hypergeometric equation defined near 1 whose ratio maps a semi-disc in the upper half-plane to a wedge-shaped region with angle $2\pi\beta$. On the other hand, we can analytically continue F_0, G_0 to this neighbourhood of 1, and their ratio gives a map which is the composite of G_1/F_1 with a Möbius map. Similarly for the point at infinity. What we conclude is that if we extend F_0, G_0 over the upper half-plane by analytic continuation and set $f = G_0/F_0$, then f maps these three semi-discs to three wedge-shaped regions bordered by segments of circlines. Now it is clear (by thinking of the analytic continuation process along the real axis) that the borders of these wedge-shaped regions must match up. That is, f maps H to a curvilinear triangle. Now the derivative of f is W/F_0^2, where $W = G_0' F_0 - G_0 F_0'$ and W satisfies a first-order ODE

$$z(1 - z) W' + (c - (1 + a + b)z) W = 0.$$

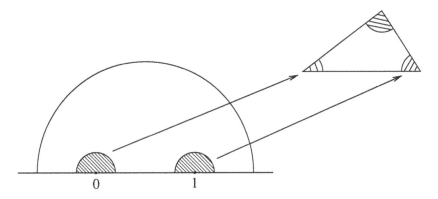

$$0 \qquad\qquad 1$$

Figure 15.2 *Map to a curvilinear triangle*

This implies that W has no zeros in the upper half-plane. Hence the derivative of f does not vanish, and a straightforward topological argument shows that f maps H bijectively to the curvilinear triangle.

We can pin down exactly which curvilinear triangle is the image of the upper half-plane under the map $f = F/G_0$. We have that $f(0) = 0$, $f(1) = r_1$, $f(\infty) = r_2 e^{i\pi\alpha}$ for r_1, r_2 real and positive. Now elementary geometry determines the ratio r_2/r_1 from the angles:

$$\frac{r_2}{r_1} = \frac{\cos \pi\alpha + \cos \pi(\alpha + \beta) - \sin \pi\gamma + \sin \pi(\alpha + \beta)}{\cos \pi\alpha(\cos \pi\gamma + \cos \pi(\alpha + \beta)) - \sin \pi\alpha(\sin \pi\gamma - \sin \pi(\alpha + \beta))}.$$

On the other hand, at $w = 1$, the power series representations of F, G_0 still converge to give (Nehari 1975, Whittaker and Watson 1996)

$$f(1) = \frac{\Gamma(2 - c)\Gamma(c - a)\Gamma(c - b)}{\Gamma(c)\Gamma(1 - a)\Gamma(1 - b)}.$$

Using this, one can write down explicitly the relation between the different local solutions of the hypergeometric equation arising from analytic continuation or, what is equivalent, the monodromy of the ODE with respect to the basis F, G_0 of local solutions. We refer to the literature for further details.

For the second point of view, we start our discussion by considering a Riemann surface S with a hyperbolic metric. Then we have a sheaf on S consisting of local Killing fields. In the upper half-plane, these are the vector fields $(a + 2bz + cz^2)(\partial/\partial)z$ for $a, b, c \in \mathbf{R}$. We call such a vector a 'translation vector field' if $ac = b^2$. This is the condition that the vector field has just one zero, which lies on the circle at infinity of the hyperbolic plane. We can define the same notion on a general hyperbolic surface: a translation vector field on some small open set $U \subset S$ is one which takes the above shape when we identify U with an open set in the upper half-plane. We may also consider 'complexified translation vector fields', for which we simply allow a, b, c to be complex with $ac = b^2$. (This complexification is a little unnatural from the hyperbolic-geometry point of view but will simplify our discussion. See the remarks on 'Möbius structures' below.)

Now suppose that we have a line bundle $L \to S$ which is a 'square root' of the tangent bundle in the sense that we have a holomorphic isomorphism $L \otimes L = TS$. We will write $L = TS^{1/2}$. (In fact, such square roots always exist, but are not generally unique.) In a local co-ordinate z, we can write sections of $TS^{1/2}$ as $f(\partial/\partial), z^{1/2}$ and such a section has a square which is a vector field. In particular, on the upper half-plane, the translation vector fields are just those which are the squares of $(\alpha + \beta z)(\partial/\partial)z^{1/2}$. So the following two conditions for a section $s = f(\partial/\partial z)^{1/2}$,

- the square of f is a translation vector field,
- $d^2 f/dz^2 = 0$,

are equivalent. But the first condition makes sense on a general hyperbolic surface, so we conclude that, given a hyperbolic surface and a square root $TS^{1/2}$, *there is a natural second-order ODE for sections of* $TS^{1/2}$. To see this more explicitly, suppose we have a local holomorphic co-ordinate w on our surface which differs from z by composing with a holomorphic map $z = \phi(w)$. Then a section $f(z)(\partial/\partial z)^{1/2}$ in one co-ordinate transforms to $f(\phi(w))\phi'(w)^{-1/2}(\partial/\partial w)^{1/2}$. One finds that the simple equation $d^2 f/dz^2 = 0$ goes over to a condition on a section $F(w)(\partial/\partial w)^{1/2}$ which is the ODE

$$\frac{d^2 F}{dw^2} - QF = 0 \qquad (15.4)$$

with

$$Q = \left(\psi^{1/2} \frac{d^2 \psi^{1/2}}{dw^2} \right),$$

where $\psi = d\phi/dw$. In turn, if we write $F = \lambda G$ for a fixed function $\lambda(w)$, this ODE transforms to the general form $d^2 G/dw^2 + \tilde{P}\, dG/dw + \tilde{Q}G = 0$ for $G(w)$.

Another point on view on this is to start with a Riemannian metric, written as $\rho^2 |dw|^2$. The Gauss curvature is $\rho^{-2}\Delta \log \rho$, so the condition that the metric is hyperbolic is $\Delta \log \rho = -\rho^2$. The uniqueness of the hyperbolic metric implies that we can find a function $z = \phi(w)$ as above, so that $\rho = |\phi'|/\mathrm{Im}\,\phi$. Then we can define Q by the above formula, and hence the ODE (15.4) for $F(z)(\partial/\partial w)^{1/2}$. The surprising thing is that this function Q is independent of the choice of ϕ and, when interpreted in terms of sections of $TS^{1/2}$, the ODE is independent of the choice of local co-ordinate.

We next introduce the concept of a hyperbolic surface with *cone singularities*. We write the metric on the hyperbolic plane in geodesic co-ordinates as $dr^2 + \sinh^2 r\, d\theta^2$. Given a *cone angle* $2\pi\alpha$, we define another metric on the punctured plane to be $dr^2 + \alpha^2 \sinh^2 r\, d\theta^2$. This has a singularity at the origin but away from the origin is hyperbolic: locally, we can just take $(r, \alpha\theta)$ as a new coordinate. We define a hyperbolic surface with cone singularities to be a smooth surface with a hyperbolic metric defined on the complement of a finite subset which agrees with this local model (for an appropriate choice of cone angle) around each of these points. Given an orientation, we get a genuine Riemann surface structure extending across the cone points. From another point of view, we could make the definition that we have a Riemann surface with a singular metric which, around the singular points, can be written in the standard form

$$\frac{\alpha^2 |z|^{2(\alpha-1)}}{(1 - |z|^{2\alpha})^2} |dz|^2,$$

which is what one obtains from the disc model of the hyperbolic plane by making the change of variable $w = z^\alpha$.

Now suppose we have a compact hyperbolic Riemann surface with cone singularities and a choice of $TS^{1/2}$. Then, away from the singularities, we get an ODE for sections of $TS^{1/2}$ as above. In a local co-ordinate around a singular point, this has the form $F'' - QF = 0$ for a section $F(w)(\partial/\partial w)^{1/2}$. Then Q is defined on a punctured disc, and the key point is that it has *at worst a double pole*. Thus the ODE has a regular singular point. This is simply a matter of checking in the explicit local model for a cone singularity.

Now return to a triangle T in the Riemann sphere and suppose that this is actually a hyperbolic triangle in the upper half-plane. We take two copies of T and glue the edges to construct a hyperbolic surface S with cone singularities. The cone angles are $2\pi\alpha, 2\pi\beta, 2\pi\gamma$. The underlying Riemann surface is homeomorphic to the sphere, so we know that it is isomorphic to the standard sphere and we can fix an isomorphism taking $0, 1, \infty$ to the three singular points. By symmetry, the isomorphism must map one half-plane to the original triangle T and, with appropriate choices of orientation, we can suppose that this the upper half-plane.

Now we define a square root of the tangent bundle of the Riemann sphere as follows. The section $\partial/\partial w$ of the tangent bundle has a double zero at the point at infinity, and we decree that local holomorphic sections of $(TCP^1)^{1/2}$ are sections $f(\partial/\partial w)^{1/2}$, where f has at worst a simple pole at infinity. So we have an ODE for sections $F(w)(\partial/\partial w)^{1/2}$ of the form $F'' - QF = 0$, where the only singularities of Q are the three points $0, 1, \infty$ and the singularities are regular, in local co-ordinates around these points. We know that any second-order ODE on the Riemann sphere with three regular singular points can be transformed to a hypergeometric equation, and it follows from this that the map from the upper half-plane to the triangle T is given by a ratio of solutions of such an equation. In more detail, the ODE must have the form $F'' = QF$, where Q has at worst double poles at $0, 1$. The regularity condition at infinity is that $Q(w) = O(w^{-1})$ as $w \to \infty$, so we must have

$$Q(w) = A(w^{-1} - (w - 1)^{-1}) + Bw^{-2} + C(w - 1)^{-2}.$$

Now, writing $F = \lambda G$ for an appropriate (multivalued) function of the form $\lambda(w) = w^p(w - 1)^q$, one can transform this ODE to the standard hypergeometric form. A local analysis around the singular points relates the cone angles to the indicial roots and hence to the parameters a, b, c, but of course we have already know how this must go from our preceding discussion.

An advantage of this second point of view is that it extends to a general 'curvilinear polygon' Π. Taking the double of Π and arguing as above, we see that the conformal map from the upper half-plane to Π is given by the ration of two solutions of an ODE $F'' + QF$, where

$$Q(w) = \sum a_i(z - z_i)^{-2} + b_i(z - z_i)^{-1},$$

and a_i, b_i, z_i are real. We also need $\sum a_i = 0$ to achieve a regular singularity at infinity. Knowledge of the angles of Π gives some constraints on the parameters, although it is not known how to determine these explicitly from Π.

The natural context for the discussion above is not really hyperbolic geometry but the more general notion of *Möbius structure*. A sign of this is the rather unnatural complexification we introduced in the discussion above A Möbius structure is given by a smooth surface with an atlas of charts which compare by elements of $PSL(2, \mathbb{C})$, acting on domains of the Riemann sphere. From this point of view, a hyperbolic structure is the special case when the maps come from $PSL(2, \mathbb{R}) \subset PSL(2, \mathbb{C})$ and the domains all lie in the upper half-plane. Given a Möbius structure, we have a subsheaf of a holomorphic vector field, modelled on those on the Riemann sphere, and the whole theory works as before.

Remark

1. Notice that cone angles of $2\pi/r$, for integer r, arise very naturally when we consider quotients. For example, the quotient of H by $PSL(2, \mathbb{Z})$ has a hyperbolic metric with two cone points having angles $\pi, 2\pi/3$, reflecting, the fact that $PSL(2, \mathbb{Z})$ does not act freely.
2. The choice of a square root of the tangent bundle is related to lifting the homomorphism $\pi_1 \to PSL(2, \mathbb{R})$, which we have for a hyperbolic surface, to a homomorphism $\pi_1 \to SL(2, \mathbb{R})$.

15.2 Periods of holomorphic forms and ordinary differential equations

15.2.1 The hypergeometric equation

Consider the function of a complex variable λ

$$F(\lambda) = 2 \int_0^\lambda \frac{dz}{\sqrt{z(z-1)(z-\lambda)}} dz.$$

We can consider this is as defined for λ in the upper half-space, say, and the integration path to be a line segment, on which we make a choice of definition of the square root. We can also regard $F(\lambda)$ as the integral around a closed loop γ on the Riemann surface associated to the equation $w^2 = z(z-1)(z-\lambda)$ of the holomorphic 1-form

$$\theta = \frac{dz}{\sqrt{z(z-1)(z-\lambda)}}.$$

From this point of view, it is clear that we can differentiate to get

$$F'(\lambda) = \int_\gamma \frac{d}{d\lambda}(z-\lambda)^{1/2}\frac{dz}{\sqrt{z(z-1)}} = \frac{1}{2}\int_\gamma (z-\lambda)^{-1}\theta,$$

and similarly

$$F''(\lambda) = \frac{3}{4}\int_\gamma (z-\lambda)^{-2}\theta.$$

For we can choose γ to avoid the point $z = \lambda$, so we do not run into any difficulty with the singularity of the integrand. Now consider the meromorphic function

$$f = 2w^{-1}(z-\lambda)^{-1}z(z-1) = 2(z-\lambda)^{-3/2}\sqrt{z(z-1)}$$

on our Riemann surface. This function has a triple pole at the point where $z = \lambda$ and no other poles. We have

$$df = f'(z)\,dz = (z-\lambda)^{-3/2}\frac{2z-1}{\sqrt{z(z-1)}} - 3(z-\lambda)^{-5/2}\sqrt{z(z-1)},$$

which gives

$$df = ((z-\lambda)(2z-1) - 3z(z-1))\frac{\theta}{(z-\lambda)^2}.$$

Write $u = z - \lambda$, so that

$$(z-\lambda)(2z-1) - 3z(z-1) = u(2u + 2\lambda - 1) - 3(u+\lambda)(u+\lambda-1)$$
$$= -(u^2 + 2(2\lambda - 1)u + 3\lambda(\lambda - 1)).$$

Then

$$-df = \left(1 + 2(2\lambda - 1)(z-\lambda)^{-1} + 3\lambda(\lambda - 1)(z-\lambda)^{-2}\right)\theta.$$

Now the integral of df around our closed loop γ, avoiding the pole, vanishes, so we have

$$F(\lambda) + 2\cdot 2(2\lambda - 1)F'(\lambda) + 3\cdot\frac{4}{3}\lambda(\lambda - 1)F''(\lambda) = 0.$$

That is,

$$\lambda(1-\lambda)F''(\lambda) + (1 - 2\lambda)F'(\lambda) - \frac{1}{4}F(\lambda) = 0.$$

This is the hypergeometric equation with parameters $a = b = 1/2, c = 1$. In fact, with a suitable sign convention, the function F is $\pi^{-1}F(\frac{1}{2}, \frac{1}{2}, 1; \lambda)$.

In the discussion above, the precise integration contour plays no role. For example, if we set

$$G(\lambda) = 2 \int_\lambda^1 \frac{dz}{\sqrt{z(z-1)(z-\lambda)}},$$

we get a second local solution of the same hypergeometric equation. The difference is that while f is defined and holomorphic in a neighbourhood of 0, the other solution G is multivalued, with a logarithmic singularity. In this final section of the book, we explain how all these phenomena fit into a general theory of *Picard–Fuchs equations*.

15.2.2   The Gauss–Manin connection

Let Δ be a finite subset of the Riemann sphere $\mathbf{CP}^1$, and write $B = \mathbf{CP}^1 \setminus \Delta$. Consider a differentiable fibre bundle $\mathcal{S} \to B$ whose fibres are smooth oriented surfaces. Each fibre Σ_b for $b \in B$ has a complex cohomology group $H^1(\Sigma_b; \mathbf{C})$ which is a complex vector space of some fixed dimension $2g$. It is clear that these form a complex vector bundle $\mathcal{V}$ over B. But, what is more, this bundle has a natural *flat connection*. That is to say, it can be given by a system of transition 'functions' which are actually constant matrices. We have a way to define the derivative of a local section of $\mathcal{V}$ and, near any point, we can find a basis of local sections whose derivatives vanish. To see this, observe that each cohomology group is a tensor product

$$H^1(\Sigma_b; \mathbf{C}) = H^1(\Sigma_b; \mathbf{Z}) \otimes \mathbf{C}.$$

For points b' close to b, we have a natural identification of the integral cohomology groups $H^1(\Sigma_b; \mathbf{Z})$, $H^1(\Sigma_{b'}; \mathbf{Z})$, and this induces an identification of their complexifications. For another point of view, fix a base point $b_0 \in B$. A flat connection is specified by a monodromy homomorphism

$$\rho_\mathcal{V} : \pi_1(B, b_0) \to GL(H^1(\Sigma_{b_0}; \mathbf{C})).$$

We have seen that the topology of the fibration gives a monodromy

$$\rho : \pi_1(B, b_0) \to \mathrm{Diff}(\Sigma_{b_0}).$$

The action of the diffeomorphisms on cohomology gives a homomorphism from $\mathrm{Diff}(\Sigma_{b_0})$ to $GL(H^1(\Sigma_{b_0}; \mathbf{C}))$, and $\rho_\mathcal{V}$ is obtained from ρ by composing with this. In fact, the image of $\rho_\mathcal{V}$ lies in the discrete subgroup of $GL(H^1(\Sigma_{b_0}; \mathbf{C})) \equiv GL(2g, \mathbf{C})$ made up of maps which preserve the integral structure and also the skew-symmetric intersection form, a copy of the group $Sp(2g, \mathbf{Z}) \subset GL(2g, \mathbf{C})$.

Now consider a family $p : \mathcal{S} \to B$ which has in addition a complex analytic structure, so that $\mathcal{S}$ is a complex surface and p is a holomorphic map. We want to give a 'complex analytic description' of the cohomology along the fibres, and for this we need a short digression.

If Σ is a compact Riemann surface, we have seen that the complex cohomology $H^1(\Sigma;\mathbf{C})$ can be expressed as a direct sum $H^{1,0} \oplus H^{0,1}$, where each of the summands has a complex analytic meaning. However, this is not what we need for our present purposes. The relation between $H^{0,1}$ and $H^1(\Sigma;\mathbf{C})$ involves differential forms of type $(0, 1)$, which are not truly 'complex analytic objects'. For example, if we are in the framework of general algebraic geometry, these differential forms do not have any meaning (that is, a meaning which extends to curves over general fields). What we do instead is to consider *meromorphic* 1-forms on Σ. A meromorphic 1-form θ has a residue at each point, which is a well-defined complex number. Suppose θ has the property that all its residues vanish. This means, of course, that any pole of θ must have order 2 or more. Such meromorphic 1-forms are called *differentials of the second kind*. Write Λ^1 for the collection of these: this is an infinite-dimensional complex vector space. Then we have a derivative

$$d : \Lambda^0 \to \Lambda^1,$$

where Λ^0 denotes the space of all meromorphic functions on Σ. Write $\mathcal{H}(\Sigma) = \Lambda^1/d\Lambda^0$. By Cauchy's formula, the integral of a form in Λ^1 around a contour does not change when the contour is moved across a pole. Similarly, the integral is unchanged if we add the derivative of a meromorphic function to the 1-form. So we get a map j from $\mathcal{H}(\Sigma)$ to $H^1(\Sigma;\mathbf{C})$. The fact we need is that this map j is an isomorphism. To see this, choose any divisor $D = \sum n_i p_i$, where p_i are distinct and $n_i > 0$. Let $\Lambda^1(D)$ denote the sheaf of meromorphic forms with at most poles of order $n_i + 1$ at the points p_i and with zero residues. Then, writing $\mathcal{O}(D)$ for the sheaf of meromorphic functions with at worst poles of order n_i at the p_i, we have an exact sequence of sheaves

$$0 \to \mathbf{C} \to \mathcal{O}(D) \to \Lambda^1(D) \to 0.$$

The long exact sequence in cohomology gives a map

$$j_D : H^0(\Lambda^1(D))/(\text{Im } d) \to H^1(\Sigma;\mathbf{C}).$$

On the other hand, the quotient space on the left-hand side above obviously injects into $\mathcal{H}(\Sigma)$. We leave it as an exercise in the definitions to check that these maps are compatible with our natural map j. Once this is done, it follows immediately that j is an injective map. Further, if we take a divisor of degree greater than $2g - 2$, the cohomology group $H^1(\Sigma, \mathcal{O}(D))$ vanishes (by the Serre duality theorem) and our long exact sequence shows that j_D is an isomorphism. This implies that j is surjective.

Now go back to our family $p : \mathcal{S} \to B$. Suppose first that α is a holomorphic 2-form on $\mathcal{S}$. Thus, at each point $x \in \mathcal{S}$, we have an element $\alpha(x)$ of the exterior power $\Lambda^2 T^* S$. In local holomorphic co-ordinates z_1, z_2 on S, we can write $\alpha = a(z_1, z_2)\, dz_1\, dz_2$ for a holomorphic function a. Let b be a point in the base B and let v be a tangent vector to B at b. Given $x \in \pi^{-1}(b)$ we lift v to $\tilde{v} \in TS_x$ and contract with α to give $i_{\tilde{v}}(\alpha(x)) \in T^* S_x$, which we can then restrict to the fibre, S_b. The crucial thing is that the result is independent of the choice of lift $\tilde{v}$. Letting x vary over the fibre, we get a holomorphic 1-form on Σ_b, which we denote by α_v. Similarly, we could start with a meromorphic 2-form on S and construct a meromorphic 1-form on Σ_b and, if this is arranged to have zero residues, it defines a class in $H^1(\Sigma_b, \mathbf{C})$ as we have seen above. Now, if we have a holomorphic vector field on B, we can apply this construction at each point to get sections of our bundle $\mathcal{V}$. If, as we can generally suppose, B is a subset of $\mathbf{C} \subset \mathbf{CP}^1$, we can take our vector field to be the standard constant vector field on $\mathbf{C}$. (Another way of expressing things is that the construction produces sections of $T^* \Sigma \otimes \mathcal{V}$.) The crucial point is that the sections which result will be holomorphic in the sense that they are given by holomorphic vector-valued functions in our local trivialisations of $\mathcal{V}$, defined by the flat structure. This amounts to the statement that if we choose a local C^∞ trivialisation of $p :$ $S \to B$, so that we can regard a local basis of the homology γ_i as being defined in all nearby fibres, the integrals

$$\int_{\gamma_i} \alpha_v$$

are holomorphic functions of the parameter on the base.

Stripped of the abstract language, the construction above becomes something we have seen many times in this book. For example, suppose we have a family of polynomials $p(z, \lambda)$ in z, of degree greater than 2, depending holomorphically on a parameter λ. So, in the case of elliptic curves above we would have $z(1 - z)(z - \lambda)$. We suppose that, for parameters λ in $B \subset \mathbf{C}$, the polynomial $p(z, \lambda)$ does not have a double root. Let Σ_λ be the compact Riemann surface of the function $\sqrt{p(z, \lambda)}$. We define

$$S = \{(x, \lambda) : \lambda \in B, x \in S_\lambda\}.$$

This is a complex surface, and we have three meromorphic functions z, w, λ on S with $w^2 = p(z, \lambda)$. Then $\alpha = d\lambda\, dz/w$ is a holomorphic 2-form on S, and if $v = \partial/\partial\lambda$, the resulting family of 1-forms is just $dz/\sqrt{p(z, \lambda)}$.

Now fix attention on a holomorphic section s of our bundle $\mathcal{V}$, for example a section obtained from the construction above. In a local flat trivialisation over a disc $N \subset B$, the section s can be viewed as a holomorphic map $s :$ $N \to \mathbf{C}^{2g}$. We differentiate $2g - 1$ times to get $2g$ vector-valued functions $s_0 = s, s_1 = ds/d\lambda, \ldots, s_{2g-1} = (d/d\lambda)^{2g-1}(s)$. At each point of N, these span a linear subspace of $\mathbf{C}^{2g}$ and the dimension of this span is semi-continuous. Suppose for simplicity that the span is the whole of $\mathbf{C}^{2g}$ on an open dense set N_0 in N, which is clearly the generic case. (If the span is everywhere smaller, one can reduce the number of derivatives and repeat the argument.) If we identify the top exterior power of $\mathbf{C}^{2g}$ with $\mathbf{C}$, then we have

$$s_0 \wedge \cdots \wedge s_{2g-1} = d(\lambda)$$

for a holomorphic function $d(\lambda)$, and N_0 is just the complement of the zeros of d. Differentiate one more time to get s_{2g}. At points of N_0, this can be expressed in terms of the basis $s_0, \ldots, s_{2g-1}$ as

$$s_{2g} = \sum_{i=0}^{2g-1} a_i s_i.$$

Take the exterior product with $s_2 \wedge \cdots \wedge s_{2g-1}$ to get

$$a_1 = d(\lambda)^{-1} s_{2g} \wedge s_2 \wedge \cdots \wedge s_{2g-1},$$

and similarly for the other a_i. (Of course, this is just the solution of a set of linear equations by the cofactor rule.) We see that the a_i are meromorphic functions, with at worst poles at the zeros of d. Thus the section s satisfies an ordinary differential equation of order $2g$, with meromorphic co-efficients. (If the span above is smaller than the generic value, we find that s satisfies an ODE of lower order.)

The discussion above was local, but we can express the conclusion in terms of the repeated derivatives $D^i s$, which are defined over all of B, and clearly the same conclusion holds. Thus, if V is a flat vector bundle of rank r, then any holomorphic section of V satisfies an ODE

$$D^p s = \sum_{j=0}^{p-1} a_j D^i s,$$

where $s \leq r$ and the co-efficients a_j are meromorphic. In particular, if α is a holomorphic 2-form on our surface S, then there is an associated ODE which is satisfied by all the periods of the resulting family of holomorphic 1-forms on the fibres.

Before going further, we can look back at our example of the periods of an elliptic curve and see that it fits into exactly this framework. Here $g = 1$, so $I(\lambda), I'(\lambda), I''(\lambda)$ must satisfy a linear relation in $H^1(\Sigma_\lambda)$. To find this relation, we compute using the description $\mathcal{H}(\Sigma_\lambda)$, which amounts to finding suitable meromorphic functions.

15.2.3   Singular points

Now suppose that our family $p : S \to B$ can be extended by adding suitable singular curves. Thus we have a compact complex surface, which we still denote by S to save notation, and a holomorphic map $p : S \to \mathbf{CP}^1$, which is a differentiable fibration over $B \subset \mathbf{CP}^1$, as before. For simplicity, we assume that the singular fibres over points of Δ have just ordinary double points. Thus p is a fibration outside a finite set of points in S, at which it is given in suitable local co-ordinates by the model $p(z_1, z_2) = z_1 z_2$. Recall that in Chapter 1 we defined a 'regular singular point' of an ODE. Our goal in this section is to show that points of Δ are regular singular points of the ODEs satisfied by the periods of meromorphic forms.

The first fact is that the bundle V naturally extends *as a holomorphic vector bundle* over the singular points $\Delta \subset \mathbf{CP}^1$. To discuss this properly, we would need rather more technical machinery than we want to assume, so we can only treat the matter informally. We fix our attention on a single point of Δ, which we may suppose is the origin $\lambda = 0$. We suppose that the fibre Σ_0 has just one singularity. Let $\tilde{\Sigma}_0$ be the normalisation of Σ_0, so we have two points $p, p' \in \tilde{\Sigma}_0$ which are identified in Σ_0. There are two cases: either $\tilde{\Sigma}_0$ has two components and the points p, p' are in different components, or $\tilde{\Sigma}_0$ has just one component. Of course, the reader who has studied the material of Chapter 14 will be quite familiar with this picture. We call these cases the 'separating' and 'non-separating' cases, respectively (the terminology being derived from the corresponding property of the vanishing cycle).

We want to define a vector space $\mathcal{H}(\Sigma_0)$ which will turn out to be the fibre of the extension of V over 0. Let $\Lambda^0(\Sigma_0)$ be the space of meromorphic functions f on the normalisation $\tilde{\Sigma}_0$ which are holomorphic at p, p' and such that $f(p) = f(p')$. Let $\Lambda^1(\Sigma_0)$ be the space of meromorphic 1-forms on $\tilde{\Sigma}_0$ which satisfy the following conditions:

- they have at worst simple poles at p, p', and the residues at these two points sum to zero;
- they have zero residues at all other points of $\tilde{\Sigma}_0$.

Then we have

$$d : \Lambda^0(\Sigma_0) \to \Lambda^1(\Sigma_0)$$

and we define

$$\mathcal{H}(\Sigma_0) = \Lambda^1(\Sigma^0)/(\mathrm{Im}\, d).$$

To see that this is a sensible candidate for the extension of V, we can check first that its dimension is correct. The discussion is slightly different in the separating and non-separating cases: we will consider the latter and leave the former as an exercise.

Let $\Lambda^0_{\perp}(\tilde{\Sigma}_0)$ denote the subspace of $\Lambda^0(\tilde{\Sigma}_0)$ consisting of functions holomorphic near to p, p', and similarly for $\Lambda^1_{\perp}(\tilde{\Sigma}_0)$. It is clear that we can compute

$\mathcal{H}(\tilde{\Sigma}_0)$ using the $\Lambda^i_\perp$ in place of Λ^i. Then we have obvious exact sequences

$$0 \to \Lambda^1_\perp(\tilde{\Sigma}_0) \to \Lambda^1(\Sigma_0) \to \mathbf{C} \to 0,$$

$$0 \to \Lambda^0(\Sigma) \to \Lambda^0_\perp(\tilde{\Sigma}_0) \to \mathbf{C} \to 0.$$

Here, in the first case, we map a form to its residue at p, and in the second we map a function f to the difference $e(f) = f(p) - f(p')$. Let K be the quotient of $\Lambda^1_\perp(\tilde{\Sigma}_0)$ by $d\lambda^0$. Then we have induced exact sequences

$$0 \to K \to \mathcal{H}(\Sigma_0) \to \mathbf{C} \to 0,$$

$$0 \to \mathcal{H}(\tilde{\Sigma}_0) \to K \to \mathbf{C} \to 0,$$

which imply that

$$\dim \mathcal{H}(\Sigma_0) = 1 + \dim K = 2 + \dim \mathcal{H}(\tilde{\Sigma}_0).$$

This is the desired relation, since we know that the genus of a nearby smooth fibre is one more than the genus of $\tilde{\Sigma}_0$.

Notice that, just as for a smooth fibre Σ_λ we have a subspace of $\mathcal{H}(\Sigma_\lambda)$ defined by the holomorphic 1-forms, so also we have a subspace of $\mathcal{H}(\Sigma_0)$ given by meromorphic 1-forms on $\tilde{\Sigma}_0$ with at worst simple poles at p, p' and with opposite residues there, but no other poles. This gives the appropriate notion of a 'holomorphic 1-form' on the singular curve.

To see further why this is a sensible definition of $\mathcal{H}(\Sigma_0)$, consider our construction starting with a meromorphic 1-form α on S, given by $a(z_1, z_2)\, dz_1\, dz_2$ in local co-ordinates. Our map is given by $\lambda = z_1 z_2$. For simplicity, suppose that a is holomorphic near the origin. With the vector field $\partial/\partial\lambda$ on the base, we take one lift at a point (z_1, z_2) with $z_1 z_2 \neq 0$ to be $\tilde{v} = z_2^{-1}\partial/\partial z_1$, which gives the 1-form $a(z_1, z_2)\, dz_2/z_2$. This construction extends to points where $z_2 \neq 0$, and we get a meromorphic 1-form on the z_2 axis given by $(a(0, z_2)/z_2)\, dz_2$. Likewise, we get a meromorphic 1-form $-(a(z_1, 0)/z_1)\, dz_1$ on the z_1 axis, using the lift $z_1^{-1}\, \partial/\partial z_2$ These forms have simple poles at the points p, p' (which of course correspond to the origins in the two axes), and the resides are $\pm a(0,0)$ and so are opposite.

A more precise statement is the following.

Proposition 53. *Let $\theta_1, \ldots, \theta_{2g}$ be meromorphic 1-forms on Σ_0 representing a basis for $\mathcal{H}(\Sigma_0)$. Then there are meromorphic 2-forms α_i defined on a neighbourhood of Σ_0 in S such that:*

- *For all small λ, the induced 1-form $\theta_i(\lambda) = p_*(\alpha_i)_\lambda$ has zero residues on Σ_λ and $p_*(\alpha_i)_0 = \theta_i$.*
- *For all small λ, the $\theta_i(\lambda)$ represent a basis for $\mathcal{H}(\Sigma_\lambda)$.*

We will not prove this, but it is easy to verify in explicit examples, which are our main focus, and we hope that the discussion above makes the assertion at least plausible. Clearly, this proposition defines an extension of the holomorphic vector bundle $\mathcal{V}$ over the origin, with fibre $\mathcal{H}(\Sigma_0)$ at this point. Notice that if we have one choice of forms $\theta_i(\lambda)$ as above, and if $G(\lambda)$ is a holomorphic matrix-valued function equal to the identity at $\lambda = 0$, then $\tilde\theta_i(\lambda) = \sum G_{ij}(\lambda)\theta_j(\lambda)$ is an equally good choice. For this, we just replace α_i by $\sum G_{ij}(\lambda)\alpha_j$.

Having understood the extension of the holomorphic vector bundle $\mathcal{V}$, we return to consider the flat connection. Recall that the monodromy of the fibre bundle around a small circle about 0 is the Dehn twist in a vanishing cycle in the fibre.

Proposition 54. *Let δ be an embedded circle in a smooth oriented surface S and let $\phi_\delta : S \to S$ be the corresponding Dehn twist. Then the action of ϕ_δ on $H_1(S)$ is given by*

$$(\phi_\delta)_* \epsilon = \epsilon + \langle \delta, \epsilon \rangle \delta,$$

where $\langle \, , \, \rangle$ denotes the intersection form on H_1.

We can safely leave the proof to the reader. The monodromy of the flat connection on our bundle $\mathcal{V}$ around 0 is given by applying this formula using the homology class of the vanishing cycle. If the vanishing cycle is not separating, then the action on homology is not trivial, the monodromy of the flat connection around 0 is not trivial, and so it is certainly not possible to extend the flat structure over 0. Our final task is to examine the nature of the singularity in this case.

Let $\gamma_3, \gamma_4, \ldots, \gamma_{2g}$ be a standard set of 1-cycles forming a basis for the homology of the normalisation $\tilde\Sigma_0$ and not meeting p, p'. The γ_i can be regarded as cycles in Σ_0 and we can obviously move them in a continuous fashion to cycles in nearby fibres, which we denote by the same symbol. Fix a small real, positive ϵ so that Σ_ϵ can be taken as the standard non-singular fibre near to Σ_0. Thus we have constructed cycles $\gamma_3, \ldots, \gamma_{2g}$ in Σ_ϵ and we want to extend this to a standard basis for the homology. We take γ_1 to be the vanishing cycle and γ_2 to be a cycle with $\langle \gamma_2, \gamma_1 \rangle = 1$ but disjoint from all the γ_i for $i \geq 3$. The vanishing cycle γ_1 is preserved by the monodromy so it makes sense to consider it as a cycle in each fibre S_λ for small non-zero λ. The monodromy acts non-trivially on γ_2—by the formula of Proposition 54, moving γ_2 around the origin (in the obvious sense) changes it to $\gamma_2 + \gamma_1$.

Now choose meromorphic 1-forms $\theta_3, \ldots, \theta_{2g}$ on $\tilde\Sigma_0$ which form a dual basis to $\gamma_3, \ldots, \gamma_{2g}$. These may also be regarded as forms on the singular fibre Σ_0. Let $\theta_1 \in \Lambda^1(\tilde\Sigma_0)$ be a form with residue $+1$ at p_1 and with 0 periods around $\gamma_3, \ldots, \gamma_{2g}$. Choose a meromorphic function f on $\tilde\Sigma_0$ with

$f(p_1) = 1/2$, $f(p_2) = -1/2$, and let $\theta_2 = df$. This is a form in $\Lambda^1(\Sigma_0)$ which also has zero periods around $\gamma_3, \ldots, \gamma_{2g}$, and the discussion above shows that $\theta_1, \ldots, \theta_{2g}$ are a basis for $\mathcal{H}(\Sigma_0)$. We apply Proposition 53 to extend these to a family of forms $\theta_i(\lambda)$, giving a basis for the cohomology of nearby fibres.

Now, for $i \neq 2$, we have cycles γ_i in the fibres, so we get functions defined initially on a small disc minus the origin

$$I(\gamma_i, \theta_j)(\lambda) = \int_{\gamma_i} \theta_j(\lambda).$$

If we work on a cut plane, we can also define $I(\gamma_2, \theta_j)$, but going once around the origin changes this by the addition of $I(\gamma_1, \theta_j)$. In other words,

$$I(\gamma_2, \theta_j) - \log \lambda I(\gamma_1, \theta_j)$$

is a holomorphic function in a punctured neighbourhood of $\lambda = 0$.

Proposition 55. *As $\lambda \to 0$, we have:*

1. *For $i, j \neq 2$, $I(\gamma_i, \theta_j)(\lambda) \to \delta_{ij}$.*
2. *$I(\gamma_i, \theta_2) \to 0$ for $i \neq 2$.*
3. *$I(\gamma_2, \theta_2) - \log \lambda I(\gamma_1, \theta_2) \to 1$.*
4. *$I(\gamma_2, \theta_i) - \log \lambda I(\gamma_1, \theta_i) \to 0$ for $i \geq 3$.*

In addition, $I(\gamma_2, \theta_1) - \log \lambda I(\gamma_1, \theta_1)$ extends to a holomorphic function over $\lambda = 0$.

Before discussing the proof of this, we give the following corollary.

Corollary 14. *By a suitable choice of $G_{ij}(\lambda)$, we can arrange that the $\tilde{\theta}_j = \sum G_{jk}\theta_k$ have periods*

$$\int_{\gamma_i} \tilde{\theta}_j = \delta_{ij}, \quad i, j \neq 2,$$

$$\int_{\gamma_i} \tilde{\theta}_2 = 0, \quad i \neq 2, \quad \int_{\gamma_2} \tilde{\theta}_2 = 1,$$

$$\int_{\gamma_2} \tilde{\theta}_i = 0, \quad i \geq 3, \quad \int_{\gamma_2} \tilde{\theta}_1 = \log \lambda.$$

This corollary is elementary. First, since we know that $I(\gamma_i, \theta_j) \to \delta_{ij}$ for $i, j \neq 2$, we can make a family of linear transformations so that, without loss of generality, $I(\gamma_i, \theta_j) = \delta_{ij}$ for $i, j \neq 2$. Since $I(\gamma_i, \theta_2)$ tends to zero, we can replace θ_2 by $\theta_2 - \sum_{i \neq 2} I(\gamma_i, \theta_2)\theta_i$ to reduce to the case when $I(\gamma_i, \theta_2) = 0$ for $i \neq 2$. This means, in particular, that $I(\gamma_2, \theta_2)$ is well defined and tends to 1 as $\lambda \to 0$. Multiplying θ_2 by $I(\gamma_2, \theta_2)^{-1}$, we can reduce to the case when $I(\gamma_2, \theta_2) = 1$. Now we know that $I(\gamma_2, \theta_1) = \log \lambda + b(\lambda)$, say, with b holomorphic. We change θ_1 to $\theta_1 - b(\lambda)\theta_2$ to reduce to the case when $b = 0$.

Now we turn to the proof of Proposition 55. We will only discuss the assertions involving the $I(\gamma_1, \theta_1), I(\gamma_2, \theta_1)$, since the other statements seem fairly obvious. We know that the extension θ_1 arises from a meromorphic 2-form $\alpha = a \, dz_1 \, dz_2$, as above. The condition on the residues that we have arranged is equivalent to saying that $a(0, 0) = 1$. For small non-zero λ, consider the part N_λ of the fibre $z_1 z_2 = \lambda$ which lies in the polydisc $|z_1|, |z_2| < 1$. We parametrise this by a single variable z using $z \mapsto (z, \lambda/z)$. Thus N_λ is identified with the annulus $A = \{z : |\lambda| < |z| < 1\}$. The vanishing cycle is a standard circle in this annulus. The cycle γ_2 does not lie in N_λ but passes through it, corresponding to a path Γ_2 in A which runs from the inner boundary to the outer boundary. The fact that γ_2 is not really well defined is just the fact that this path can be changed by adding multiples of γ_1.

Recall from our preceding discussion that for each λ the induced 1-form $\theta_1(\lambda)$ is given in our coordinate z above by

$$\theta_1 = a(z, \lambda/z) \, dz/z.$$

Write $a(z_1, z_2) = \sum a_{pq} z_1^p z_2^q$. Then

$$\theta_1 = \left(\sum_{pq} a_{pq} \lambda^q z^{p-q-1} \right) dz.$$

Thus

$$\int_{\gamma_1} \theta_1 = \sum_{p \geq 0} a_{pp} \lambda^p,$$

which converges to $a_{00} = a(0, 0) = 1$ as $\lambda \to 0$, as required.

The other assertion is that $I(\gamma_2, \theta_1) - \log \lambda I(\gamma_1, \theta_1)$ is holomorphic across $\lambda = 0$. By the Riemann Removal-of-Singularities Theorem, it suffices to prove that this function is bounded in the punctured disc. We write

$$\int_{\gamma_2} \theta_1 = \int_{\Gamma_2} \theta_1 + \int_{\Gamma_2'} \theta_1,$$

where Γ_2' is the part of Γ_2 outside N. Clearly, we can choose our set-up so that the integral over Γ_2' is bounded. We can also suppose that when $\lambda = |\lambda|$, the contour Γ_2 runs along the positive real axis from $|\lambda|$ to 1. When $\lambda = e^{i\phi}|\lambda|$ for $0 \leq \phi < 2\pi$, the contour $\Gamma_2(\phi)$, say, should be taken to be a path from $e^{i\phi}|\lambda|$ to 1, as in Figure 15.3.

Thus, if we put $\phi = 2\pi$, we get a different contour from $\Gamma_2(0)$, and this phenomenon is precisely the Dehn twist monodromy. Now, with this understanding about the integration contour, we have

$$\int_{\Gamma_2} \theta(\lambda) = \int_{\lambda}^{1} \sum a_{pq} \lambda^q z^{p-q-1} \, dz,$$

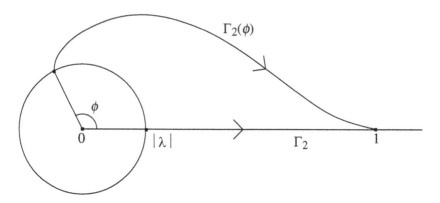

Figure 15.3 *The integration contour*

which is

$$\left(\sum_p a_{pp}\lambda^p\right)\log\lambda + \sum_{p\neq q}\frac{a_{pq}}{p-q}(\lambda^q - \lambda^p),$$

and hence equal to $I(\gamma_1,\theta_1)\log\lambda + f(\lambda)$, where

$$f(\lambda) = \sum_{p\neq q}\frac{a_{pq}}{p-q}(\lambda^q - \lambda^p),$$

which is holomorphic across $\lambda = 0$.

We can express Corollary 14 in a different way. Consider a general holomorphic section corresponding to some other family of forms $\theta(\lambda) = \sum f_i\tilde\theta_i$ of our bundle V around $\lambda = 0$. The condition that the periods of the section are constant is

$$\frac{\partial f_i}{\partial\lambda} = 0, \ i \neq 2, \quad \frac{\partial f_2}{\partial\lambda} + \lambda^{-1}f_1 = 0.$$

In other words, in this basis of sections,

$$D\underline{f} = \frac{\partial \underline{f}}{\partial\lambda} + \lambda^{-1}A\underline{f},$$

where A is the matrix with $A_{21} = 1$ and all other entries zero. In particular, $\lambda D\underline{f}$ is again a holomorphic section. So we get $2g$ holomorphic sections

$$\underline{f},(\lambda D)\underline{f},\ldots,(\lambda D)^{2g-1}\underline{f}.$$

We suppose for simplicity that these form a basis for the fibre of zero, and hence for all nearby fibres. Then, just as before, we can write

$$(\lambda D)^{2g}\underline{f} = \sum_{p=0}^{2g-1} b_p(\lambda)(\lambda D)^p(\underline{f})$$

for holomorphic functions $b_p(\lambda)$. We now interpret this equation away from the origin, where we can use a flat trivialisation. The equation says that any period $I(\lambda)$ of our family of forms $\theta(\lambda)$ satisfies an equation

$$\left(\lambda \frac{d}{d\lambda}\right)^{2g} I = \sum_{p=0}^{2g-1} b_p(\lambda) \left(\lambda \frac{d}{d\lambda}\right)^{p} I. \tag{15.5}$$

Now we have the following lemma.

Lemma 44. *Equation* (15.5) *can be written in the form*

$$\left(\frac{d}{d\lambda}\right)^{2g} I = \sum_{p=0}^{2g-1} a_p(\lambda) \left(\lambda \frac{d}{d\lambda}\right)^{p} I,$$

where a_p has at worst a pole of order $2g - p$ at 0.

The proof, by induction on g, is elementary. This achieves our goal of showing that the ODE satisfied by the periods has a regular singular point (looking back at the definition in Chapter 1).

To illustrate this theory, consider the Riemann surface S_λ associated to the function $w = \sqrt{(z - \lambda)z(z^4 - 1)}$ and the holomorphic form $\theta = dz/w$. Then, computing as before, we have

$$\theta' = \frac{1}{2}\theta_1, \quad \theta'' = \frac{3 \cdot 1}{2 \cdot 2}\theta_2, \quad \theta''' = \frac{5 \cdot 3 \cdot 1}{2 \cdot 2 \cdot 2}\theta_3, \quad \theta^{(iv)} = \frac{7 \cdot 5 \cdot 3 \cdot 1}{2 \cdot 2 \cdot 2 \cdot 2}\theta_4,$$

where $\theta_p = \theta(z - \lambda)^{-p}$. The Riemann surface S_λ has genus 2, so $H^1(S_\lambda, \mathbf{C})$ has dimension 4 and there must be a meromorphic function F and scalars a_p such that $\partial F = \sum_{p=0}^{4} a_p\theta_p$. Our task is to find these scalars. Let x be the point in S_λ lying over $z = \lambda$. Thus any sum $\sum_{p=0}^{4} a_p\theta_p$ gives a meromorphic 1-form that has a pole of order at most 8 at x, and is otherwise holomorphic. So we need to consider functions with poles of order up to 7 at x. The function we need must be reversed by the involution $w \mapsto -w$ of S_λ, and it is easy to check that the only possibility is a linear combination of $f_1 = w(z - \lambda)^{-3}$ and $f_2 = w(z - \lambda)^{-4}$. Now, if we write $Q = z(z^4 - 1)$, we have

$$f_1' = \frac{1}{2\sqrt{(z - \lambda)Q}} \left(\frac{-5Q + (z - \lambda)Q'}{(z - \lambda)^3}\right),$$

where we write f_1' for $\partial f_1/\partial z$. Similarly,

$$f_2' = \frac{1}{2\sqrt{(z - \lambda)Q}} \left(\frac{-7Q + (z - \lambda)Q'}{(z - \lambda)^4}\right).$$

Consider a linear combination $f = af_1 + f_2$ for $a \in \mathbf{C}$. Then

$$f' = \frac{1}{2(z - \lambda)^4}\sqrt{(z - \lambda)Q}P,$$

where

$$P(z) = ((z - \lambda)Q' - 7Q) + a((z - \lambda)Q' - 5Q)).$$

At first sight, P is a polynomial of degree 6, but on closer inspection one sees that it has degree at most 5, owing to cancellation. Expanding out, one sees that the co-efficient of z^5 is $5a\lambda - 2$, so if we choose $a = 2/5\lambda$, the polynomial P has degree 4. If we expand about the point $z = \lambda$ as $P(z) = \sum_{p=0}^{4} b_p(z - \lambda)^p$ for $b_p \in \mathbf{C}$, then we have $2\partial f = \sum_{p=0}^{4} b_p\theta_p$. In other words, $\sum b_p\theta_p = 0$ in $\mathcal{H}(\Sigma_\lambda)$. This is the desired linear relation, which translates into an ODE for the periods. We leave the reader to check that this ODE is

$$(\lambda^5 - \lambda)\frac{2^4}{15}I^{(iv)} + (7\lambda^4 - 1)\frac{2^5}{15}I''' + \left(42\lambda^3 + \frac{8}{5\lambda}\right)\frac{4}{3}I'' + 56I' + 7I = 0, \quad (15.6)$$

where $I = I(\lambda)$ is an integral of θ around a closed path in S_λ.

The indicial equation at $\lambda = 0$ is

$$v(v - 1)(v - 2)(v - 3) - 2v(v - 1)(v - 2) + 2v(v - 1) = 0,$$

which has roots $v = 0, 1, 2, 3, 4$. The root $v = 0$ is realised by the 'vanishing cycle', the double cover of a short path from 0 to λ. This gives

$$I_0(\lambda) = \int_0^\lambda \frac{dz}{\sqrt{z(z - \lambda)(z^4 - 1)}} = \int_0^1 \frac{du}{\sqrt{u(1 - u)(1 - \lambda^4 u^4)}},$$

which clearly has the non-zero limit

$$I_0(0) = \int_0^1 \frac{du}{\sqrt{u(1 - u)}} = \pi.$$

Clearly I_0 is a function of λ^4. Let $\gamma_1 \in H_1(S_\lambda, \mathbf{C})$ correspond to the sum

$$\left(\int_1^i + i\int_i^{-1} + i^2\int_{-1}^{-i} + i^3\int_{-i}^1\right).$$

Then $I_1 = \int_{\gamma_1} \theta$ satisfies $I_1(i\lambda) = iI_1(\lambda)$, so $I_1(\lambda)/\lambda$ is a function of λ^4 and I_1 realises the indicial root $v = 1$. Similarly, replacing i by $-i$, we get a cycle realising the indicial root $v = 3$. The remaining indicial root $v = 4$ corresponds to the logarithmic solution, integrating θ around a contour which intersects the vanishing cycle.

Remark In this chapter, we have just discussed the simplest case of a theory which extends much further. In one direction, one can consider more compli-cated singularities: in another direction, one can consider complex manifolds of higher dimension. In some situations, the fact that such 'periods' satisfy an ODE can be used to derive complete power series expansions for them.

A particularly topical application is to mirror symmetry on 'Calabi–Yau' manifolds (Candelas *et al.* 1991).

Exercises

1. Show that a curvilinear triangle with angles $\pi\alpha, \pi\beta, \pi\gamma$ can be transformed by a Möbius map to a hyperbolic triangle if and only if $\alpha + \beta + \gamma < 1$.
2. Let b_i be real and $\alpha_i \in (0,1)$. Show that solutions of the first-order equation

$$\frac{df}{dz} = \left(\sum \frac{\alpha_i}{z - b_i}\right) f$$

   map the upper half-plane to a Euclidean polygon. (This is the Schwarz–Christoffel transformation.) How does this fit in with the hypergeometric solution in the case of triangles?
3. Verify that, as asserted in Subsection 15.2.1, $F = \pi^{-1} F(\frac{1}{2}, \frac{1}{2}, 1; \lambda)$ by integrating the power series for $(z - \lambda)^{-1/2}$.
4. Show that one solution of equation ((15.6)), for small λ, is $J_0(\lambda^4)$, where

$$J_0(\mu) = \pi + \frac{\pi}{2} \sum_{k=1}^{\infty} \mu^k \left(\frac{1 \cdot 3 \cdots (8k-1) \cdot 1 \cdot 3 \cdots (2k+1)}{2^{5k}(4k)!k!}\right).$$

   (Use the Taylor expansion of $(1 - \mu u^2)^{-1/2}$ and substitute $u = \sin^2\theta$.) Find a similar formula for the solutions corresponding to the indicial roots $v = 1, 3$. (Apply the residue theorem on the Riemann surface of $\sqrt{z^4 - 1}$.)

References

Candelas, P., De La Ossa, X., Green, P. and Parkes, L. (1991). A pair of Calabi–Yau manifolds as an exactly soluble superconformal theory. *Nucl. Phys. B* **359**, 21–74.

Cassels, J. W. S. (1986). *Local Fields*, London Mathematical Society Student Texts, No. 3. Cambridge University Press, Cambridge.

Cohn, P. M. (1991). *Algebraic Numbers and Algebraic Functions*, Chapman and Hall Mathematics Series. Chapman & Hall, London.

Dold, A. and Thom, R. (1958). Quasifaserungen und unendliche symmetrische Produkte. *Ann. Math.* (2) **67**, 239–281.

Griffiths, P. and Harris, J. (1994). *Principles of Algebraic Geometry*, Wiley Classics Library. Reprint of the 1978 original. Wiley, New York.

Hatcher, A. (2002). *Algebraic Topology*. Cambridge University Press, Cambridge.

Hirsch, M. W. (1994). *Differential Topology*, Graduate Texts in Mathematics, No. 33. Corrected reprint of the 1976 original. Springer, New York.

Hubbard, J. H. (2006). *Teichmüller Theory and Applications to Geometry, Topology, and Dynamics*. Vol. 1, *Teichmüller Theory*. Matrix Editions, Ithaca, NY.

Hudson, R. W. H. T. (1990). *Kummer's Quartic Surface*, Cambridge Mathematical Library. With a foreword by W. Barth. Revised reprint of the 1905 original. Cambridge University Press, Cambridge.

Hurwitz, A. (1891). Über Riemann'sche Flächen mit gegebenen Verzweigungspunkten. *Math. Ann.* **39**, No. 1, 1–60.

Kirwan, F. (1992). *Complex Algebraic Curves*, London Mathematical Society Student Texts, No. 23. Cambridge University Press, Cambridge.

Levy, S. (ed.) (1999). *The Eightfold Way: The Beauty of Klein's Quartic Curve*, Mathematical Sciences Research Institute Publications, No. 35. Cambridge University Press, Cambridge.

Lickorish, W. B. R. (1962). A representation of orientable combinatorial 3-manifolds. *Ann. Math.* (2) **76**, 531–540.

Milnor, J. (1956). On manifolds homeomorphic to the 7-sphere. *Ann. Math.* (2) **64**, 399–405.

Mumford, D. (1975). *Curves and Their Jacobians*. University of Michigan Press, Ann Arbor, MI.

Nehari, Z. (1975). *Conformal Mapping*. Reprint of the 1952 edition. Dover Publications, New York.

Spivak, M. (1979). *A Comprehensive Introduction to Differential Geometry*, Vol. 1, 2nd edition. Publish or Perish, Inc., Willington, DE.

Stein, E. M. (1970). *Singular Integrals and Differentiability Properties of Functions*, Princeton Mathematical Series, No. 30. Princeton University Press, Princeton, NJ.

Whittaker, E. T. and Watson, G. N. (1996). *A course of Modern Analysis*, Cambridge Mathematical Library. Reprint of the 4th (1927) edition. Cambridge University Press, Cambridge.

Index